Essentials of Geophysics

Volume I

Essentials of Geophysics
Volume I

Edited by **Karl Seibert**

R CALLISTO REFERENCE

New York

Published by Callisto Reference,
106 Park Avenue, Suite 200,
New York, NY 10016, USA
www.callistoreference.com

Essentials of Geophysics: Volume I
Edited by Karl Seibert

International Standard Book Number: 978-1-63239-329-6 (Hardback)

Printed in the United States of America.

Contents

Preface

Geophysics is a branch of science which studies about the physics and environment of earth in space. It covers important aspects regarding earth by using quantitative physical methods, which happen to be an integral component in this field. In the 19th century, this branch of science was recognized as a unique and separate discipline. However, the first magnetic compasses were lodestones, which can be traced to fourth century in China, India and Greece. Geophysical scientists study the Earth using gravity, magnetic, electrical and seismic methods. It is a highly interdisciplinary area of expertise and geophysicists contribute significantly to each and every area of the Earth sciences. This discipline has also played a highly crucial role in developing researches and practical theories of plate tectonics. Other applications of Geophysics include analysing potential petroleum reservoirs, assessing sites for environmental remediation and locating groundwater among others.

Sometimes, the term geophysics also refers to only for geological applications such as Earth's shape, Earth's internal structure and composition etc. On the other hand, in modern geophysics the definition is broader and also encompasses the hydrological cycle which includes snow and ice on earth, fluid dynamics of seas and atmospheric fluids, electricity and magnetism in the layers of atmosphere and also the relations and problems related to moon and other planets. Geophysics in twentieth century had developed methods for remote exploration of solid earth and natural water.

The aim of this book has been to provide geophysics professionals with a solid foundation on this important area of research. I wish to thank all the authors for their efforts and time that they have given to this project. Last but not the least, I wish to thank my family and friends, who have supported me in my life through everything.

Editor

Crosscorrelation of Earthquake Data Using Stationary Phase Evaluation: Insight into Reflection Structures of Oceanic Crust Surface in the Nankai Trough

Shohei Minato,[1] Takeshi Tsuji,[1, 2] Toshifumi Matsuoka,[1] and Koichiro Obana[3]

[1] *Graduate School of Engineering, Kyoto University, C1 Kyoto-Daigaku Katsura, Nishikyo-ku, Kyoto 6158540, Japan*
[2] *International Institute for Carbon-Neutral Energy Research (I2CNER), Kyushu University, Ito Campus, 744 Motooka, Nishi-Ku, Fukuoka 819-0395, Japan*
[3] *Japan Agency for Marine-Earth Science and Technology, Institute for Research on Earth Evolution, 3173-25 Showa-machi, Kanazawa-ku, Yokohama 236-0001, Japan*

Correspondence should be addressed to Shohei Minato, s_minato@earth.kumst.kyoto-u.ac.jp

Academic Editor: Yun-tai Chen

Seismic interferometry (SI) has been recently employed to retrieve the reflection response from natural earthquakes. We perform experimental study to apply SI to Ocean Bottom Seismogram (OBS) records in the Nankai Trough, southwest Japan in order to reveal the relatively shallow geological boundaries including surface of oceanic crust. Although the local earthquakes with short raypath we use to retrieve reflection response are expected to contain the higher-frequency components to detect fine-scale structures by SI, they cannot be assumed as plane waves and are inhomogeneously distributed. Since the condition of inhomogeneous source distribution violates the assumption of SI, the conventional processing yields to the deteriorated subsurface images. Here we adopt the raypath calculation for stationary phase evaluation of SI in order to overcome this problem. To find stationary phase, we estimate the raypaths of two reflections: (1) sea-surface P-wave reflection and (2) sea-surface multiple P-wave reflection. From the estimated raypath, we choose the crosscorrelation traces which are expected to produce objective reflections considering the stationary phase points. We use the numerical-modeling data and field data with 6 localized earthquakes and show that choosing the crosscorrelation traces by stationary phase evaluation improves the quality of the reflections of the oceanic crust surface.

1. Introduction

Among various seismic exploration methods using the body-wave of natural earthquakes [1, 2], seismic interferometry (SI) has been recently employed to retrieve the reflection response. Although the receiver function method [1] has been broadly used to image the Moho and mantle discontinuities, there is a study claiming that retrieving and migrating the reflection response using SI is superior to the receiver function method [3].

SI retrieves Green's function between receivers by crosscorrelating wavefield [4, 5]. This theory requires the physical sources homogeneously distributed along the enclosed surface which surrounds the receivers [4]. There are several successful applications of SI to natural earthquakes. Abe

et al. [3] and Tonegawa et al. [6] retrieved the crustal reflection response in central Japan using P coda and S coda, respectively. Ruigrok et al. [7] used P wave to retrieve reflection response using Laramie array in USA. Abe et al. [3] further showed the comparison of migrated images of SI and those using receiver function analysis.

These applications focused on the teleseismic wavefields in which the epicentral distance is much longer than the length of receiver array. Consequently the wavefields can be assumed as plane wave. Ruigrok et al. [7] replaced the integral for source position of SI into a one for ray parameter. These teleseismic events are suitable to SI processing since the most earthquakes are generated at the numerous plate boundaries in the world, and consequently the teleseismic records contain the earthquakes propagating from

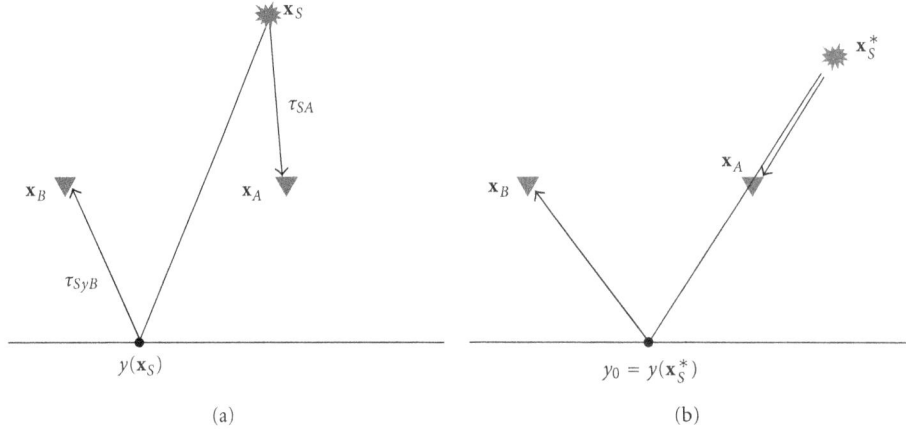

FIGURE 1: (a) Receiver $\mathbf{x}_A$ observes direct wave with the travel time τ_{SA} and $\mathbf{x}_B$ observes reflected wave with τ_{SyB}. The specular reflect position $y(\mathbf{x}_S)$ varies with the source position $\mathbf{x}_S$. (b) When the source position $\mathbf{x}_S$ satisfies the stationary phase position $\mathbf{x}_S^*$, two events have common raypath between $\mathbf{x}_S^*$ and $\mathbf{x}_A$.

the various directions. This can moderately enable us to assume that their source distribution is homogeneous.

Here, we performed experimental study to apply SI to the localized earthquake records acquired by Ocean Bottom Seismogram (OBS) in the Nankai Trough, southwest Japan in order to reveal the relatively shallow geological boundaries including surface of oceanic crust (i.e., plate boundary). There is no application of SI to the natural earthquake recorded by OBS. The OBS we used in this study was deployed to observe the local earthquake for earthquake observation in subduction zone [11, 12]. Those local earthquakes cannot be assumed as a teleseismic wavefield since the source-receiver distance is smaller than the length of receiver array. Furthermore, those localized earthquakes are usually inhomogeneously distributed and violate the assumption of SI. Therefore, the conventional SI processing (crosscorrelation and summation) yields to the deteriorated subsurface images. However, focusing on the local earthquakes with shorter raypath can give us the advantages over using the teleseismic wavefields because the teleseismic events usually have long propagating path and lose their higher-frequency components, which leads to detect only large-scale structure such as Moho reflections. On the other hand, the local earthquakes which have shorter raypath are expected to contain higher-frequency components and to resolve more fine-scale structures by using SI. In this study, we discuss a method to retrieve the reflection response using localized natural earthquakes observed by OBS; we adopt the raypath calculation for stationary phase evaluation in SI analysis in order to overcome the noise originated from the violation of homogeneous source distribution.

The physical interpretation of the condition to be posed on the SI can be explained by the stationary phase approximation [5, 13, 14]. In this approximation, it explains that the dominant contribution of the retrieval of the Green's function comes from the crosscorrelation of the records from the physical source located at the stationary phase position (stationary phase source). Therefore, in the localized source distribution, crosscorrelation pairs with stationary phase positions have physical meaning. The other crosscorrelation traces will produce noise and deteriorate the quality of the imaging results. In this study, we identify crosscorrelation pairs with stationary phase position using the estimated raypath information of two reflections: (1) sea-surface P-wave reflection and (2) sea-surface multiple P-wave reflection.

Note that Chaput and Bostock [15] successfully retrieved reflection response using the subsurface noise sources located from 10 km to 60 km depth which is similar source depth to our study. They found that the source illumination is imperfect from the discussion using the stationary phase approximation. Our study is different from them in the point that we focus on the natural earthquakes.

2. Estimation of Stationary Phase Records Using Raypath Calculation

2.1. Stationary Phase Approximation of Seismic Interferometry. Seismic interferometry by crosscorrelation (e.g., [4]) is written as

$$
\begin{aligned}
&\hat{G}(\mathbf{x}_B, \mathbf{x}_A, \omega) + \hat{G}^*(\mathbf{x}_B, \mathbf{x}_A, \omega) \\
&\quad = \frac{2k}{\omega\rho} \oint_{\mathbb{S}_{src}} \hat{G}(\mathbf{x}_B, \mathbf{x}_S, \omega) \hat{G}^*(\mathbf{x}_A, \mathbf{x}_S, \omega) d^2\mathbf{x}_S,
\end{aligned}
\tag{1}
$$

where $\hat{G}(\mathbf{x}_A, \mathbf{x}_S, \omega)$ and $\hat{G}(\mathbf{x}_B, \mathbf{x}_S, \omega)$ are the observed wavefield from the sources $\mathbf{x}_S$ along the closed surface $\mathbb{S}_{src}$. $\hat{G}(\mathbf{x}_B, \mathbf{x}_A, \omega)$ is a Green's function between two receiver positions.

We assume that the primary reflection is retrieved by crosscorrelating a direct wave and a specular reflection from the physical sources (Figure 1(a)).

$$
\begin{aligned}
\hat{G}^d(\mathbf{x}_A, \mathbf{x}_S, \omega) &= e^{i\omega\tau_{SA}}, \\
\hat{G}^r(\mathbf{x}_B, \mathbf{x}_S, \omega) &= e^{i\omega\tau_{SyB}},
\end{aligned}
\tag{2}
$$

Crosscorrelation of Earthquake Data Using Stationary Phase Evaluation: Insight into Reflection Structures of Oceanic Crust Surface in the Nankai Trough

3

where τ_{SA} denotes a travel time of a direct wave from $\mathbf{x}_S$ to $\mathbf{x}_A$. τ_{SyB} denotes a travel time of a specular reflection from the source position $\mathbf{x}_S$ to the receiver position $\mathbf{x}_B$ through specular reflection point y (Figure 1(a)). Superscript d or r indicates that we only consider a direct wave for $\mathbf{x}_A$ and reflected wave for $\mathbf{x}_B$. Note that the amplitudes of these wavefields are assumed to be normalized in (2). Substituting these waves into (1) and applying stationary phase approximation (e.g., [5, 16]) yields the following equation:

$$\hat{G}(\mathbf{x}_B, \mathbf{x}_A, \omega) \approx e^{i\omega\tau_{Ay_0B}} \oint_S e^{i\omega(\tau_{SyB} - \tau_{SA} - \tau_{Ay_0B})} d^2\mathbf{x}_S \tag{3}$$

$$\approx \alpha e^{i\omega\tau_{Ay_0B}},$$

where τ_{Ay_0B} denotes a travel time of the specular reflection between two receiver positions through its specular reflection point y_0 (Figure 1(b)). α denotes a coefficient for stationary phase approximation. Note that we removed $\hat{G}^*(\mathbf{x}_B, \mathbf{x}_A, \omega)$ in (1) since we only consider causal part of Green's function. The acausal part is obtained by considering a direct wave for $\mathbf{x}_B$ and reflected wave for $\mathbf{x}_A$. The integral in (3) has a stationary point at $\mathbf{x}_S = \mathbf{x}_S^*$, and the objective primary reflection is retrieved [5]. Furthermore, in this stationary point, the following relation is satisfied:

$$\tau_{S^*yB} - \tau_{S^*A} - \tau_{Ay_0B} = 0. \tag{4}$$

This relation states that, for the stationary phase source position, the two events (direct wave and reflected wave in this case) have same raypath from the source position to the receiver position (Figure 1(b)). This corresponds to the fact that the crosscorrelation processing subtracts the travel times, cancels the common raypath, and produces the traveltime of the objective reflection event between receivers.

Note that the amplitudes derived from the crosscorrelation of the nonstationary phase sources are cancelled after the summation of the homogeneously distributed sources along the enclosed surface. In the case of the localized source distribution, the cancellation is insufficient and the unwanted amplitudes remain.

2.2. Selection of Receiver Pairs by Stationary Phase Evaluation. When the physical sources are widely distributed, the stationary phase sources effectively produce the objective reflection events. On the other hand, when the source distribution is localized, only the reflection events with the stationary phase sources are retrieved. Therefore, we evaluate the crosscorrelation traces for the existence of the stationary phase source and exclude those without the stationary phase sources in the summation of the crosscorrelation traces. We assume that we can estimate the raypath propagating from the source position to the two receivers. When these two raypaths have common pathway, we define that the crosscorrelated trace using these receivers contains objective reflections.

In order to evaluate the existence of the stationary phase sources, the raypaths for the direct waves and the arbitrary multiple reflected waves are needed to be estimated. We adopted a method developed by Tamagawa et al. [17] for

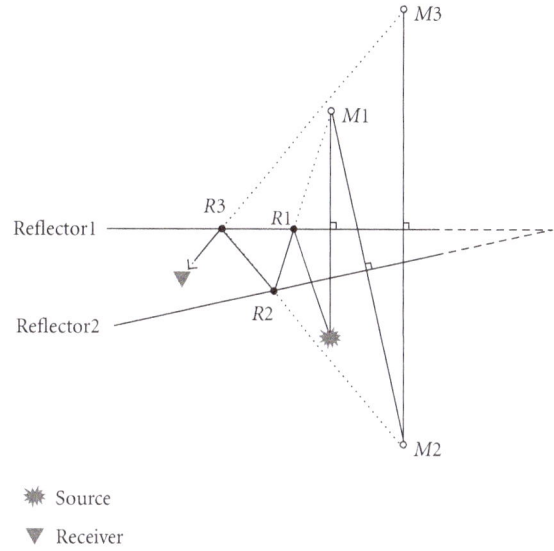

Figure 2: Raypath calculation for multiple reflections.

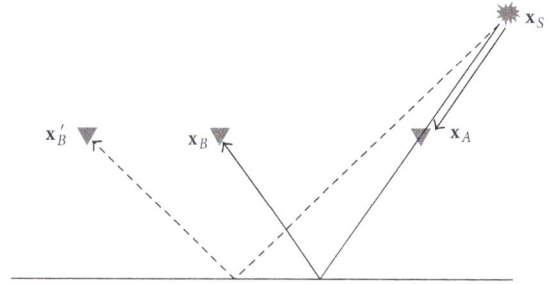

Figure 3: Source $\mathbf{x}_S$ is a stationary source for the crosscorrelation pair $\mathbf{x}_A - \mathbf{x}_B$, but not for the pair $\mathbf{x}_A - \mathbf{x}_B'$.

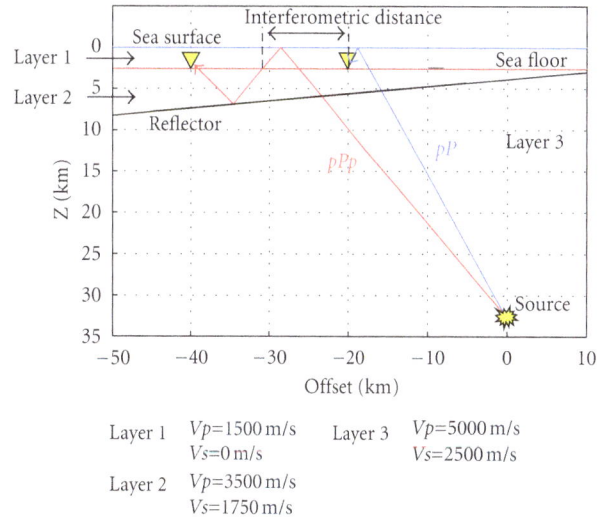

Figure 4: Velocity model for the numerical simulation. pP and pPp denote the sea-surface reflection and the sea-surface multiple reflection, respectively.

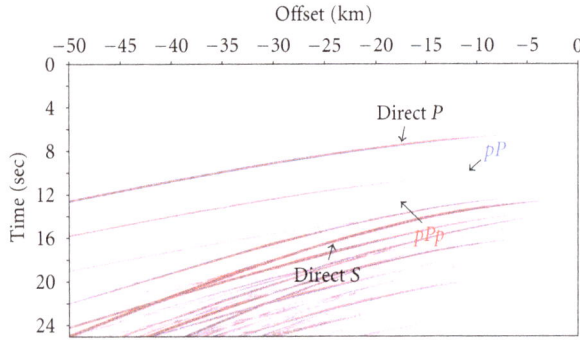

FIGURE 5: Vertical component of the modeled seismic wavefields. Arrows indicate the dominated events.

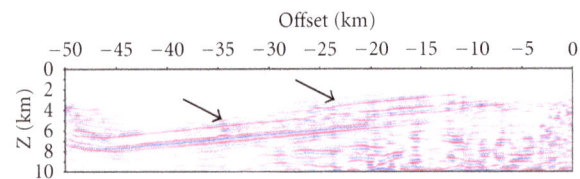

FIGURE 6: PSDM-imaging result using all crosscorrelation traces. Arrows indicate the artifact events.

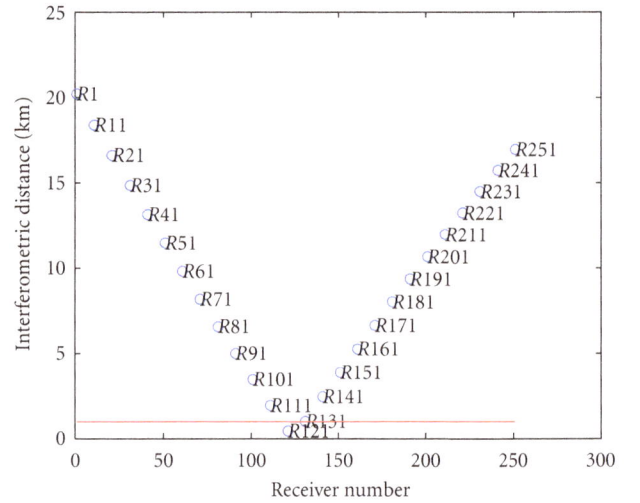

FIGURE 7: Example of the interferometric distance for different R^i_{pPp} with the fixed R_{pP}. The receiver R^i_{pPp} with the interferometric distance less than the threshold value (1 km) is defined as the stationary phase pair.

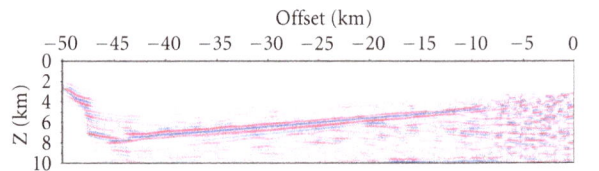

FIGURE 8: PSDM imaging result using selected crosscorrelation traces by stationary phase evaluation. This result much improves compared with Figure 6.

raypath calculation. This method geometrically calculates the raypath for arbitrary multiple reflections with given 3 dimensionally deviated structures assuming straight raypath (Figure 2). When we have two reflector planes, the method calculates the mirror point of the source position for the reflector planes ($M1 \rightarrow M2 \rightarrow M3$ in Figure 2). A mirror point is defined as a projection of the position with plane symmetry. The arbitrary multiple reflection raypath can be derived by connecting those mirror points to the receiver position ($R3 \rightarrow R2 \rightarrow R1$ in Figure 2). This method is simple and fast to calculate raypaths.

We evaluate the stationary phase using these raypaths. For example, we assume that the receiver $\mathbf{x}_A$ observes direct wave, and two different receivers $\mathbf{x}_B$ and $\mathbf{x}'_B$ observe reflected wave (Figure 3). In this model, the candidate for crosscorrelation processing is $\mathbf{x}_A - \mathbf{x}_B$ and $\mathbf{x}_A - \mathbf{x}'_B$. Stationary phase evaluation using raypath can give us the information that only the crosscorrelation of $\mathbf{x}_A - \mathbf{x}_B$ has a stationary phase source and contain the objective reflection event. By evaluating this procedure for all source position and receiver combination, we remove the crosscorrelation traces which do not have stationary sources for objective reflections. Note that we need some knowledge of the position of reflector to evaluate raypaths before imaging. However, a rough estimate of the position of reflector can be sufficient since we do not need precise information about the location of the stationary points.

3. Numerical Modeling Results

We numerically simulate the wavefield with localized source distribution in order to evaluate the processing considering stationary phase point. For the simulation study, we consider the local earthquakes recorded by Ocean Bottom Seismogram (OBS) on the seafloor. The objective reflection events are assumed to be those generated from the oceanic crust surface. We use only one physical source for the simulation study because we evaluate our proposed method for localized source distribution. The velocity model is a three-layer structure including sea water (Figure 4). The dipping structure in the velocity model simulates a oceanic crust surface (i.e., plate interface) in the Nankai Trough.

We define the two events which produce the objective reflection event for the stationary phase evaluation. In our source-receiver configuration, the source is located in the subsurface, and the sea-surface (modeled as a free-surface) expects to produce the strong downpropagating reflections. This is the difference of the OBS recorded wavefields with those by the surface receivers. Therefore, we assume that the sea-surface P-wave reflections (denoted as pP in Figure 4) and the sea-surface multiple P-wave reflections (pPp in Figure 4) produce the reflection events from the oceanic crust surface after crosscorrelation. Although other events may contribute to produce the objective reflection event (e.g., the crosscorrelation of the direct P-wave and the multiple P-wave reflections; Abe et al. [3]), we consider the crosscorrelation of these two events (pP and pPp) since they

Crosscorrelation of Earthquake Data Using Stationary Phase Evaluation: Insight into Reflection Structures of Oceanic
Crust Surface in the Nankai Trough

5

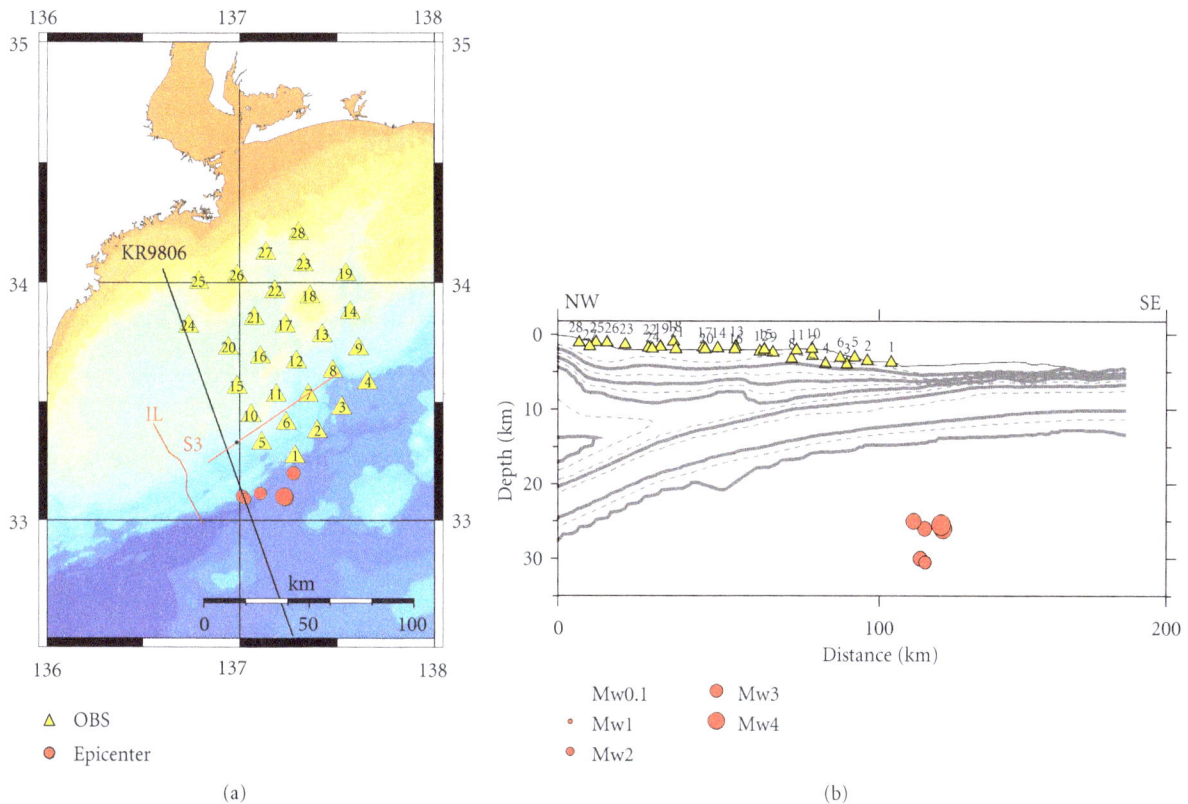

FIGURE 9: (a) Survey area in the Nankai Trough. Three seismic survey lines, KR9806, IL, and S3 indicate those from Figure 9(b) [8], Figure 11(a) [9], and Figure 13(c) [10], respectively. (b) The depth of OBS and earthquakes projected to the line KR9806. Background velocity contour is derived from refraction tomography [8]. The hypocenters associated with 6 earthquakes are localized around the trough axis.

propagate with high energy and are easy to be separated from S-waves.

We numerically modeled the seismic wavefield with a one subsurface source using the 2-dimensional staggered grid method [18]. The subsurface source is installed at the 30 km depth (Figure 4). We simulate the earthquake by giving horizontal stress as a source and using a Ricker wavelet with the central frequency of 5 Hz. The length of receiver array is 50 km with 200 m intervals.

Figure 5 shows the calculated wavefield (vertical component) at the receiver array. One can see that the sea-surface reflections (pP) and the sea-surface multiple reflections (pPp) are observed as well as the direct P-wave and direct S-wave (Figure 5). Since we focus on only P-waves in this study, we muted the amplitude of the direct S-wave and subsequent wavefields. Because we have 251 receivers, the total number of the crosscorrelation traces is $_{251}C_2$ which is the number of 2 combinations from 251 elements. Figure 6 shows the result of subsurface image derived from all crosscorrelation traces using prestack depth migration (PSDM). The target-dipping structure was appeared on the image. However, the signal-to-noise ratio at the near offset is low, and the artifact-dipping events are appeared (arrows in Figure 6). These artifacts are caused by the crosscorrelation traces which do not have a stationary phase source for objective reflections as pointed out by, for example, Snieder et al. [19].

We evaluate the stationary phase source in order to remove the unwanted crosscorrelation traces. We estimate the raypath of pP and pPp and compare their raypaths. We fix the receiver which observe pP (denoted as R_{pP}) and calculate the raypath of pPp for all receivers (R_{pPp}^i). We calculate the horizontal distance of the receiver R_{pP} and the point where pPp passes through the seafloor to a downward direction (Figure 7). We refer to the distance as an interferometric distance (Figure 4). The interferometric distance of zero indicates that the two events have the common raypath between the buried source to the receiver R_{pP}. Due to the receiver spacing, the interferometric distance is not always zero (Figure 7). Therefore, we define the threshold for the interferometric distance. The receivers R_{pPp}^i with the interferometric distance less than the threshold value (1 km in this case) are assumed to have the common raypath. This threshold corresponds to the determination of the size of the first Fresnel zone around the stationary points. Note that the interferometric distance projected from the first Fresnel zone is dependent on the position of reflector and the receiver geometry. We iterate this procedure by changing the fixed receiver R_{pP} and evaluate all combination of crosscorrelation traces.

We remove the crosscorrelation traces which do not have a stationary phase source for objective reflections and apply PSDM using 2489 traces. Figure 8 shows the result of

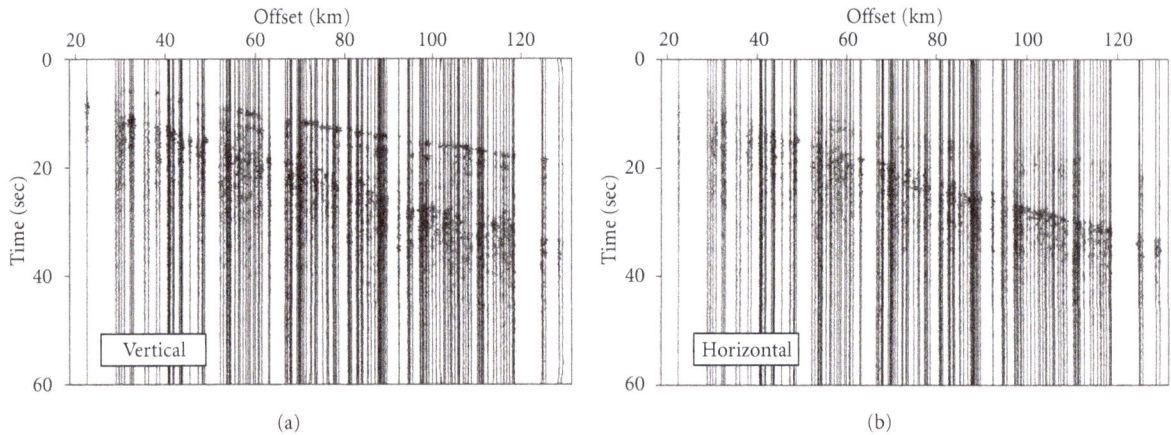

FIGURE 10: Recorded signal of the 168 traces from the 6 events aligned by the epicentral distance after correcting the origin time as the time of the earthquakes.

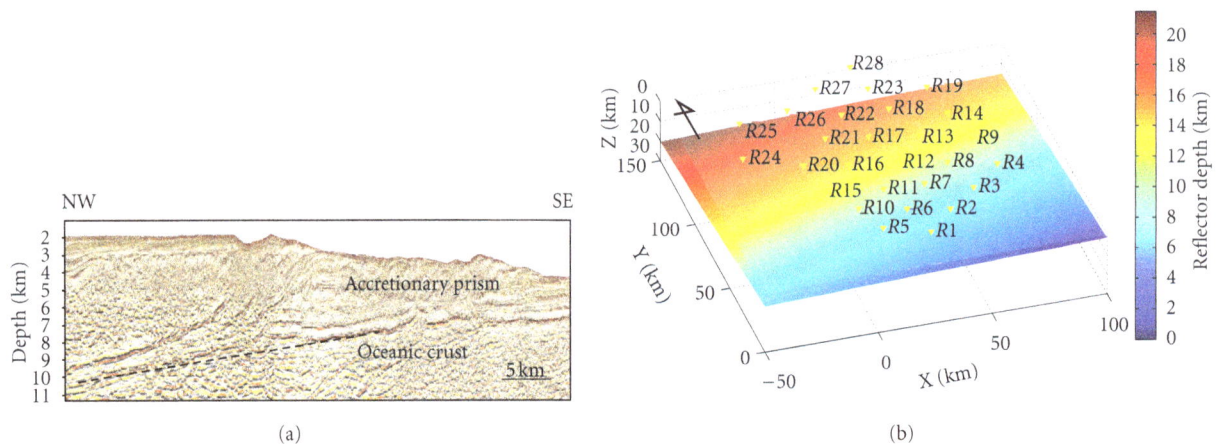

FIGURE 11: (a) 2D reflection profile across the Nankai Trough (Line IL in Figure 9(a), modified after Moore et al. [9]). Dashed line shows the angle of 5.5 degree. (b) Constructed 3D dipping structure for the raypath calculation.

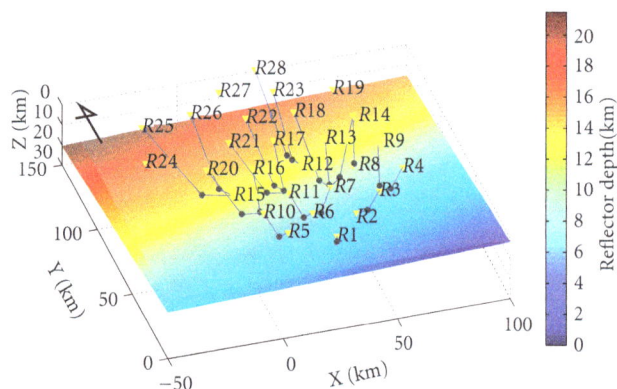

FIGURE 12: Reflection points estimated by a raypath calculation (black dots).

subsurface image considering stationary phase. One can see that the signal-to-noise ratio at the near offset improved, and the artifact events are suppressed. We conclude that evaluating stationary phase by raypath calculation and removing

unwanted crosscorrelation traces improve reflection images in SI.

4. Extracting Reflected Waves of Oceanic Crust Surface from OBS Records in the Nankai Trough

We apply this analysis to the field passive-seismic data. The field data consists of the OBS records deployed in the Nankai Trough to observe local earthquakes. This dataset was originally obtained by JAMSTEC [11, 12] for the earthquake observation as well as for aftershocks of the September 2004 intraplate earthquake ruptured around this survey area [20]. The 28 OBS are 3 dimensionally deployed at the approximately 100 km-squared area (Figure 9(a)). The recorded period was for approximately 3 months from March 2005, and we used the data during 12 days in March. We extract the reflection events generated at the oceanic crust surface using crosscorrelation. Within the 653 local earthquakes detected in this duration, we extract 6

Crosscorrelation of Earthquake Data Using Stationary Phase Evaluation: Insight into Reflection Structures of Oceanic
Crust Surface in the Nankai Trough

7

(a)

(b)

(c)

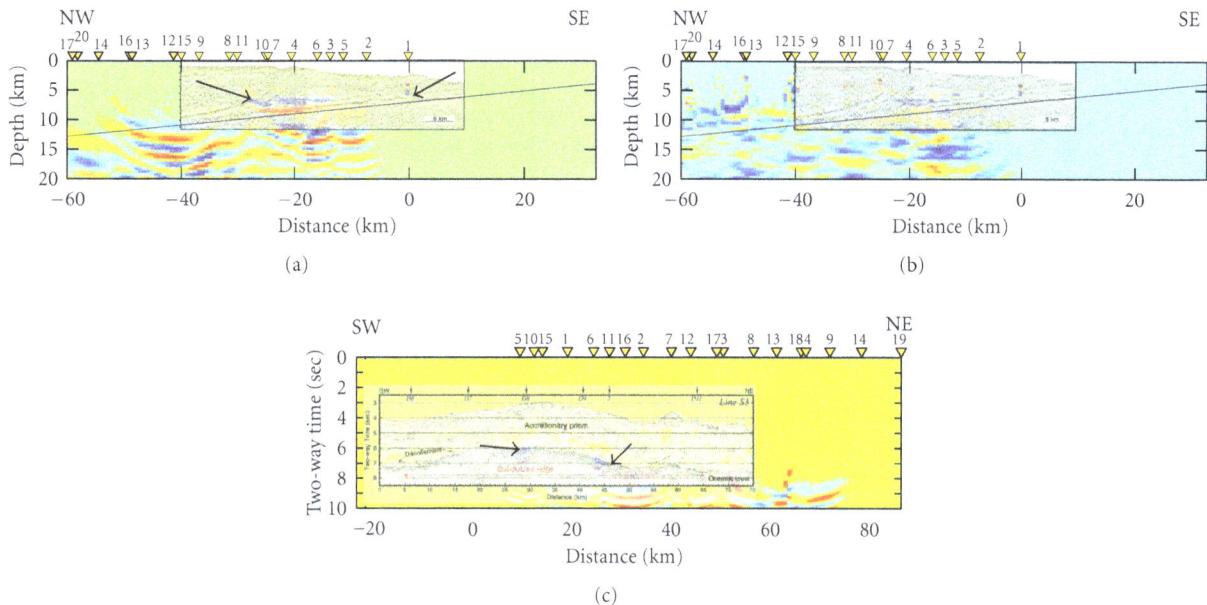

FIGURE 13: PSDM-imaging results. (a) Stacked profile projected to Line KR9806 (Figure 9(a)) running perpendicular to the trough axis.
The seismic profile from Moore et al. [9] is overwrapped in this profile. Arrows show the imaged oceanic crust surface. (b) Stacked profile
projected to Line KR9806 using all crosscorrelation traces. (c) Stacked profile projected to Line S3 (Figure 9(a)) running parallel to the
trough axis. The seismic profile from Park et al. [10] is overwrapped.

earthquakes whose magnitudes are larger than 3.0 with
>20 km in depth (Figure 9(b)) because these earthquakes
contain strong energy. We choose deep earthquakes since the
P-wave energy is dominated in the vertical component of the
records, and the arrival time of S-waves is late enough to be
separated from P-waves. These earthquakes are localized at
south-west of the survey area (Figure 9(a)).

We show the 168 traces from the 6 events aligned by the
epicentral distance and corrected origin time with the earthquake nucleation time (Figure 10). The dominant frequency
was ~5 Hz. One can see two events appearing with different
propagating velocity on the seismic record. Since the first
arrival is dominated in the vertical component (Figure 10),
we assumed them as the P-waves and the second arrivals as
the direct S-waves. We muted the amplitude of the S-waves
as well as subsequent records in order to analyze only P-wave
events.

Figure 11(a) shows the reflection profile across the
Nankai Trough (Moore et al. [9]; Line IL in Figure 9(a)).
The oceanic crust surface in this area is located from
7 km to 10 km in depth and dipping landward direction
(dashed line in Figure 11(a)). We extended this structure
(landward-dipping structure with the angle of 5.5 degree)
perpendicular to the 2D cross-line (Line IL in Figure 9(a))
and obtain the 3D dipping structure used for raypath
calculation (Figure 11(b)).

We estimate the raypaths of the sea-surface reflection
(pP) and the sea-surface multiple reflection (pPp) as in the
case of simulation study. To account for the difference of
each elevation of OBS, we define the interferometric distance

as the distance between the receiver R_{pP} and the point
where pPp passes through the horizontal plane constructed
by the receiver R_{pP} to a downward direction. We defined
the threshold of the interferometric distance as 5 km. We
removed the crosscorrelation traces without the stationary
phase sources and obtained 55 crosscorrelation traces. Since
we calculate the raypath, we can estimate the reflection
point at the given structure (Figure 12). Due to the source
localization, only a part of combination of the receivers is
selected.

We applied the 3D PSDM to the crosscorrelation traces.
We extended the tomographic velocity estimated by Nakanishi et al. [8] perpendicular to its survey line (Line KR9806
in Figure 9(a)) and used it as a velocity model for PSDM.
The expected reflection points are sparsely distributed in the
3D region (Figure 12). Therefore, we spatially stacked the 3D
imaging result and obtained the pseudo-2D profile.

We show the pseudo-2D profile projected to Line KR9806
(Figure 13(a)) running perpendicular to the trough axis. The
result shows that the oceanic crust surface is imaged at the
same depth of the active-source reflection profile of Line IL
(arrows in Figure 13(a)). For a comparison, we show the
PSDM result using all crosscorrelation traces projected to
Line KR9806 (Figure 13(b)). Since it is difficult to detect the
oceanic crust surface in the profile (Figure 13(b)), stationary
phase evaluation improves the quality of the imaging result
(Figure 13(a)).

We show the pseudo-2D profile projected to Line S3 (Figure 13(c)) running parallel to the trough axis. We convert the
depth axis of our imaging result to the time axis using the

migration velocity model. The imaged oceanic crust surface (arrows in Figure 13(c)) is discontinuous around the horizontal distance of 20 km due to the migration aperture and the sparse distribution of the reflection points (Figure 12). However, we can observe the dominated amplitudes at the same two-way time of the oceanic crust surface in Line S3 [10] as well as the local bulge of the oceanic crust surface.

5. Conclusion

We use the stationary phase interpretation to obtain high-quality imaging results in the localized source distribution. We estimate the raypath of two reflection events which are sea-surface P-wave reflection and sea-surface multiple P-wave reflection. We choose the crosscorrelation traces which is expected to produce objective reflections due to the stationary phase sources using the estimated raypath. We show the numerical modeling result to check the validity of this method. Furthermore, we use Ocean Bottom Seismogram (OBS) records which observe localized earthquakes. We show that choosing the crosscorrelation traces by stationary phase evaluation improves the quality of the imaged reflection boundary of the oceanic crust surface. This processing technique has a possibility to monitor the Nankai seismogenic fault without the active sources and higher resolution than using teleseismic records.

Acknowledgments

The OBS data were acquired by JAMSTEC. S. Minato is grateful for a support by JSPS Fellows (212666). T. Tsuji is grateful for a support by Grant-in-Aid for Scientific Research on Innovative Areas (21107003). This work is partially supported by JST/JICA, SATREPS and Kyoto University GCOE.

References

[1] C. A. Langston, "Corvallis, oregon, crustal and upper mantle receiver structure from teleseismic p and s waves," *Bulletin of the Seismological Society of America*, vol. 67, pp. 713–724, 1977.

[2] J. Shragge, B. Artman, and C. Wilson, "Teleseismic shot-profile migration," *Geophysics*, vol. 71, no. 4, pp. SI221–SI229, 2006.

[3] S. Abe, E. Kurashimo, H. Sato, N. Hirata, T. Iwasaki, and T. Kawanaka, "Interferometric seismic imaging of crustal structure using scattered teleseismic waves," *Geophysical Research Letters*, vol. 34, no. 19, Article ID L19305, 2007.

[4] K. Wapenaar and J. Fokkema, "Green's function representations for seismic interferometry," *Geophysics*, vol. 71, no. 4, pp. SI33–SI46, 2006.

[5] G. Schuster, *Seismic Interferometry*, Cambridge University Press, 2009.

[6] T. Tonegawa, K. Nishida, T. Watanabe, and K. Shiomi, "Seismic interferometry of teleseicmic S-wave coda for retrieval of body waves: an application to the Philippine Sea slab underneath the Japanese Islands," *Geophysical Journal International*, vol. 178, no. 3, pp. 1574–1586, 2009.

[7] E. Ruigrok, X. Campman, D. Draganov, and K. Wapenaar, "High-resolution lithospheric imaging with seismic interferometry," *Geophysical Journal International*, vol. 183, no. 1, pp. 339–357, 2010.

[8] A. Nakanishi, N. Takahashi, J. O. Park et al., "Crustal structure across the coseismic rupture zone of the 1944 Tonankai earthquake, the Central Nankai trough seismogenic zone," *Journal of Geophysical Research B*, vol. 107, no. 1, pp. 1–21, 2002.

[9] G. Moore, J. Park, N. Bangs et al., "Structural and seismic stratigraphic framework of the nantroseize stage 1 transect," *Proceedings of the Integrated Ocean Drilling Program*, 2009.

[10] J. O. Park, G. F. Moore, T. Tsuru, S. Kodaira, and Y. Kaneda, "A subducted oceanic ridge influencing the Nankai megathrust earthquake rupture," *Earth and Planetary Science Letters*, vol. 217, no. 1-2, pp. 77–84, 2004.

[11] K. Obana, S. Kodaira, and Y. Kaneda, "Microseismicity around rupture area of the 1944 Tonankai earthquake from ocean bottom seismograph observations," *Earth and Planetary Science Letters*, vol. 222, no. 2, pp. 561–572, 2004.

[12] K. Obana, S. Kodaira, and Y. Kaneda, "Seismicity in the incoming/subducting Philippine Sea plate off the Kii Peninsula, central Nankai trough," *Journal of Geophysical Research B*, vol. 110, no. 11, Article ID B11311, pp. 1–13, 2005.

[13] G. T. Schuster, J. Yu, J. Sheng, and J. Rickett, "Interferometric/daylight seismic imaging," *Geophysical Journal International*, vol. 157, no. 2, pp. 838–852, 2004.

[14] R. Snieder, "Extracting the Green's function from the correlation of coda waves: a derivation based on stationary phase," *Physical Review E*, vol. 69, no. 4, article 46610, 2004.

[15] J. A. Chaput and M. G. Bostock, "Seismic interferometry using non-volcanic tremor in Cascadia," *Geophysical Research Letters*, vol. 34, no. 7, Article ID L07304, 2007.

[16] M. Bath, *Mathematical Aspects of Seismology*, Elsevier, 1968.

[17] T. Tamagawa, T. Matsuoka, and A. Sakai, "Multiple reflections and ghost for tilted sea bottom," *Butsuri-Tansa*, vol. 50, pp. 477–485, 1997.

[18] J. Virieux, "P- SV wave propagation in heterogeneous media: velocity- stress finite-difference method.," *Geophysics*, vol. 51, no. 4, pp. 889–901, 1986.

[19] R. Snieder, K. Wapenaar, and K. Larner, "Spurious multiples in seismic interferometry of primaries," *Geophysics*, vol. 71, no. 4, pp. SI111–SI124, 2006.

[20] T. Tsuji, J. O. Park, G. Moore et al., "Intraoceanic thrusts in the Nankai Trough off the Kii Peninsula: Implications for intraplate earthquakes," *Geophysical Research Letters*, vol. 36, no. 6, Article ID L06303, 2009.

2

Latitudinal and Local Time Variation of Ionospheric Turbulence Parameters during the Conjugate Point Equatorial Experiment in Brazil

Charles S. Carrano, Cesar E. Valladares, and Keith M. Groves

Boston College, Chestnut Hill, MA 02467, USA

Correspondence should be addressed to Charles S. Carrano, charles.carrano@bc.edu

Academic Editor: Y. Sahai

Previous authors have reported on the morphology of GPS scintillations and irregularity zonal drift during the 2002 Conjugate Point Equatorial Experiment (COPEX) in Brazil. In this paper, we characterize the turbulent ionospheric medium that produced these scintillations. Using 10 Hz GPS carrier-to-noise measurements at Boa Vista (2.9°N, 60.7°W), Alta Floresta (9.9°S, 56.1°W), and Campo Grande (20.5°S, 54.7°W), we report on the variation of turbulent intensity, phase spectral index, and irregularity zonal drift as a function of latitude and local time for the evening of 1-2 November 2002. The method of analysis is new and, unlike analytical theories of scintillation based on the Born or Rytov approximations, it is valid when the scintillation index saturates due to multiple-scatter effects. Our principal findings are that (1) the strength of turbulence tended to be largest near the crests of the equatorial anomaly and at early postsunset local times, (2) the turbulent intensity was generally stronger and lasted two hours longer at Campo Grande than at Boa Vista, (3) the phase spectral index was similar at the three stations but increased from 2.5 to 4.5 with local time, and (4) our estimates of zonal irregularity drift are consistent with those provided by the spaced-receiver technique.

1. Introduction

The distribution of free electrons in the ionosphere is dictated by production from solar radiation, transport, and loss through chemical recombination. It is also subject to instability mechanisms that generate large-scale depletions and irregularities in the ambient electron density over a wide range of spatial scales (plasma turbulence). Radio waves that propagate through these irregularities experience scattering and diffraction, causing random fluctuations in amplitude and phase referred to as scintillations. The scintillation of satellite signals has been shown to severely degrade the performance of satellite communications systems such as AFSATCOM [1, 2], satellite global navigation satellite systems (GNSS) such as the Global Positioning System (GPS) [3–5], and space radars used to conduct cloud-free, day-and-night observations of the Earth's surface [6–8]. Scintillations associated with irregularities in the equatorial ionosphere are generally the most intense encountered worldwide, and the occurrence morphology depends on local time, season, longitude, solar cycle, magnetic activity, and exhibits a high degree of night-to-night variability [9]. Ionospheric irregularities and scintillations constitute one of the most important space weather threats to modern technological systems which increasingly rely on transionospheric radio propagation. While a multitude of plasma instabilities may operate in the equatorial ionosphere after sunset, interchange instabilities which generate large-scale depletions in the ambient electron density, commonly referred to as equatorial plasma bubbles (EPB), are believed to be the dominant source of irregularities that cause scintillation of the GPS satellite signals at L band frequencies [2–5].

The Conjugate Point Equatorial Experiment (COPEX) was conducted in Brazil from October to December 2002. Its purpose was to explore the occurrence morphology of scintillations and the physical processes by which plasma instabilities occur in the equatorial ionosphere. In this experiment, a multitude of ionospheric monitoring instruments, including ionosondes, optical imagers, VHF receivers, and GPS receivers, were deployed to three sites nearly along the same magnetic meridian, one at the magnetic equator and the other two at magnetically conjugate points.

Several authors have reported on the morphology of GPS scintillations and irregularity zonal drift during this experiment, for example, Batista et al. [10], Muella et al. [11], Abdu et al. [12], Sobral et al. [13], and de Paula et al. [14]. In this paper, we characterize the turbulent ionospheric medium that produced these scintillations. For this purpose, we developed a new technique called Iterative Parameter Estimation (IPE) for inferring ionospheric turbulence parameters from a time series of scintillating intensity measurements resulting from radio propagation through the low latitude ionosphere. Unlike analytical theories of scintillation based on the Born or Rytov approximations, the IPE technique is valid for strong scatter when the scintillation index (S_4) saturates due to multiple-scatter effects. This generality is crucial for proper analysis of GPS scintillations during the COPEX campaign, since a large percentage of these observations were in the strong scatter regime, with S_4 frequently approaching or exceeding unity [11]. Weak scatter theory cannot be applied satisfactorily to analyze these observations. It is also beneficial that the IPE technique requires only the intensity fluctuations to infer the turbulence parameters. While the GPS phase measurements from the COPEX campaign are also available, their analysis is complicated by frequent cycle slips and loss of phase lock events which occur when the scintillation is strong. IPE employs the phase screen approximation [15–18] and a numerical inversion technique to estimate the parameters of the screen which are consistent with the ionospheric turbulence model employed and the scintillations observed. Due to the analytic intractability of the 4th moment equation governing fluctuations in the field we must determine the optimal parameters by numerical iteration, and therefore we cannot guarantee the uniqueness of this parametrization.

The purpose of this paper is to explore the latitudinal and local time variation of ionospheric turbulence characteristics during the COPEX campaign in Brazil. The IPE technique was developed to investigate these variations using ground-based observations rather than *in situ* measurements of the ionospheric turbulence, which are limited in several respects. First, the temporal coverage of *in situ* measurements within any geographic region is sparse, whereas ground-based observations of scintillation are continuous. Second, *in situ* observations of the density are usually made in the topside ionosphere, whereas the scattering of L band signals is generally believed to be concentrated near the F region peak where the density is highest. It is not obvious how to infer the level of density fluctuations near the F region peak from *in situ* measurements made in the topside (and it is clearly not possible to do so when the irregularities have not risen to the orbital altitude of the satellite making the *in situ* observations). An advantage of using ground-based GPS scintillation observations to characterize the morphology of ionospheric turbulence is that irregularities at all ionospheric altitudes are sampled by the satellite signals.

The organization of this paper is as follows. Section 2 describes the equipment and the measurements used in our analysis. Section 3 presents the methodology used to analyze the data. The exposition of the methodology is given in three parts. First, we describe a model for the correlation of phase fluctuations after the GPS satellite signals have penetrated the ionosphere. Second, we describe a technique that uses this correlation function to model the spectrum of intensity fluctuations on the ground. Third, we describe an iterative technique used to infer the turbulence parameters by fitting this model spectrum to the spectrum of measured intensity fluctuations. In Section 4, we present the results of our analysis and describe their significance. A summary of these results is provided in Section 5.

2. Experimental Techniques

During the COPEX experiment in 2002, Air Force Research Laboratory operated Ashtech μZ-CGRS model dual frequency GPS receivers at three locations in Brazil: Boa Vista (2.9°N, 60.7°W), Alta Floresta (9.9°S, 56.1°W), and Campo Grande (20.5°S, 54.7°W). Boa Vista is located near the northern crest of the equatorial anomaly, Alta Floresta is located near the magnetic equator, and Campo Grande is located near the southern crest of the equatorial anomaly. Figure 1 shows the location of the three stations and the local magnetic field geometry, while Table 1 provides their geographic and geomagnetic coordinates. All three stations lie on approximately the same magnetic meridian. The three AFRL GPS receivers recorded the carrier-to-noise ratio (C/No) of the C/A code on the L1 frequency of 1575.42 MHz at a rate of 10 samples per second. The data was collected in October through December of 2002, a period of high solar activity. Additional details of the COPEX experiment are summarized in [10–14] and the references therein. For validation purposes, we compare our results with the zonal irregularity drift measurements calculated by Muella et al. [11] using the GPS spaced-receiver technique, and also the 4-channel VHF receivers which were colocated with each of the GPS receivers. The spaced VHF receiver measurements of the zonal irregularity drift provided to us for this study were obtained by simple cross-correlation of the intensity fluctuations obtained along geostationary links from two receivers separated in the magnetic east-west direction. The geometrical corrections described in [19], which account for the look angles to the geostationary satellites and the magnetic dip angles at the penetration point locations, were not applied. The GPS spaced receiver measurements provided by Muella et al. [11] did account for these geometrical effects. Unfortunately, neither the VHF nor GPS spaced-receiver measurements of the zonal drift were provided to us with error bars. Table 2 gives the locations of the 350 km ionospheric penetration points and look angles for the VHF west links (receiver channels 1-2) and east links (receiver channels 3-4) for the three COPEX stations.

In this paper, we consider measurements collected during the evening of 1-2 November 2002, since during this

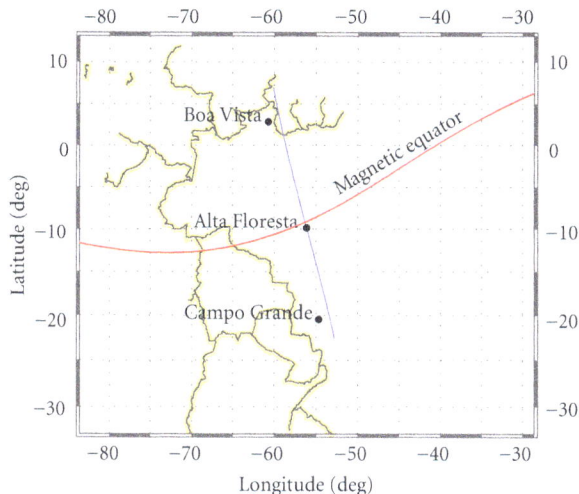

Figure 1: Locations of the three GPS receivers operated by AFRL during the COPEX campaign in Brazil. The magnetic equator is shown in red, and a magnetic meridian passing through Alta Floresta is shown in blue.

Table 1: Geographic coordinates for the three COPEX stations equipped with Ashtech μZ-CGRS receivers.

Station	Geographic		Geomagnetic		
	Lat.	Lon.	Dip Angle	Inclination	Declination
Boa Vista	2.8°N	60.7°W	22.1°N	22.1°	−13.0°
Alta Floresta	9.9°S	56.1°W	3.38°S	−3.2°	−14.6°
Campo Grande	20.5°S	54.7°W	22.3°S	−21.8°	−13.8°

Table 2: Locations of the 350 km ionospheric penetration points and lookangles for the west and east VHF links from the three COPEX stations.

Station	Channels 1-2 (West)			Channels 3-4 (East)		
	Lat.	Lon.	Elevation	Lat.	Lon.	Elevation
Boa Vista	2.7°N	63.7°W	44°	2.7°N	57.9°W	46°
Alta Floresta	9.2°S	59.7°W	38°	9.2°S	53.7°W	49°
Campo Grande	19.0°S	58.9°W	34°	19.0°S	52.1°W	47°

period all three AFRL GPS receivers were operating, all three VHF spaced receivers were operating, and the spaced GPS receiver estimates of the zonal drift velocity were available for validation. This evening was typical of others during the COPEX experiment, with moderate magnetic activity (Kp ranged from 2° to 5⁻) and the 10.7 cm solar flux was 160 W/m²/Hz. Unfortunately, there were only a few other evenings during the COPEX campaign when all of these instruments were operating simultaneously, and most of these evenings were more magnetically active than the evening of 1-2 November 2002. Following [14], we restricted our analysis to evenings with low-to-moderate magnetic activity, since it is easier to compare different techniques for

estimating the zonal irregularity drift when the actual drift is regular and changing slowly in the absence of storm-time perturbation electric fields.

For our analysis, we use the receiver reported C/No as a proxy for the signal intensity. From the time series of raw intensity fluctuations, we selected data segments which are approximately stationary and for which the scintillation intensity index, calculated as the standard deviation of normalized (by the mean) signal intensity, exceeded 0.3. This threshold criterion was applied to minimize the contribution of receiver noise to the scintillation statistics. A satellite elevation cutoff of 30° was used to avoid multipath and to reduce possibility that the radio wave may have propagated through multiple plasma bubbles with different turbulence characteristics. The stationary requirement is enforced to ensure the time series admits a spectral representation. We chose to use 4-minute data segments for the analysis, rather than 1-minute data segments which have become rather customary in scintillation studies, for example, [3]. The reason for this choice is that IPE analysis provides the most accurate results when all frequencies that contribute to the power spectrum of intensity fluctuations are resolved, including frequencies somewhat smaller than the Fresnel scale. Resolving smaller frequencies requires the analysis of a longer data segment. Over these relatively longer segment durations, however, the stationary requirement becomes more important to enforce. To confirm the (approximate) stationarity of the data, we calculated the S_4 index on a 1-minute cadence and also on a 4-minute cadence. Stationary segments were identified as segments for which 4 consecutive 1-minute S_4 values differed from the corresponding 4-minute S_4 by less than 0.1. The mean intensity was observed to vary slowly over a 4-minute timescale for satellites viewed at elevation angles above the 30° cutoff. The time to 50% decorrelation of intensity (τ_m) was calculated from the segmented time series using the FFT technique. We computed the temporal spectrum of intensity fluctuations $I_m(f)$ from the measured time series using Welch's method [20], whereby each 4 minute data segment is subdivided into 1-minute subsegments. We apply a Hamming window to each of these 1-minute subsegments prior to FFT analysis and then average these spectra with 50% overlap. The averaging is performed to minimize noise in the spectra due to spectral leakage [20]. The resulting power spectral densities range from the minimum resolvable frequency of 1/60 sec = 0.0167 Hz to the Nyquist frequency of 10 Hz/2 = 5 Hz.

3. Method of Analysis

3.1. Ionospheric Turbulence Model. Following the development by Rino [16], a continuously displaced coordinate system is chosen in which the measurement plane follows the principal propagation direction. The propagation geometry is shown in Figure 2. The coordinate system is defined with origin at the ionospheric penetration point (IPP), (x_p, y_p, and z_p), which is located at the center of the ionospheric layer. At each point along the propagation path the x, y, and z axes point toward geomagnetic north, geomagnetic

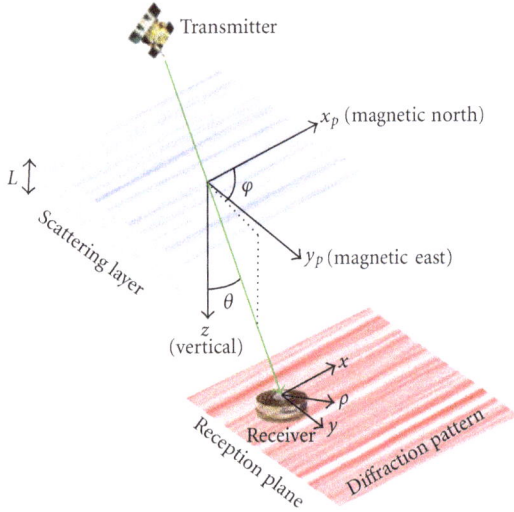

FIGURE 2: Geometry of propagation through a layer of field-aligned irregularities in the equatorial ionosphere. The coordinate system is defined with the origin at the ionospheric penetration point which is located at the center of the scattering layer. The continuously displaced coordinate system follows the principal propagation direction. At each point along the propagation path the x, y, and z axes point toward geomagnetic north, geomagnetic east, and downward, respectively.

east, and downward, respectively. The thickness of the ionospheric layer, assumed to contain homogenously distributed irregularities, is L. The angle θ is the propagation (nadir) angle at the IPP, and φ is the magnetic meridian angle of the propagation vector.

As in [16], the 3D spatial spectrum of electron density fluctuations (δN) is modeled as a power law with an outer scale as follows:

$$\Phi_{\delta N}(q) = \frac{C_s}{\left(q_0^2 + q^2\right)^{(\nu + 1/2)}}. \qquad (1)$$

In (1), C_s is the strength of turbulence, ν is the irregularity spectral index, and q is the isotropic (radial) wavenumber. The outer scale wavenumber, q_0, is related to the outer scale of the turbulence, L_0, as $q_0 = 2\pi/L_0$. In this model, $\nu > 1$ is required for the variance of electron density fluctuations to remain finite. For irregularity scales such that $q \gg q_0$ this spectrum reduces to the pure power law form $\Phi_{\delta N}(q) = C_s q^{-(2\nu+1)}$. In the equatorial region, electron density irregularities are highly elongated and aligned with the geomagnetic field. To account for irregularity anisotropy, Rino [16] employed a scaling and rotation of the coordinate system; this transformation is described in considerably more detail in [18]. Application of this transformation to (1), followed by straight-line integration through the ionospheric layer along the line of sight, followed by Fourier transformation, gives an expression for the correlation function of phase fluctuations just beneath the screen:

$$R_{\delta\phi}(\xi) = r_e^2 \lambda^2 \sec\theta \, G C_s L \left| \frac{\xi}{2q_0} \right|^{\nu - 1/2} \frac{K_{\nu-1/2}(q_0\xi)}{2\pi\Gamma(\nu+1/2)}. \qquad (2)$$

In (2), r_e is the classical electron radius, λ is the radio wavelength, ξ is the separation distance in the transverse plane, $K_{\nu-1/2}$ is the modified Bessel function of order $\nu - 1/2$, and Γ is Euler's gamma function. Equation (2) above was first derived in [16] and appears in that paper as (11). G is a geometric enhancement factor which is given by

$$G = \frac{ab}{\sqrt{AC - B^2/4} \cos\theta}. \qquad (3)$$

Equation (3) also appeared in [16] but with a typographical error; the factor $\cos\theta$ should be outside the radical as shown in (3). The scaling factors a and b in (3) elongate contours of constant phase correlation along and transverse to the magnetic field, respectively. The coefficients A, B, and C depend on the direction of propagation and the orientation of the irregularity axes [16]:

$$A = C_{11} + C_{33}\tan^2\theta\cos^2\varphi - 2C_{13}\tan\theta\cos\varphi,$$
$$B = 2[C_{12} + C_{33}\tan^2\theta\sin\varphi\cos\varphi$$
$$\qquad - \tan\theta(C_{13}\sin\varphi + C_{23}\cos\varphi)], \qquad (4)$$
$$C = C_{22} + C_{33}\tan^2\theta\sin^2\varphi - 2C_{23}\tan\theta\sin\varphi,$$

where

$$C_{11} = a^2\cos^2\psi + \sin^2\psi\left(b^2\sin^2\delta + \cos^2\delta\right),$$
$$C_{22} = b^2\cos^2\delta + \sin^2\delta,$$
$$C_{33} = a^2\sin^2\psi + \cos^2\psi(b^2\sin^2\delta + \cos^2\delta),$$
$$C_{12} = (b^2 - 1)\sin\psi\sin\delta\cos\delta, \qquad (5)$$
$$C_{13} = (a^2 - b^2\sin^2\delta - \cos^2\delta)\sin\psi\cos\psi,$$
$$C_{23} = -(b^2 - 1)\cos\psi\sin\delta\cos\delta.$$

In (5), ψ is the magnetic inclination angle and δ is the angle at which irregularities are inclined from the xz plane (which we take to be zero). We use the International Geomagnetic Reference Field (IGRF) 2000 [21] to compute these parameters at the location of the ionospheric penetration points, using an assumed ionospheric shell height of 350 km.

It is convenient, at this point, to introduce the vertically integrated strength of turbulence at the 1 km scale [22]:

$$C_k L = \left(\frac{1000}{2\pi}\right)^{2\nu+1} C_s L. \qquad (6)$$

It is common to evaluate the turbulence strength in this form since the layer thickness (L) and turbulent intensity appear together as a product and need not be known independently [5, 22–24]. Characterizing the strength of scatter using $C_k L$, rather than S_4, is advantageous because the former is a property of the random medium alone, whereas the latter depends on the random medium, propagation geometry, and frequency of the radio wave.

Up to this point, the ionospheric turbulence model involves purely spatial quantities, whereas a time series of

Latitudinal and Local Time Variation of Ionospheric Turbulence Parameters during the Conjugate Point Equatorial
Experiment in Brazil

13

field fluctuations is measured by the receiver during an experiment. Assuming the random medium is invariant over the measurement interval (this is the Taylor hypothesis of frozen-in flow), spatial fluctuations and temporal fluctuations can be related by a model-dependent effective scan velocity. The effective scan velocity consistent with the Rino spectral model is [16, 18]

$$v_{\text{eff}} = \left[\frac{CV_{sx}^2 - BV_{sx}V_{sy} + AV_{sy}^2}{AC - B^2/4} \right]^{1/2}. \tag{7}$$

Assuming the irregularity drift is purely in the zonal direction, the scan velocity in the continuously displaced coordinate system is given by

$$
\begin{aligned}
V_{sx} &= -\left(V_{px} - \tan(\theta)\cos(\varphi)V_{pz} \right), \\
V_{sy} &= V_D - \left(V_{py} - \tan(\theta)\sin(\varphi)V_{pz} \right),
\end{aligned}
\tag{8}
$$

where V_D is the zonal irregularity drift velocity, and V_{px}, V_{py}, V_{pz} are the velocity components of the ionospheric penetration point (IPP) in the magnetic north, magnetic east, and down directions, respectively. In (8), the terms involving $\tan\theta$ account for the horizontal translation of the continuously displaced coordinate system as it follows the propagation ray path. We note that the effective scan velocity has a particularly simple interpretation for the special case of normal propagation through infinitely long irregularities; in this case v_{eff} is equal to the zonal irregularity drift velocity minus the zonal component of the IPP velocity. The general case of oblique propagation and finite irregularity axial ratio is used for the calculations in this paper. Following the approach implemented by the Wideband Scintillation Model (WBMOD) [22] and also [25], the component of IPP velocity parallel to the line of sight has been removed prior to the calculation (8) in order to reduce spectral smearing. We compute the GPS satellite locations and velocities using the SGP4/SDP4 satellite propagator from Spacetrack [26].

In the next section, we use the model for the correlation function of phase fluctuations given in (2) and phase screen theory to compute the model spectrum of intensity fluctuations in the reception plane.

3.2. Calculation of the Model Intensity Spectrum. A widely employed simplifying assumption in transionospheric radio propagation problems is the phase screen approximation. In the phase screen approximation, the interaction of the wave with an extended irregularity layer is replaced by that with an equivalent thin phase-changing screen [15–18, 27]. This approximation neglects diffraction effects that develop within the irregularity layer, which can be considerable when the scattering is strong. Nevertheless, several authors (e.g., see [27] and the references therein) have demonstrated that an equivalent phase screen accurately reproduces the amplitude and phase scintillation predicted by a more complex formulation that accommodates diffraction within the scattering layer provided that the height of the phase screen is appropriately chosen. In this paper, we employ the phase screen approximation so that the ionospheric turbulence

may be characterized using a small number of parameters (i.e., those parameters which specify the equivalent phase screen).

When a plane wave is incident on a thin phase-changing screen, the spectrum of intensity fluctuations at the reception plane, at some vertical distance z beyond the screen, may be calculated using the algorithm proposed by Booker and MajidiAhi [28]. This algorithm provides an intensity spectrum that satisfies the differential equation governing the 4th moment of the field in the thin screen approximation (e.g., see Ishimaru [15, Section 20–14]). The Booker and MajidiAhi approach is not limited to weak scatter conditions, as are the analytical results relating the screen parameters to S_4 given in [16]. These analytical results are implemented in the Wideband Model [22, 29], which applies an empirical correction for strong scatter based on the questionable assumption of Rician statistics. By solving the 4th moment equation directly there is no need to assume Rician statistics when the scatter is strong. While the 4th moment equation has been solved in various other forms (e.g., [30–32]), Booker et al. [27] explains that the Booker and MajidiAhi formulation [28] provides the most accurate results for the same level of computational effort. In this section, we adapt the Booker and MajidiAhi approach so that it may be used with Rino's model for the correlation function of phase fluctuations (2).

We begin by defining the Fresnel parameter, F, in terms of the signal wavelength and the slant propagation distance, $z \sec\theta$:

$$F = \left(\frac{\lambda z \sec\theta}{2\pi} \right)^{1/2}. \tag{9}$$

Next, we form two functions that depend on the separation distance in the transverse plane, ξ, and the spatial wavenumber, k:

$$
\begin{aligned}
f(\xi, k) &= 2R_{\delta\phi}(\xi) - R_{\delta\phi}(\xi - kF^2) - R_{\delta\phi}(\xi + kF^2), \\
g(\xi, k) &= \exp\{-[f(0, k) - f(\xi, k)]\} - \exp\{-f(0, k)\}.
\end{aligned}
\tag{10}
$$

Finally, the spectrum of intensity fluctuations is given by

$$I(k) = 2 \int_0^\infty g(\xi, k)\cos(k\xi)d\xi. \tag{11}$$

The factor of 2 appears in (11), instead of 4 as in the Booker and MajidiAhi paper [28], in order to change from a one-sided spectrum to a two-sided spectrum for consistency with Rino [16]. The integral in (11) is oscillatory and challenging to evaluate. This is especially true when the scatter is weak and Fresnel ringing is present in the spectrum. We use an adaptive bisection algorithm from QUADPACK [33] to evaluate (11). Once the intensity spectrum has been computed, the scintillation index may be calculated by integrating the spectrum over all wavenumbers:

$$S_4^2 = \frac{1}{2\pi} \int_0^\infty I(k)dk. \tag{12}$$

Once again, we have a formulation that involves spatial quantities, while a time series of intensity fluctuations are measured during an experiment. It can be shown by a change of variables that when temporal frequencies (f) in the time series of intensity fluctuations and spatial wavenumbers (k) are related as

$$f = v_{\text{eff}}\left(\frac{k}{2\pi}\right), \tag{13}$$

then temporal power spectral densities $I(f)$ and spatial power spectral densities $I(k)$ are related as

$$I(f) = \frac{I(k)}{v_{\text{eff}}}. \tag{14}$$

Once the temporal intensity spectrum is known, the decorrelation time (τ) corresponding to the model spectrum can be evaluated by Fourier transforming $I(f)$ to obtain the intensity correlation function and then determining the time lag to 50% decorrelation.

3.3. Iterative Parameter Estimation. The preceding sections have described a technique to compute the temporal spectrum of intensity fluctuations corresponding to a given ionospheric screen. The free parameters of the model are the altitude of the phase screen (H_p), the anisotropy parameters (a, and b), the outer scale (L_0), the turbulence strength (C_kL), irregularity spectral index (ν), and the zonal irregularity drift velocity (V_D). All the other parameters in the model can be computed from the signal frequency and propagation geometry. In this paper we assume $a = 50$, $b = 1$, and $L_0 = 10$ km. The results of a sensitivity study revealed that the intensity spectrum so computed is insensitive to the value of a once the ratio a/b exceeds approximately 20. In general, we cannot unambiguously measure L_0 from the GPS observations alone, because L_0 is much larger than the Fresnel break scale. Scales larger than the Fresnel break scale are naturally filtered out by diffraction and therefore contribute little to the intensity fluctuations on the ground (except in the case of strong focusing). In a second sensitivity study it was determined that the phase spectral index, $p = 2\nu$, as determined using the IPE technique, is insensitive to the value chosen for L_0 so long as the latter is much larger than the Fresnel scale. The effective scan velocity, v_{eff}, and therefore also our estimates of the zonal irregularity drift velocity (V_D), depend on the assumed altitude of the phase screen. We assume the fixed value $H_p = 350$ km for the calculations in this paper. With these assumptions, the remaining independent parameters are C_kL, p, and V_D; these parameters specify the ionospheric screen and will be estimated from the scintillation observations by fitting the model intensity spectrum to the observed intensity spectrum as described next.

We define a metric to measure the difference between the modeled intensity spectrum (I_m) and measured intensity spectrum (I) in the log-log domain:

$$\chi^2 = \frac{2}{f_{\max} - f_{\min}} \int_{f_{\min}}^{f_{\max}} [\log I(f) - \log I_m(f)]^2 d\log(f). \tag{15}$$

Given an initial guess for the screen parameters C_kL, p, and V_D, we determine the optimal values of these parameters which provide the best fit of the model intensity spectrum to the measured intensity spectrum by minimizing χ^2. We use the downhill simplex method [34] to perform the multidimensional minimization. The factor 2 in (15) is to account for both positive and negative frequencies, and the frequency difference in the denominator is a normalization factor. The integration over frequency spans from $f_{\min}$ to $f_{\max}$. We chose $f_{\min}$ to be the smallest nonzero frequency after averaging the spectrum according to Welch's method [20], which corresponds to 0.0167 Hz as explained in Section 2. The frequency $f_{\max}$ is chosen as the maximum useable frequency before the noise floor in the PSD is reached. We estimate this frequency from the measured intensity spectrum by identifying frequency samples for which $I_m(f) < -35$ dB, and then taking $f_{\max}$ as the frequency that separates the first 5% of these samples from the remaining 95%. The -35 dB cutoff is receiver specific and was determined by numerical experimentation.

We note that IPE analysis, as described above, is a computationally intensive procedure. The calculations shown in this paper required more than three days (wall clock time) to produce on a Pentium-class Quad Core workstation. A faster approach is possible, however, where the forward propagation calculation described in Section 3.2 is replaced by an approximate calculation, as follows. First, numerical realizations of the screen phase are generated which are characterized by the autocorrelation function given in (2). Next, a plane wave is propagated through the screen and through free space down to the ground (using, for example, the phase screen technique described in [8, 35] or [24]). Finally, this process is repeated many times and the results ensemble averaged to produce a smooth intensity spectrum, which is needed for least-squares fitting the measured intensity spectrum.

4. Results and Discussion

4.1. Example Application of the IPE Technique. The methodology described in Section 3 is a technique for inferring the parameters of the ionospheric screen from a time series of intensity fluctuations due to transionospheric propagation in the equatorial ionosphere. Figure 3 shows two example applications of the IPE technique using GPS measurements collected at Campo Grande on 1-2 November 2002. The time series of intensity fluctuations in Figure 3(a) correspond to a case of relatively weak scatter measured late in the evening (26:00 UT). The C/No fluctuates a few dB about the background level and the measured scintillation intensity index, $S_{4m} = 0.34$, is relatively small. The rate of intensity fluctuations, quantified by the decorrelation time, $\tau_m = 1.2$ sec, is moderate. Accounting for the propagation geometry, magnetic field configuration, and velocity of the GPS satellites at the time the measurement was taken, as described in Section 3, the IPE technique gives the ionospheric screen parameters as $C_kL = 1.63 \times 10^{35}$, $p = 3.26$, and $V_D = 113.8$ m/s. The corresponding effective scan velocity in this case was $v_{\text{eff}} = 69$ m/s. The χ^2 for the spectral fit was

Latitudinal and Local Time Variation of Ionospheric Turbulence Parameters during the Conjugate Point Equatorial Experiment in Brazil

15

PRN 08

S_{4m} : 0.34

τ_m : 1.2

C/N_0 (dB-Hz)

Time (s)

(a)

f_o f_b $1/\tau$

PSD intensity (dB)

$C_k L$: 1.63×10^{35}
$p = 3.26$
V_D: 113.8
χ^2: 0.052

Frequency

(b)

PRN 02

S_{4m} : 0.93

τ_m : 0.46

C/N_0 (dB-Hz)

Time (s)

(c)

f_o f_b $1/\tau$

PSD intensity (dB)

$C_k L$: 5.8×10^{35}
$p = 3.39$
V_D: 147.4
χ^2: 0.026

Frequency

(d)

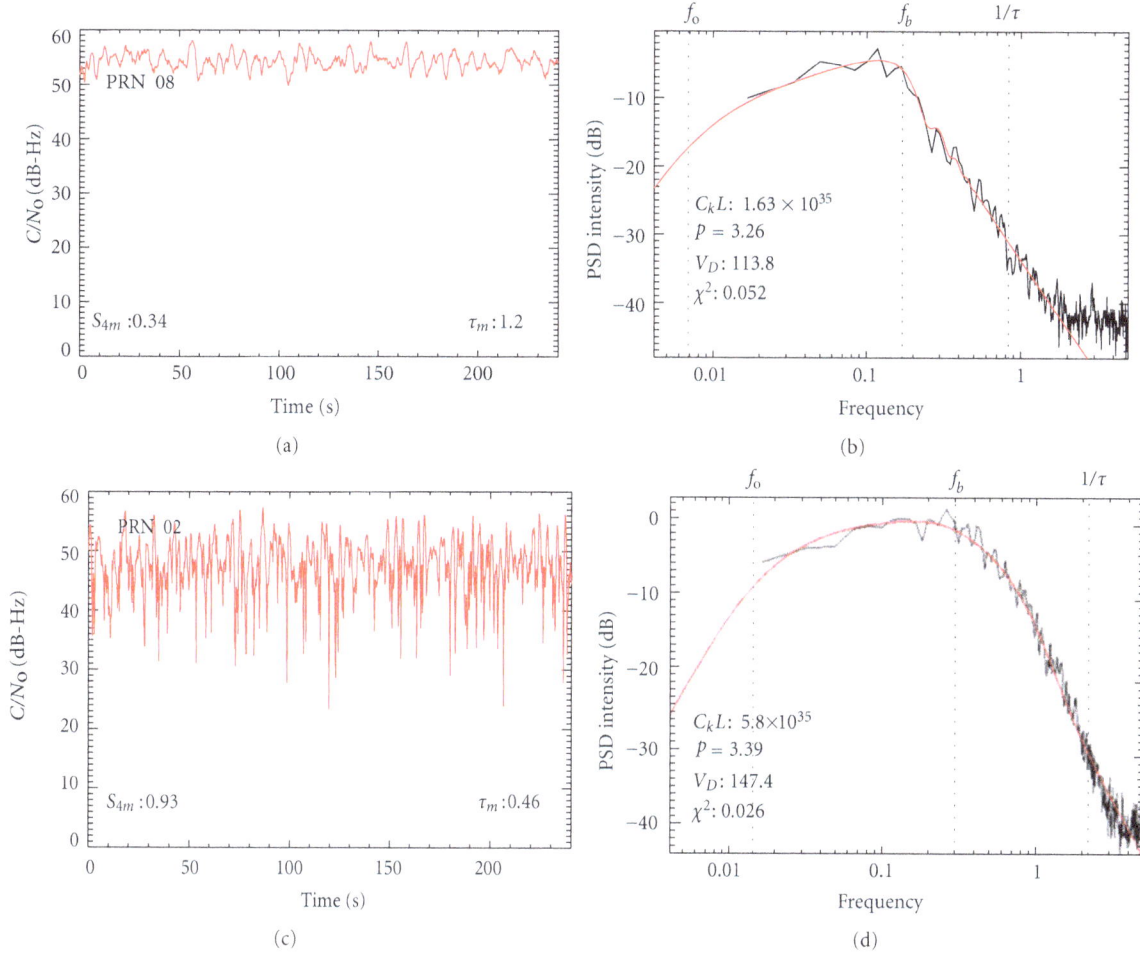

FIGURE 3: Time series of intensity fluctuations at Campo Grande during weak scatter (a) and during strong scatter (c). The corresponding temporal power spectral densities (PSD) are shown in (b) and (d), respectively. The PSD for the data are shown in black, while the PSD from the model are shown in red. The frequencies corresponding to the outer scale (f_0), break scale according to weak scatter theory (f_b), and inverse decorrelation time ($1/\tau$) are shown. The shaded regions indicate the portions of the spectra ($f_{\min} \leq f \leq f_{\max}$) over which the least-squares fits to the spectra are calculated.

0.052. Both the model intensity spectrum (Figure 3(b) red curve) and the measured spectrum (Figure 3(b) black curve) show evidence of Fresnel rings, which occur in weak scatter when the ionospheric scattering layer is relatively thin. Since the scatter is weak and the phase spectral index is relatively shallow, frequencies smaller than the Fresnel break frequency, $f_b = v_{eff}/(2z \sec\theta)^{1/2}$, are suppressed by diffraction during free-space propagation [28]. For frequencies larger than f_b, the intensity spectrum follows a power law with slope equal to the spectral index of the screen. The time series of intensity fluctuations in Figure 3(c) correspond to a case of strong scatter encountered earlier in the evening (23:34 UT). In this case, the C/No undergoes deep fades of up to 20 dB, and the scintillation index is $S_{4m} = 0.93$, which is close to the saturation value of 1.0. The fades occur more rapidly with a decorrelation time of $\tau_m = 0.46$ sec, which is partly due to multiple scatter effects which generate small scale features at the reception plane [28, 36], and also partly due to a faster effective scan velocity ($v_{eff} = 146$ m/s) in this case [5]. Frequencies smaller than the Fresnel break scale are

suppressed less effectively than in the weak scatter case due to the influence of refractive scatter [28]. For frequencies larger than f_b but less than $1/\tau_m$, the intensity spectrum deviates substantially from a power law by broadening and steepening under the influence of refractive scatter. For frequencies larger than about $1/\tau_m$, the power law behavior of the intensity spectrum is restored, but at these frequencies the spectral power is often smaller than the noise floor of the receiver. In this case, it is not possible to measure the spectral index of the ionospheric screen directly from the slope of the intensity fluctuations when the scatter is strong. The IPE technique provides a way to circumvent this problem; the phase spectral index (and therefore also the irregularity spectral index) can be retrieved even when the measured intensity spectrum does not manifest a power law regime.

4.2. Directly Measured Parameters. We begin by presenting the latitudinal and local time morphology of those parameters which are measured directly from the GPS observations (i.e., without IPE analysis). Figure 4 shows the variation of vertical-equivalent total electron content (TEC), the scintillation index (S_{4m}), and the decorrelation time (τ_m). The latitudes and local times for which the measurements have been assigned correspond to the latitudes and local times of the 350 km ionospheric penetration points. The total electron content was computed from the GPS observations using the two-frequency technique as described in [37]. For the TEC, all available data samples (with 60 second cadence) are shown in Figure 4(a). For the scintillation index and decorrelation time, the plots (Figures 4(b) and 4(c)) show data samples only for the statistically stationary segments of the intensity time series for which $S_4 > 0.3$, as described in Section 2. Because of this S_4 threshold, fewer samples are shown for the equatorial station Alta Floresta than for the anomaly stations Boa Vista and Campo Grande, since the scintillations at Alta Floresta were much weaker.

We observe, from Figure 4(a), that a well-developed equatorial anomaly was present on this evening (1-2 November 2002). The northern crest of the anomaly was located a few degrees northward of Boa Vista. The southern crest of the anomaly was located a few degrees southward of Campo Grande. Multiple gaps in the TEC data are evident prior to local midnight when the scintillation activity caused loss of lock on either the GPS L1 or L2 carriers (or both). These TEC gaps preclude a direct quantitative correlation between TEC and S_{4m} during strong scintillations, but it is clear from Figure 4 that the largest S_{4m} values generally occur at or near the crests of the equatorial anomaly and during early postsunset local times. The decorrelation time tends to be short ($\tau_m < 1 \sec$) when the scintillation is intense (i.e., S_{4m} larger than 0.5-0.6), but there can be exceptions to this rule. In general, the decorrelation time depends on both the strength of scatter and also the effective scan velocity [5, 36].

At this point we can address the validity of the IPE approach, as applied to this specific dataset. Booker et al. [27] state that the approach we have used to solve the 4th moment equation is limited to small-angle scatter such that $\lambda/(2\pi L_c) \ll 1$, where λ is the signal wavelength (0.19 m for the GPS L1 signal) and L_c is the correlation length of intensity fluctuations on the ground. The correlation length can be computed from the effective scan velocity v_{eff} and decorrelation time τ as $L_c = v_{\mathrm{eff}}\tau$. Since we have computed both v_{eff} and τ for every case in which the IPE technique has been applied, we can calculate the correlation lengths explicitly. The minimum correlation length we encountered in this dataset was 40 m. This corresponds to a maximum scattering angle of 8×10^{-4} radians, which is much smaller than 1, and therefore our propagation calculations using the IPE are valid even for the most strongly disturbed conditions that were observed during the experiment.

4.3. Parameters Inferred by IPE Analysis. As described in Section 3, application of the IPE technique provides the parameters of the ionospheric screen that result in a model intensity spectrum that best matches the measured intensity spectrum in a least-squares sense. Figure 5 shows the χ^2

values from these least-squares fits for all applications of the IPE technique presented in this paper (from all three stations combined). We chose to discard the IPE results if the χ^2 of the fit exceeded a threshold of 0.5, considering these as poor fits between the model and the measurements. As can be seen from the histogram, however, the number of IPE results rejected on this basis was very small. Figure 6 shows a comparison between the scintillation index and decorrelation time calculated from the model intensity spectrum with those calculated directly from the measurements. The close agreement between the measured and modeled scintillation index and decorrelation time indicate the accuracy of the least-squares fitting, and the ability of the ionospheric turbulence model to reproduce the observations.

Figure 7 shows the ionospheric screen parameters inferred from the IPE analysis of the GPS observations on this evening (1-2 November 2002). The calculated values for the vertically integrated turbulent strength ($C_k L$) span roughly 2 decades, ranging from 1×10^{34} to 1×10^{36}. Comparing Figures 7(a) and 4(a) reveals that the largest values of $C_k L$ were generally encountered close to the crests of the anomaly, northward of Boa Vista and southward of Campo Grande. The largest values of $C_k L$ were observed by the receiver at Campo Grande, and the scintillations persisted 2 hours longer at Campo Grande than at Boa Vista. This is consistent with the results reported by de Paula et al. [14], which suggests the scintillations may be more intense at Campo Grande than at Boa Vista due to a combination of the effects of a larger postsunset vertical plasma drift and delayed collapse in peak electron density (over Campo Grande). Since the geometrical aspects of the scattering are not reflected in $C_k L$ (it is a property of the random medium alone), it is conjectured that this parameter may correlate better with the vertical TEC in the region than would the scintillation index S_4, which depends on both the medium and the propagation geometry. We were not able to confirm this hypothesis in a quantitative manner using the GPS observations alone, due to the numerous loss of lock events caused by the scintillation itself (which prevented us from measuring the TEC). The difficulties associated with TEC estimation using satellite links that are strongly scintillating are widely known, for example, [38].

The values for the phase spectral index of the ionospheric screen (p), as inferred from the IPE analysis, generally ranged from 2.5 to 4.5. No significant differences in the phase spectral index were observed between the three COPEX stations. Figure 7(b) shows a clear increase in the spectral index with increasing local time, which the authors believe is a new result. The increase of p with increasing local time may suggest the erosion (decay) of small scale features in the turbulence as it evolves in time. On the other hand, the values of $C_k L$ (Figure 7(a)) or S_4 (Figure 4(b)) on this evening do not show much evidence of decay with increasing local time, except that after 25 LT there are no longer any data samples which meet the requirements of $S_4 > 0.3$ (and stationarity). The observation that the spectral index is steepening while the scintillation strength is unchanging may suggest that while the small scale features are eroding away, the large scales features near the Fresnel scale (on the order of

Latitudinal and Local Time Variation of Ionospheric Turbulence Parameters during the Conjugate Point Equatorial
Experiment in Brazil

17

(a)

(b)

(c)

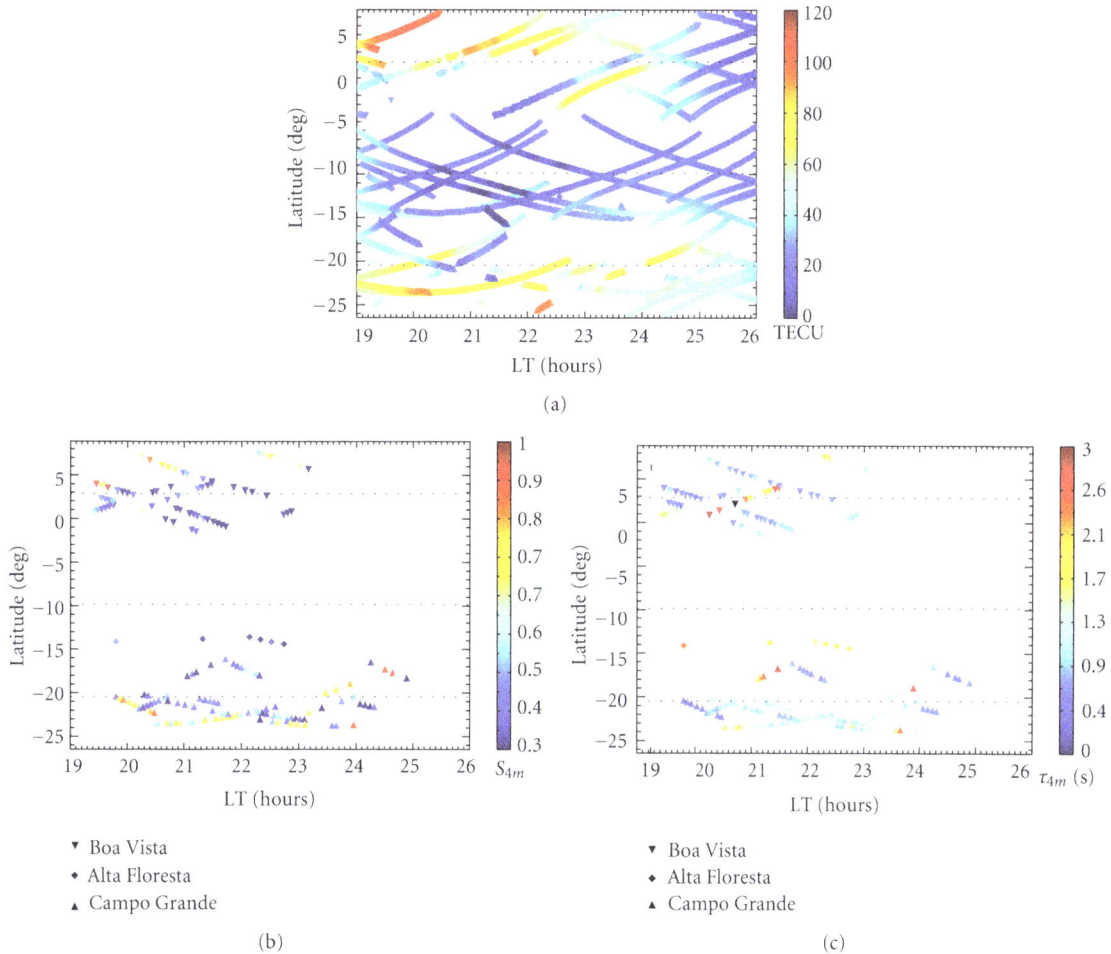

FIGURE 4: Directly measured parameters from the AFRL GPS receivers at the three COPEX stations on 1-2 November 2002: (a) vertical equivalent TEC, (b) scintillation intensity index (S_{4m}), and (c) decorrelation time (τ_m). Dotted lines demark the latitudes of the three stations.

400 m) which contribute most strongly to the scintillations at L band, have been only modestly affected.

Figure 7(b) shows the irregularity zonal drift velocities (V_D) inferred from the IPE analysis. The irregularity zonal drift velocities were roughly the same at all three COPEX stations, generally ranging from 100 to 200 m/s, with the exception of those values computed for early local times. Prior to about 20:30 LT, the estimates of V_D were anomalously small at Boa Vista (less than 100 m/s) and anomalously large at Campo Grande (larger than 200 m/s). At early local times the equatorial plasma bubbles that generate the plasma turbulence are still evolving and have a vertical component of velocity in addition to a zonal component. As pointed out by Rino et al. [16–18], the ionospheric turbulence model described in Section 2 is valid only for fully developed irregularities that do not evolve as they translate. Furthermore, we have assumed the irregularity motion to be purely zonal, which is clearly incorrect at early local times. Later in this paper we will show that after 20:30 LT the IPE inferred zonal irregularity drifts compare favorably with measurements of the zonal drift made using the spaced-receiver technique. It is interesting to note that

it is possible, using the IPE technique, to infer the zonal drift using only a single receiver. This capability could provide new opportunities to measure the irregularity zonal drift velocity at sites where only a single receiver is available.

4.4. Characterization of Phase Scintillations. The parameters $C_k L$, p, and V_D, along with the assumed values we have used for the altitude H_p, outer scale L_0, and anisotropy ratio $(a : b)$, completely specify the ionospheric screen and its translational velocity. For the sake of completeness, we also show the latitudinal and local time variation of two additional parameters which characterize phase scintillations and can be calculated from these others, namely, the strength of the spectrum of phase fluctuations and absolute phase variance. Rino [16] shows that when the ionospheric screen is described as in Section 2, and if diffraction effects on the phase are neglected, the 1D temporal spectrum of phase fluctuations can be expressed as

$$P(f) = \frac{T}{[f_0^2 + f^2]^{p/2}}, \tag{16}$$

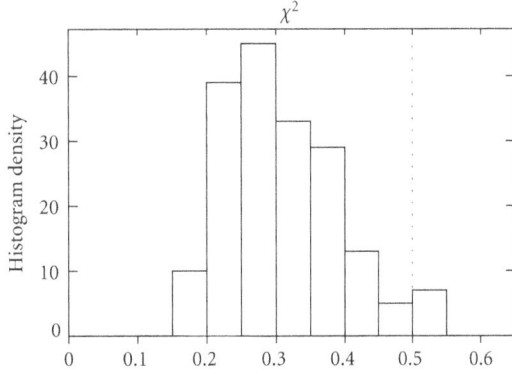

FIGURE 5: Histogram of the χ^2 values from the least-squares fits of the model intensity PSD to the measured intensity PSD, for all applications of the IPE technique presented in this paper (from all three stations combined). A threshold of $\chi^2 = 0.5$ (shown as the dotted vertical line) was chosen to identify good and bad fits between the model and the measurements.

where T is the strength of the temporal spectrum of phase at a scale of 1 Hz. The parameter T can be expressed in terms of other parameters of the screen and the geometry which are known after the IPE analysis:

$$T = r_e^2 \lambda^2 G \sec\theta \left(\frac{2\pi}{1000} \right)^{p+1} C_k L \frac{\sqrt{\pi}\,\Gamma(p/2)}{(2\pi)^{p+1}\Gamma[(p+1)/2]} v_{\text{eff}}^{p-1}. \tag{17}$$

Integrating $P(f)$ over all frequencies gives an expression for the absolute phase variance:

$$\sigma_\varphi^2 = r_e^2 \lambda^2 G \sec\theta \left(\frac{2\pi}{1000} \right)^{p+1} C_k L \frac{q_0^{1-p}\Gamma((p-1)/2)}{4\pi\Gamma((p+1)/2)}. \tag{18}$$

Note that the phase variance depends on the outer scale wavenumber, which we do not measure. Hence the values for the phase variance we report, like $C_k L$, must be considered as relative to our specific choice of outer scale (10 km). It is also important to note that the expressions (16) and (17) were derived while neglecting the effects of diffraction on the measured phase, which become important when the scatter is strong [8]. Nevertheless, the parameters T and σ_φ^2 have been used extensively in the literature for characterizing the effect of scintillation on GPS tracking loop performance [39, 40] so we report these findings here.

Figure 8 shows the latitudinal and local time variations of the parameters T and σ_φ^2 during the evening of 1-2 November 2002. The strength of the phase spectrum (T) ranged from -60 dB (weak) to -15 dB (very strong), with the largest values encountered at early local times at the anomaly stations Boa Vista and Campo Grande. The absolute phase variance reached up to 500 radians squared, which is well into the strong multiple-scatter regime [28]. The largest values of the absolute phase variance occurred near the anomaly crests, northward of Boa Vista and southward of Campo Grande. We note that using the IPE technique, it is possible to estimate the parameters T and σ_φ^2 even when the GPS phase observations cannot be analyzed directly due

to loss of phase lock. This may be helpful when studying the influence of scintillation on loss of lock and GPS positioning accuracy, where one of the principal challenges is to characterize the conditions under which loss of phase lock occurs in spite of the fact that loss of phase lock precludes direct measurement of T and σ_φ^2 [5].

4.5. Local Time Variation at the Three Stations. The plots shown in Figures 4, 7, and 8 allow examination of the parameters as a function of latitude and local time. This format has been chosen in order to compare the parameters to the structure of the equatorial anomaly, which varies with latitude and local time. Nevertheless, the local time dependence of the screen parameters is perhaps more clearly shown in Figure 9. The plots shown in Figure 9 include all data, irrespective of the latitude at which they were measured. Figure 9(a) shows the dependence of $C_k L$ with local time. The largest values of $C_k L$ were encountered at the anomaly stations Boa Vista and Campo Grande. A clear latitudinal asymmetry in the turbulent intensity is evident in Figure 9(a), as the $C_k L$ values were generally larger at Campo Grande than they were at Boa Vista. There was little change in $C_k L$ with local time during this particular evening, but the scintillations exceeding our threshold of ($S_4 > 0.3$) persisted 2 hours longer at Campo Grande than at Boa Vista. As commented earlier, the phase spectral index (Figure 9(b)) showed a marked change from 2.5 just after local sunset increasing to as large as 4.5 after local midnight. Again, this steepening of the spectrum may suggest that the small scale features of the turbulence are being eroded away. Figure 9(c) reveals that the strength of the phase fluctuation spectrum decreases with increasing local time. From (17), it is easy to show that if $C_k L$ and all other parameters except p are held constant, then T will decrease approximately linearly with increasing p on a logarithmic scale. Under these conditions the strength of phase scintillations decrease with increasing local time. We note that it would be very useful, for both academic interest and also modeling purposes, to know how the strength and spectral index of the plasma turbulence within the same equatorial plasma bubble change as a function of time. Since the measurements from the COPEX campaign are from receivers located along a fixed magnetic meridian, we cannot follow the evolution of a single bubble as it drifts zonally. It would be interesting to repeat this experiment using a chain of receivers distributed zonally to infer how the screen parameters vary in a frame of reference that follows the bubbles as they drift.

4.6. Comparison with Spaced-Receiver Measurements of the Irregularity Zonal Drift. The IPE analysis provides estimates for the parameters of the ionospheric screen which yield a temporal spectrum of intensity fluctuations that most closely match the observations in a least-squares sense. One of these screen parameters is its translational velocity in the zonal direction, V_D. This is a model-inferred irregularity drift velocity, not a direct measurement of the rate at which the density irregularities sweep past two points in space as is provided by the spaced-receiver technique. Both the

(a)

(b)

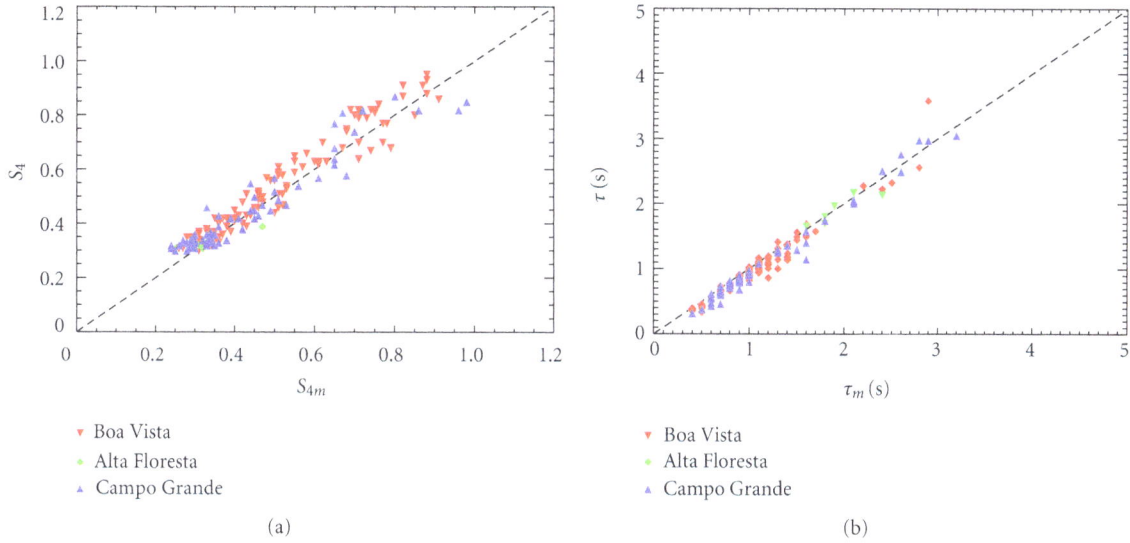

Figure 6: Comparison between parameters inferred from the model intensity spectra and those calculated directly from the measurements: (a) measured scintillation intensity index (S_{4m}) and model scintillation intensity index (S_4); (b) measured intensity decorrelation time (τ_m) and model intensity decorrelation time (τ).

spaced-receiver technique and also IPE analysis are subject to errors if the irregularities evolve temporally as they drift. An excellent discussion of this issue and other potential sources of error when measuring the zonal irregularity drift using spaced GPS receivers may be found in [19]. A detailed analysis of zonal irregularity drift velocity using the spaced GPS receiver technique during the COPEX campaign is provided by Muella et al. [11] and de Paula et al. [14].

In this section, we compare the zonal drift estimates provided by the IPE technique with the estimates provided by the colocated 4-channel VHF receivers at Boa Vista, Alta Floresta, and Campo Grande, and also with the GPS spaced-receiver measurements provided by Muella et al., as described in [11]. The estimates from Muella were measured at Boa Vista, Cachimbo, and Campo Grande. We will compare our measurements at Alta Floresta (9.9°S, 56.1°W) to Muella's measurements at Cachimbo (9.5°S, 54.8°W), since these stations are close to one another. For both the GPS spaced-receiver drift estimates and the IPE drift estimates, the geometry of propagation with respect to the magnetic field has been taken into account, and also the motion of the ionospheric penetration points. For the VHF spaced-receiver measurements, these geometrical considerations have not been accounted for, but these corrections are smaller when the satellites are viewed at high elevation angles. Therefore we only show results for the east links which are visible at higher elevation angles than the west links.

Figure 10 shows a comparison of the zonal drift velocity estimated using the IPE technique with estimates provided by the VHF east links (channels 3-4), and the GPS spaced-receiver measurements provided by Muella et al. [11]. We observe that after approximately 20:30 LT, the IPE estimates compare favorably with both the VHF spaced-receiver estimates and the GPS spaced-receiver estimates. All three techniques show considerable scatter at early local times,

presumably because the plasma bubbles are still evolving and have significant vertical components of velocity.

The fact that the IPE inferred drifts agree with the VHF and GPS spaced-receiver drift estimates is encouraging and may lend confidence to the IPE estimates of C_kL and p, parameters for which no independent source of validation data is readily available. We plan to repeat the IPE analysis for the remaining evenings of the COPEX campaign in order to validate the technique and also to study the average seasonal behavior of the ionospheric screen parameters during the duration of the experiment.

We should emphasize that the IPE technique is not intended as a replacement for the spaced-receiver technique. The spaced-receiver technique provides a direct measurement of the zonal irregularity drift once the necessary geometric corrections [19] have been applied, whereas IPE analysis provides a model-inferred drift estimate. For example, the IPE drift estimates would change if a different model for the irregularities were used (e.g., a two-component spectral model). As such, the IPE technique cannot be expected to provide as reliable estimates of the zonal drift as the spaced-receiver technique. Furthermore, the nonlinear nature of the inversion process makes it unclear how to provide error bounds for these drift estimates. Despite these limitations, however, the IPE technique should be useful for ground stations where zonal drift estimates are desired but only a single receiver is available.

5. Remarks and Conclusions

In this report, we introduce a new technique called Iterative Parameter Estimation (IPE) for inferring ionospheric turbulence parameters from a time series of scintillating intensity measurements resulting from propagation through the low latitude ionosphere. The method of analysis is new and,

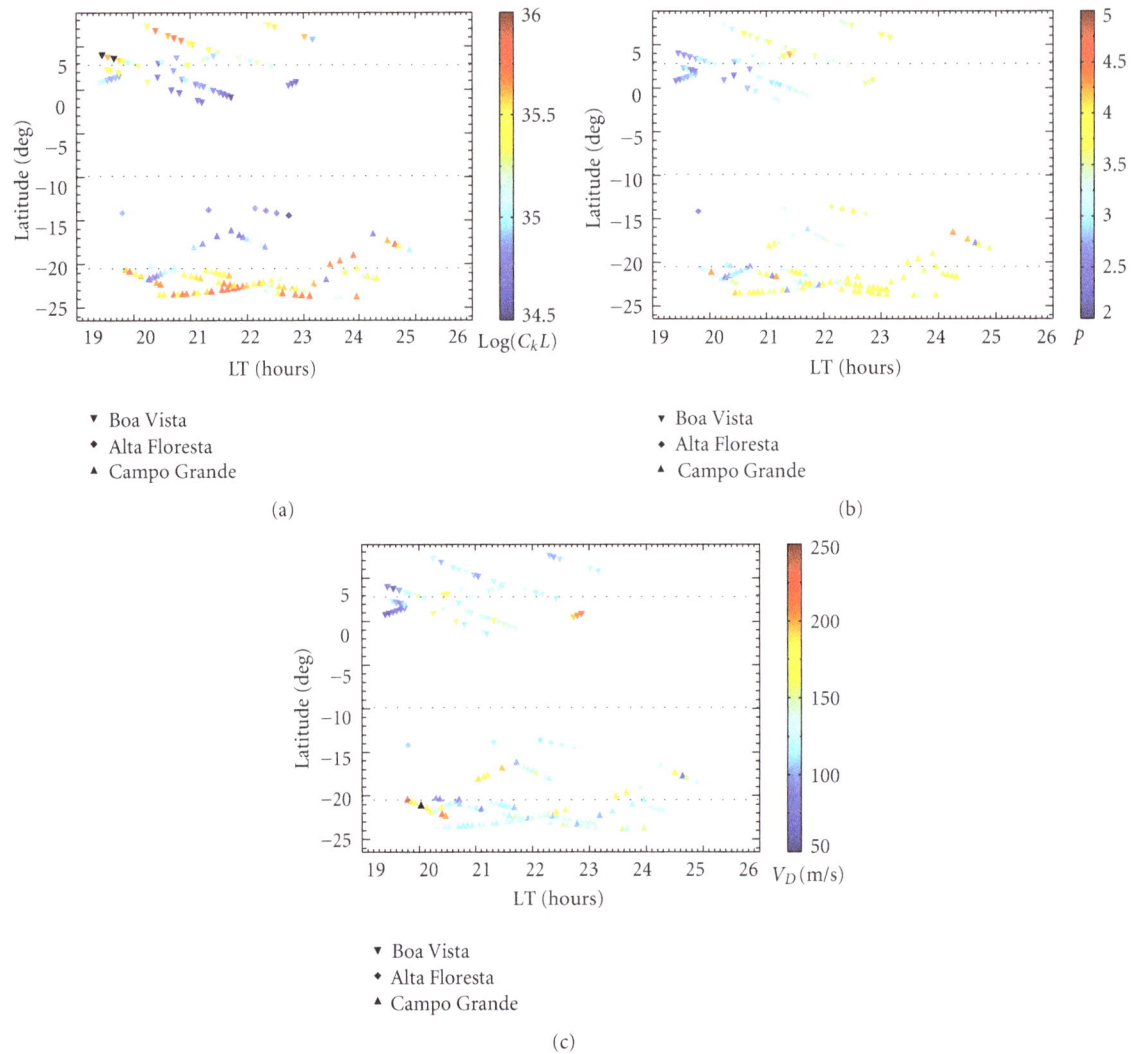

FIGURE 7: Parameters inferred from IPE analysis: (a) log of the vertically integrated turbulence strength at the 1 km scale ($C_k L$), (b) phase spectral index (p), and (c) zonal irregularity drift velocity (V_D). Dotted lines demark the latitudes of the three stations.

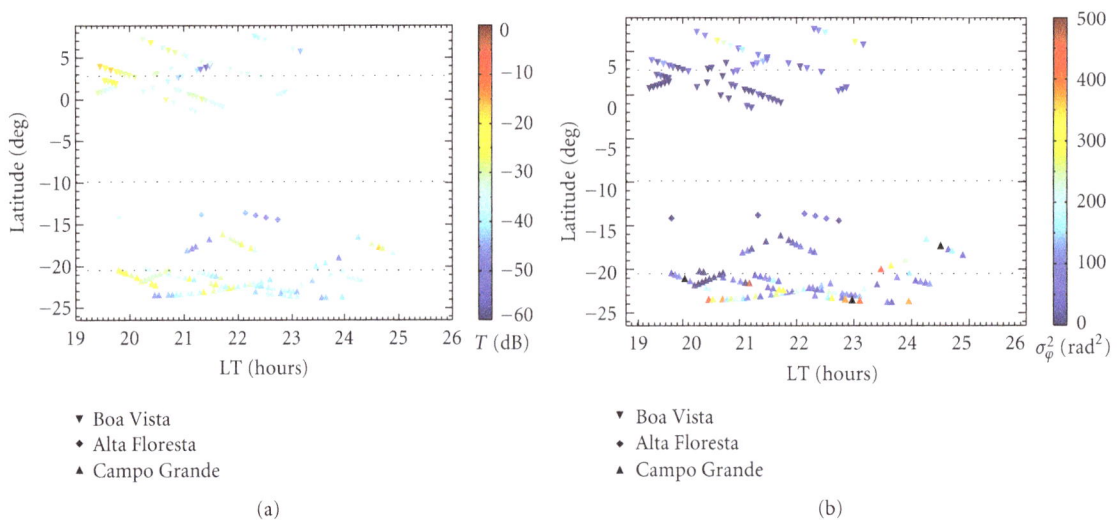

FIGURE 8: Phase scintillation parameters infered from IPE analysis and the ionospheric turbulence model: (a) strength of the phase temporal spectrum at the 1 Hz scale (T) and (b) absolute phase variance (σ_φ^2). Dotted lines demark the latitudes of the three stations.

Latitudinal and Local Time Variation of Ionospheric Turbulence Parameters during the Conjugate Point Equatorial Experiment in Brazil

21

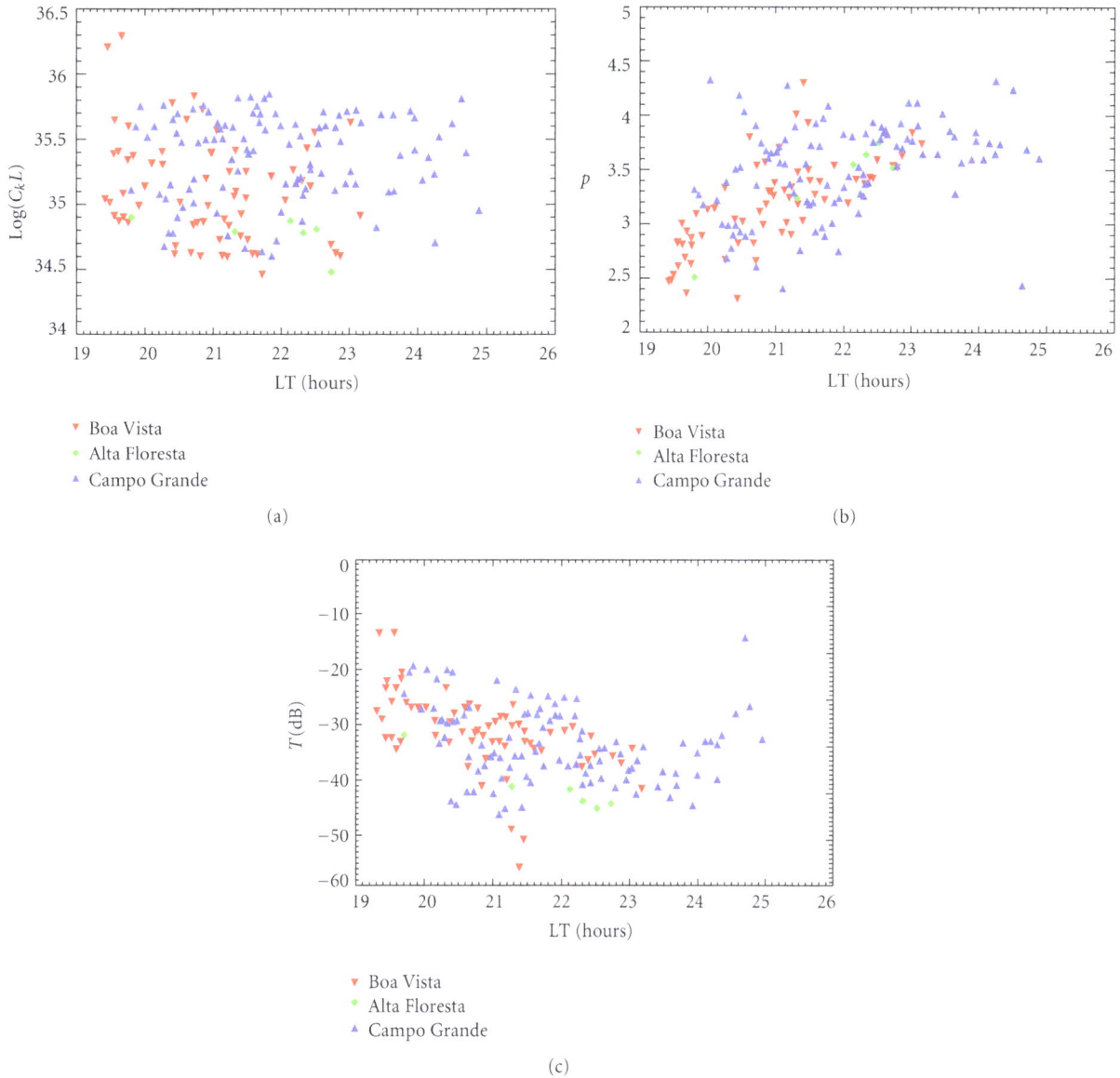

FIGURE 9: Local time variation of the parameters: (a) vertically integrated strength of turbulence at the 1 km scale (C_kL), (b) phase spectral index (p), and (c) strength of the phase temporal spectrum at the 1 Hz scale (T).

unlike analytical theories of scintillation based on the Born or Rytov approximations, it is valid for strong scatter when the scintillation index (S_4) saturates due to multiple-scatter effects. IPE employs the phase screen approximation and a numerical inversion technique to estimate the parameters of an ionospheric screen which are consistent with the ionospheric turbulence model employed and the scintillations observed. More specifically, IPE provides a set of screen parameters that produces a model intensity spectrum which best matches the measured intensity spectrum in a least-squares sense. Due to the analytic intractability of the 4th moment equation governing fluctuations in the field we must determine the optimal parameters by numerical iteration, and therefore we cannot guarantee the uniqueness of this parametrization.

The underlying ionospheric turbulence model we have used to fit the scintillation observations described in this paper is that developed by Rino [16]. This model assumes the electron density irregularities can be characterized by a slab of homogenous, anisotropic fluctuations with a single-slope power law spectral density function in three dimensions. Actual turbulence in the equatorial ionosphere, by contrast, consists of large-scale deterministic structure (equatorial plasma bubbles) with small-scale random structure embedded within these bubbles [17, 24]. Furthermore, there is evidence to suggest that a model based on a two-component power law is better able to characterize the plasma turbulence under some circumstances [41]. Despite these limitations, the Wideband Scintillation Model which is based on this formulation has been shown to produce a very satisfactory

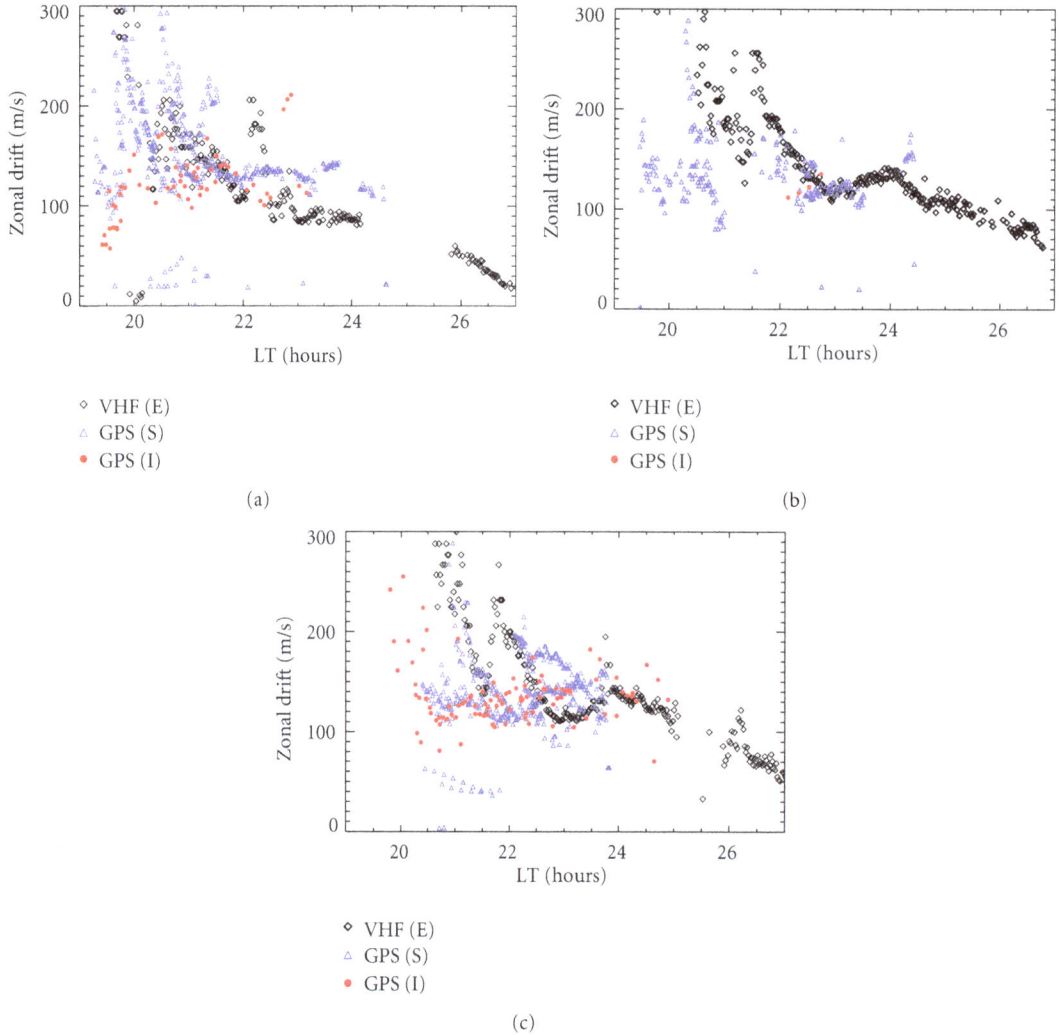

FIGURE 10: Comparison of the zonal irregularity drift velocity calculated using the VHF east link (E), GPS spaced-receivers (S), and AFRL GPS receivers using the IPE technique (I). (a) is for Boa Vista, (b) is for Alta Floresta, and (c) is for Campo Grande.

description of the scintillation of transionospheric signals at frequencies ranging from VHF to L band (e.g., see [22] and the references therein). The IPE technique itself is not constrained to use any particular turbulence model, however, and it would certainly be possible to extend the approach to accommodate a two-component spectrum, for example.

We use the IPE technique to investigate the latitudinal and local time variation of the plasma turbulence parameters from 10 Hz GPS C/No measurements during one evening (1-2 November 2002) of the Conjugate Point Equatorial Experiment (COPEX) in Brazil. We find that the strength of turbulence tends to be largest near the crests of the equatorial anomaly and at early postsunset local times. We note that the results of Muella et al. [42] suggest that the most intense scintillations do not occur exactly at the crests of the anomaly, but instead at the edges of the crests where TEC gradients are largest. Due to the limited number of observations considered in our analysis, and the gaps in our TEC estimates due to scintillation induced loss of lock, we can neither corroborate nor refute this claim. During the

evening we considered, the strength of turbulence showed little variation as function of local time but was generally stronger and lasted two hours longer at Campo Grande than at Boa Vista. The phase spectral index was similar at the three stations but increased significantly from 2.5 to 4.5 with increasing local time. We hypothesize that this change in spectral slope may have occurred as small scale features of the turbulence were gradually eroded. The larger-scale sizes in the turbulence (near the Fresnel break scale), which contributed most to the intensity scintillations, may have decayed more slowly so that the turbulent intensity decreased more gradually. An alternative interpretation for the local time dependence of the spectral index is that the physical mechanisms which generated these irregularities may have differed depending on the time at which the bubbles were generated. With GPS receivers operating from fixed locations on the ground, it is not possible to distinguish between old bubbles which have been decaying for a long period of time and new bubbles which have developed at later local times. Additional work needs to be performed in order to

Latitudinal and Local Time Variation of Ionospheric Turbulence Parameters during the Conjugate Point Equatorial
Experiment in Brazil

23

resolve this issue. It is hoped that either *in situ* density observations or perhaps a longitudinally distributed chain of GPS receivers may enable us to determine how the spectral index varies as a function of the local time at which the bubbles developed. Even though we examined only the GPS intensity fluctuations and not the GPS phase measurements, we were able to infer the strength of phase fluctuations and the absolute phase variance based on theoretical considerations. From this analysis, we inferred that the absolute phase variance reached up to $500 \, \text{rad}^2$, which is well into the strong multiple-scatter regime. The strength of phase scintillations was inferred to decrease with increasing local time. We found that estimates of zonal irregularity drift using the IPE technique agreed favorably with those made using the VHF and GPS spaced-receiver techniques. This encouraging result suggests that IPE analysis may provide useful estimates of zonal irregularity drift at scintillation monitoring sites which are equipped with only a single receiver, so that the spaced-receiver technique cannot be employed.

Since the COPEX campaign was conducted back in 2002, the number of GPS receivers capable of providing high rate (10 Hz or faster) scintillation observations in South America has increased dramatically. The Low-Latitude Ionosphere Sensor Network (LISN) [43, 44] includes more than 35 such receivers and also will include five ionosondes distributed along a magnetic field line in a fashion similar to COPEX. The application of IPE analysis to GPS scintillation observations collected along this field line could enable routine and systematic investigation of the latitudinal and local time morphology of ionospheric turbulence that produces radio wave scintillations at low-latitudes. It may also be instructive to use a longitudinally distributed chain of GPS receivers to follow the evolution of individual plasma bubbles to determine how the turbulence evolves in a reference frame that follows the bubbles. Such an investigation might help better explain why the spectral slope was observed to increase significantly with local time.

Acknowledgments

The authors would like to thank Robert Livingston for providing the VHF spaced-receiver irregularity drift estimates and Marcio Muella for providing the GPS spaced-receiver irregularity drift estimates during the COPEX experiment in Brazil. This work was supported by AFRL contract FA8718-09-C0041 and also funding from AFOSR.

References

[1] A. Basu, E. MacKenzie, and S. Basu, "Ionospheric constraints on VHF/UHF communications links during solar maximum and minimum periods," *Radio Science*, vol. 23, no. 3, pp. 363–378, 1988.

[2] K. Groves, S. Basu, E. Weber et al., "Equatorial scintillation and systems support," *Radio Science*, vol. 32, no. 5, pp. 2047–2064, 1997.

[3] K. M. Groves, S. Basu, J. M. Quinn et al., "A comparison of GPS performance in a scintillating environment at Ascension Island," in *Proceedings of the 13th International Technical Meeting of the Satellite Division of The Institute of Navigation (ION GPS '00)*, pp. 672–679, Salt Palace Convention Center, Salt Lake City, Utah, USA, September 2000.

[4] S. Datta-Barua, P. H. Doherty, S. H. Delay, T. Dehel, and J. A. Klobuchar, "Ionospheric scintillation effects on single and dual frequency GPS positioning," in *Proceedings of the 16th International Technical Meeting of the Satellite Division of The Institute of Navigation (ION GPS/GNSS '03)*, pp. 336–346, Portland, OR, USA, September 2003.

[5] C. S. Carrano and K. M. Groves, "Temporal decorrelation of GPS satellite signals due to multiple scattering from ionospheric irregularities," in *Proceedings of the 23rd International Technical Meeting of The Satellite Division of the Institute of Navigation (ION GNSS '10)*, pp. 361–374, Oregon Convention Center, Portland, OR, USA, September 2010.

[6] Z.-W. Xu, J. Wu, and Z.-S. Wu, "A survey of ionospheric effects on space-based radar," *Waves in Random Media*, vol. 14, no. 2, pp. S189–S273, 2004.

[7] N. Rogers and P. Cannon, "The synthetic aperture radar trans-ionospheric radio propagation simulator (SAR-TIRPS)," in *Proceedings of the 11th International Conference on Ionospheric Radio Systems and Techniques (IRST '09)*, pp. 112–116, Edinburgh, UK, April 2009.

[8] C. S. Carrano, K. M. Groves, and R. G. Caton, "Accuracy of phase screen models using deterministic screens derived from GPS and ALTAIR measurements," in *Proceedings of the Ionospheric Effects Symposium*, Alexandria, Va, USA, May 2011.

[9] J. Aarons, "Global morphology of ionospheric scintillations," *Proceedings of the IEEE*, vol. 70, no. 4, pp. 360–378, 1982.

[10] I. S. Batista, M. A. Abdu, A. J. Carrasco et al., "Equatorial spread F and sporadic E-layer connections during the Brazilian conjugate point equatorial experiment (COPEX)," *Journal of Atmospheric and Solar-Terrestrial Physics*, vol. 70, no. 8-9, pp. 1133–1143, 2008.

[11] M. T. A. H. Muella, E. R. de Paula, I. J. Kantor et al., "GPS L-band scintillations and ionospheric irregularity zonal drifts inferred at equatorial and low-latitude regions," *Journal of Atmospheric and Solar-Terrestrial Physics*, vol. 70, no. 10, pp. 1261–1272, 2008.

[12] M. A. Abdu, I. S. Batista, B. W. Reinisch et al., "Conjugate point equatorial experiment (COPEX) campaign in Brazil: electrodynamics highlights on spread F development conditions and day-to-day variability," *Journal of Geophysical Research A*, vol. 114, no. 4, Article ID A04308, 21 pages, 2009.

[13] J. H. A. Sobral, M. A. Abdu, T. R. Pedersen et al., "Ionospheric zonal velocities at conjugate points over Brazil during the COPEX campaign: experimental observations and theoretical validations," *Journal of Geophysical Research A*, vol. 114, no. 4, Article ID A04309, 24 pages, 2009.

[14] E. R. de Paula, M. T. A. H. Muella, J. H. A. Sobral et al., "Magnetic conjugate point observations of kilometer and hundred-meter scale irregularities and zonal drifts," *Journal of Geophysical Research A*, vol. 115, no. 8, Article ID A08307, 15 pages, 2010.

[15] A. Ishimaru, *Wave Propagation and Scattering in Random Media*, IEEE Press and Oxford University, New York, NY, USA, 1997.

[16] C. L. Rino, "Power law phase screen model for ionospheric scintillation, 1 weak scatter," *Radio Science*, vol. 14, no. 6, pp. 1135–1145, 1979.

[17] C. L. Rino and C. S. Carrano, "The application of numerical simulations in Beacon scintillation analysis and modeling,"

Radio Science, vol. 46, no. 3, Article ID RS0D02, 10 pages, 2011.

[18] C. L. Rino, *The Theory of Scintillation With Applications in Remote Sensing,* John Wiley, New York, NY, USA, 2011.

[19] B. M. Ledvina, P. M. Kintner, and E. R. de Paula, "Understanding spaced-receiver zonal velocity estimation," *Journal of Geophysical Research A,* vol. 109, no. 10, Article ID A10306, 12 pages, 2004.

[20] P. D. Welch, "The use of fast Fourier transform for the estimation of power spectra: a method based on time averaging over short, modified periodograms," *IEEE Transactions on Audio and Electroacoustics,* vol. 15, no. 2, pp. 70–73, 1967.

[21] N. Olsen, T. Sabaka, and L. Tøffner-Clausen, "Determination of the IGRF 2000 model," *Earth, Planets and Space,* vol. 52, no. 12, pp. 1175–1182, 2000.

[22] J. A. Secan, R. M. Bussey, E. J. Fremouw, and S. Basu, "Improved model of equatorial scintillation," *Radio Science,* vol. 30, no. 3, pp. 607–617, 1995.

[23] M. A. Cervera and R. M. Thomas, "Latitudinal and temporal variation of equatorial ionospheric irregularities determined from GPS scintillation observations," *Annales Geophysicae,* vol. 24, no. 12, pp. 3329–3341, 2006.

[24] C. S. Carrano, K. M. Groves, R. G. Caton, C. L. Rino, and P. R. Straus, "Multiple phase screen modeling of ionospheric scintillation along radio occultation raypaths," *Radio Science,* vol. 46, no. 3, Article ID RS0D07, 14 pages, 2011.

[25] N. C. Rogers, P. S. Cannon, and K. M. Groves, "Measurements and simulation of ionospheric scattering on VHF and UHF radar signals: channel scattering function," *Radio science,* vol. 44, Article ID RS0A07, 10 pages, 2009.

[26] F. R. Hoots and R. L. Roehrich, *Spacetrack Report #3, Models for Propagation of the NORAD Element Sets,* United States Air Force, Colorado Springs, Colo, USA, 1980.

[27] H. Booker, J. A. Ferguson, and H. O. Vats, "Comparison between the extended-medium and the phase-screen scintillation theories," *Journal of Atmospheric and Terrestrial Physics,* vol. 47, no. 4, pp. 381–399, 1985.

[28] H. Booker and G. MajidiAhi, "Theory of refractive scattering in scintillation phenomena," *Journal of Atmospheric and Terrestrial Physics,* vol. 43, no. 11, pp. 1199–1214, 1981.

[29] E. J. Fremouw and J. A. Secan, "Modeling and scientific application of scintillation results," *Radio Science,* vol. 19, no. 3, pp. 687–694, 1984.

[30] K. C. Yeh and C. H. Liu, "Radio wave scintillations in the ionosphere," *Proceedings of the IEEE,* vol. 70, no. 4, pp. 324–360, 1982.

[31] A. Bhattacharyya, K. C. Yeh, and S. J. Franke, "Deducing turbulence parameters from transionospheric scintillation measurements," *Space Science Reviews,* vol. 61, no. 3-4, pp. 335–386, 1992.

[32] C. S. Carrano and C. L. Rino, "Split-step solution of the 4th moment equation for propagation through intense ionospheric disturbances," in *Proceedings of the International Conference on Electromagnetics in Advanced Applications (ICEAA '11),* Turin, Italy, 2011.

[33] R. Piessens, E. deDoncker-Kapenga, C. Uberhuber, and D. Kahaner, *Quadpack: A Subroutine Package for Automatic Integration,* vol. 1 of *Computational Mathematics,* Springer, 1983.

[34] J. A. Nelder and R. Mead, "A simplex method for function minimization," *The Computer Journal,* vol. 7, no. 4, pp. 308–313, 1965.

[35] R. G. Caton, C. S. Carrano, C. M. Alcala, K. M. Groves, T. Beach, and D. Sponseller, "Simulating the effects of scintillation on transionospheric signals with a two-way phase screen

constructed from ALTAIR phase-derived TEC," *Radio Science,* vol. 44, no. 4, Article ID RS0A12, 9 pages, 2009.

[36] C. L. Rino and J. Owen, "The time structure of transionospheric radio wave scintillation," *Radio Science,* vol. 15, no. 3, pp. 479–489, 1980.

[37] C. S. Carrano, A. Anghel, R. A. Quinn, and K. M. Groves, "Kaiman filter estimation of plasmaspheric total electron content using GPS," *Radio Science,* vol. 44, no. 4, Article ID RS0Al0, 14 pages, 2009.

[38] C. S. Carrano and K. Groves, "The GPS segment of the AFRL-SCINDA global network and the challenges of real-time TEC estimation in the equatorial ionosphere," in *Institute of Navigation, National Technical Meeting (NTM '06),* pp. 1036–1047, January 2006.

[39] M. Knight and A. Finn, "The effects of ionospheric scintillation on GPS," in *Proceedings of the Institute of Navigation (ION) GPS meeting,* Nashville, Tenn, USA, 1998.

[40] R. S. Conker, M. B. El-Arini, C. J. Hegarty, and T. Hsiao, "Modeling the effects of ionospheric scintillation on GPS/Satellite-based augmentation system availability," *Radio Science,* vol. 38, no. 1, article 1001, 23 pages, 2003.

[41] S. J. Franke and C. H. Liu, "Modeling of equatorial multifrequency scintillation," *Radio Science,* vol. 20, no. 3, pp. 403–415, 1985.

[42] M. T. A. H. Muella, E. A. Kherani, E. R. de Paula et al., "Scintillation-producing Fresnel-scale irregularities associated with the regions of steepest TEC gradients adjacent to the equatorial ionization anomaly," *Journal of Geophysical Research A,* vol. 115, no. 3, Article ID A03301, 19 pages, 2010.

[43] C. E. Valladares and P. H. Doherty, "The low-latitude ionosphere sensor network (LISN)," in *Proceedings of the International Technical Meeting of The Institute of Navigation,* pp. 16–24, Anaheim, Calif, USA, January 2009.

[44] C. Valladares, "The low-latitude ionosphere sensor network (LISN): initial results," in *Proceedings of the 2011 Ionospheric Effects Symposium,* Alexandria, Va, USA, May 2011.

3

Numerical Modeling of the Influence of Solar Activity on the Global Circulation in the Earth's Mesosphere and Lower Thermosphere

Igor Mingalev and Victor Mingalev

Kola Scientific Center of the Russian Academy of Sciences, Polar Geophysical Institute, Fersman Street 14, Apatity 184209, Russia

Correspondence should be addressed to Victor Mingalev, mingalev@pgia.ru

Academic Editor: M. V. Klimenko

The nonhydrostatic model of the global neutral wind system of the earth's atmosphere, developed earlier in the Polar Geophysical Institute, is utilized to investigate how solar activity affects the formation of the large-scale global circulation of the mesosphere and lower thermosphere. The peculiarity of the utilized model consists in that the internal energy equation for the neutral gas is not solved in the model calculations. Instead, the global temperature field is assumed to be a given distribution, that is, the input parameter of the model. Moreover, in the model calculations, not only the horizontal components but also the vertical component of the neutral wind velocity is obtained by means of a numerical solution of a generalized Navier-Stokes equation for compressible gas, so the hydrostatic equation is not applied. The simulation results indicate that solar activity ought to influence considerably on the formation of global neutral wind system in the mesosphere and lower thermosphere. The influence is conditioned by the vertical transport of air from the lower thermosphere to the mesosphere and stratosphere. This transport may be rather different under distinct solar activity conditions.

1. Introduction

During the last three decades, several general circulation models of the lower and middle atmosphere have been developed (e.g., see [1–11]). It can be noticed that the existing general circulation models of the lower and middle atmosphere may be successfully utilized for simulation of the slow climate changes. Unfortunately, these models cannot produce the vertical atmospheric wind with an acceptable accuracy. The fact is that the momentum equation for the vertical velocity is omitted in commonly used general circulation models, and the vertical velocity is obtained with the help of simple hydrostatic equation. Unfortunately, these models can not produce the vertical atmospheric wind with an acceptable accuracy. As a consequence of the simplification, such models do not provide an ability to simulate the large-scale global circulation in all regimes, in particular, under disturbed conditions. As is well known, the global models just mentioned produce the vertical component of the wind

velocity having the values of several centimeters per second at levels of the lower thermosphere. While the observed vertical velocity is known to achieve up to some tens m/s at levels of the mesosphere and lower thermosphere in high-latitude regions (see [12–16]). Thus, it is necessary to use more complex general circulation models, describing the vertical transport more adequately, for simulation the transient large-scale global circulation of the middle atmosphere and lower thermosphere.

In the Polar Geophysical Institute (PGI), the nonhydrostatic model of the global neutral wind system in the earth's atmosphere has been developed not long ago [17, 18]. This model enables to calculate three-dimensional global distributions of the zonal, meridional, and vertical components of the neutral wind at levels of the troposphere, stratosphere, mesosphere, and lower thermosphere, with whatever restrictions on the vertical transport of the neutral gas being absent. This model has been utilized in order to simulate the global circulation of the middle atmosphere for conditions

(a)

(b)

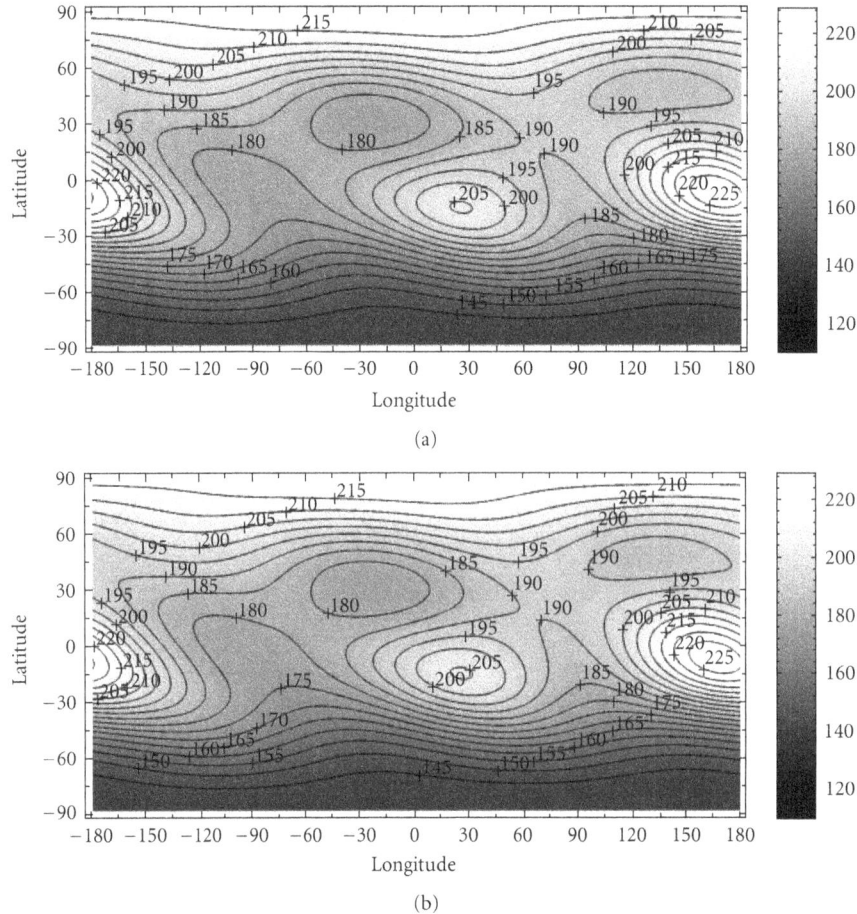

FIGURE 1: The global distributions of the atmospheric temperature (K) at 90 km altitude, obtained from the NRLMSISE-00 empirical model for 16 January, UT = 10.30 and calculated for two distinct values of solar activity: $F_{10.7} = 101$ (a) and $F_{10.7} = 230$ (b).

corresponding to summer and winter in the northern hemisphere (see [17–19]). It has been found in these studies that the global distributions of the neutral wind, calculated for summer and winter periods in the northern hemisphere, in particular, the large-scale circumpolar vortices, are consistent with the planetary circulations of the atmosphere, obtained from observations.

The purpose of the present work is to continue these studies and to investigate numerically, using the nonhydrostatic model of the global neutral wind system [17, 18], how solar activity affects the formation of the large-scale global circulation of the mesosphere and lower thermosphere.

2. Mathematical Model

The nonhydrostatic model of the global neutral wind system in the earth's atmosphere, developed earlier in the PGI [17, 18] is utilized in the present study. The utilized model produces three-dimensional global distributions of the zonal, meridional, and vertical components of the neutral wind velocity and neutral gas density at the levels of the troposphere, stratosphere, mesosphere, and lower thermosphere. The characteristic feature of the model is that the vertical component of the neutral wind velocity, as well as horizontal

components of the neutral wind, is obtained by means of a numerical solution of the appropriate momentum equation for a viscous gas without any simplifications of this equation, with the hydrostatic equation being not used. Moreover, the model does not include the internal energy equation for the neutral gas. Instead, the global temperature field is assumed to be a given distribution. It is known that the atmospheric temperature distributions, calculated by using the existing global circulation models containing the internal energy equation, as a rule, differ from observed distributions of the atmospheric temperature. These differences are conditioned by uncertainty and complexity in various chemical-radiational heating and cooling rates. Therefore, there is no reason to expect an exact correspondence between the calculated and measured temperatures of the neutral gas. On the other hand, over the last years empirical models of the global atmospheric temperature field have been successfully developed which have a satisfactory accuracy. In the present study, we take the global temperature distribution from the NRLMSISE-00 empirical model [20] and consider it to be an input parameter of the utilized model.

However, the utilized model is not able to simulate the gravity waves origin and propagation because of the time independence of the temperature in the model calculations.

Numerical Modeling of the Influene of Solar Activity on the Global Circulation in the Earth's Mesosphere and Lower Thermosphere

27

(a)

(b)

(c)

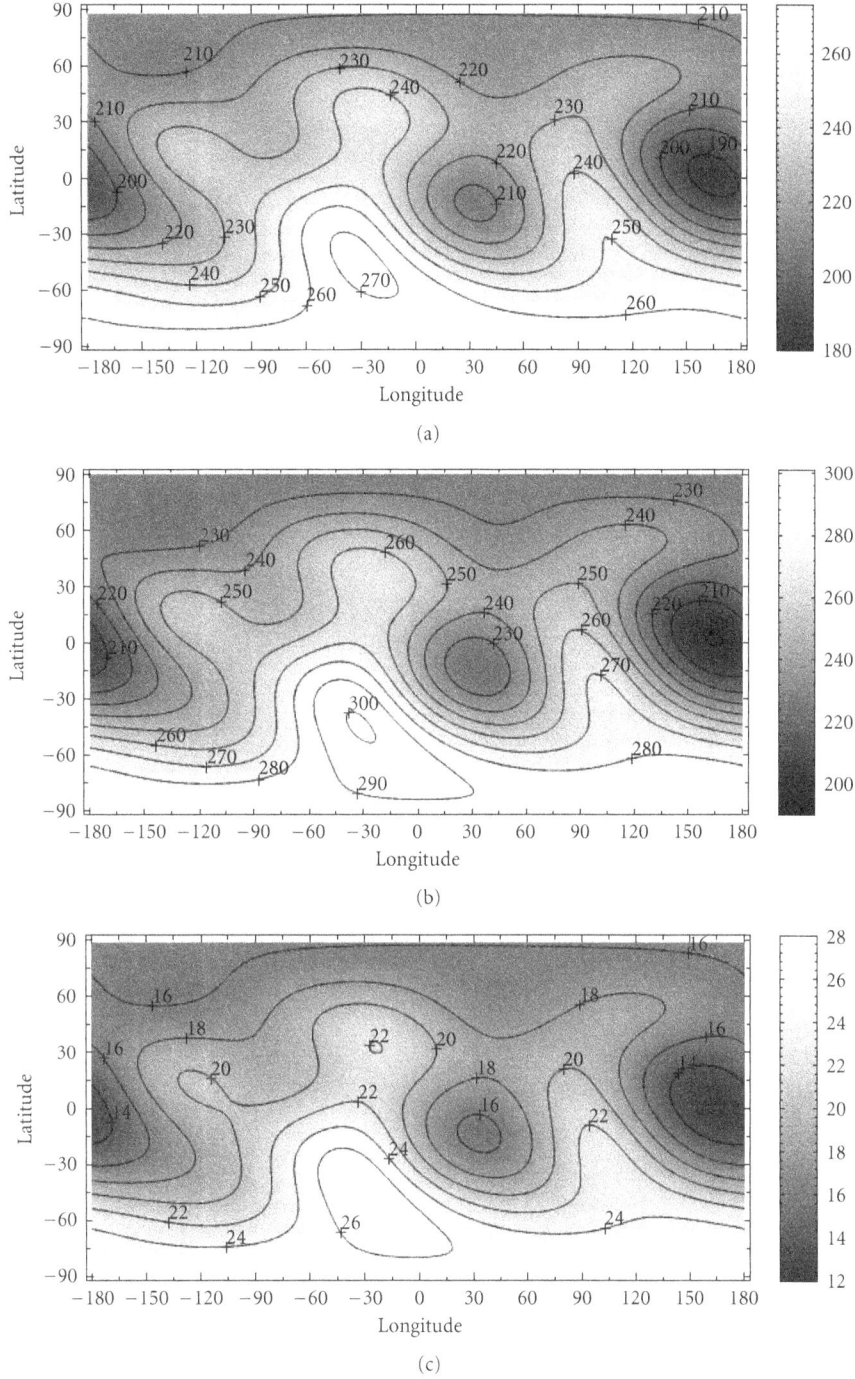

FIGURE 2: The global distributions of the atmospheric temperature (K) at 110 km altitude, obtained from the NRLMSISE-00 empirical model for 16 January, UT = 10.30 and calculated for two distinct values of solar activity: $F_{10.7} = 101$ (a) and $F_{10.7} = 230$ (b). Also the difference between the latter and the former distributions (c).

The model has the potential to describe the global neutral wind system under disturbed conditions when the vertical component of the neutral wind velocity at the levels of the lower thermosphere can be as large as several tens of meters per second [12–16].

The mathematical model, utilized in the present study, is based on the numerical solution of the system of equations containing the dynamical equation and continuity equation for the neutral gas. For solving the system of equations, the finite-difference method is applied. The dynamical equation for the neutral gas in vectorial form can be written as

$$\rho \left(\frac{\partial \vec{V}}{\partial t} + \left(\vec{V}, \nabla \right) \vec{V} \right) = \rho \vec{F} + \nabla \cdot \widehat{P}, \qquad (1)$$

(a)

(b)

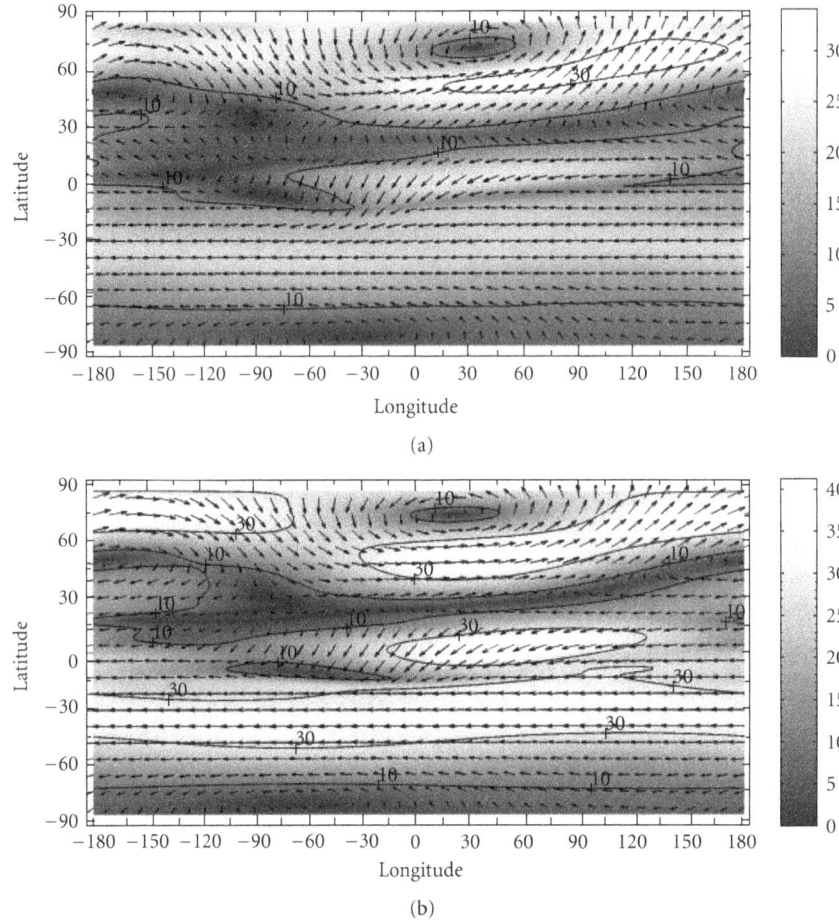

FIGURE 3: The global distributions of the vector of the simulated horizontal component of the neutral wind velocity at the altitude of 30 km, obtained for 16 January and calculated for two distinct values of solar activity: $F_{10.7} = 101$ (a) and $F_{10.7} = 230$ (b). The velocities are given in m/s.

where ρ is the neutral gas density, $\vec{V}$ is the neutral wind velocity, $\vec{F}$ is the acceleration comprising the gravity acceleration, Coriolis acceleration, acceleration of translation, and acceleration due to elastic collisions with the ion gas, and $\widehat{P}$ is the total stress tensor. The latter tensor can be decomposed as follows:

$$\widehat{P} = -p\,\widehat{I} + \widehat{\tau}, \tag{2}$$

where p is the pressure, $\widehat{I}$ is the unit tensor, and $\widehat{\tau}$ is the extra stress tensor whose components are given by the rheological equation of state or the law of viscous friction. A spherical coordinate system rotatable together with the earth is utilized in model calculations. Therefore, from the dynamical equation (1), momentum equations for the zonal, meridional, and vertical components of the neutral gas velocity may be derived. These equations include not only the pressure gradients but also partial derivatives of components of the extra stress tensor, $\widehat{\tau}$. The latter tensor is composed of a Newtonian part, $\widehat{\tau}_0$, and a complementary part, $\widehat{\tau}_1$, namely,

$$\widehat{\tau} = \widehat{\tau}_0 + \widehat{\tau}_1. \tag{3}$$

The former tensor, $\widehat{\tau}_0$, is given by the well-known Newton's law of viscous friction,

$$\widehat{\tau}_0 = 2\mu\widehat{\varepsilon}, \tag{4}$$

where μ is the coefficient of molecular viscosity, whose dependence on the temperature is assumed to obey the Sutherland's law, and $\widehat{\varepsilon}$ is the tensor defined as

$$\widehat{\varepsilon} = \widehat{D}_0 - \frac{1}{3}\widehat{I}\,\mathrm{Tr}\left(\widehat{D}_0\right), \tag{5}$$

where $\widehat{D}_0$ is the strain rate tensor and $\mathrm{Tr}(\ )$ denotes the trace of a tensor. The complementary stress tensor, $\widehat{\tau}_1$, is supposed to be conditioned by a small-scale turbulence having the scales equal and less than the steps of the finite-difference approximations. It is assumed that this tensor represents the effect of the turbulence on the mean flow and is given by an expression, analogous to the Newton's law of viscous friction (4), with the scalar coefficient of viscosity, μ, being replaced by three distinct coefficients describing the eddy viscosities in the directions of the basis vectors of the utilized spherical coordinate system. For computing the eddy viscosities, the turbulence theory of Obukhov [21] is applied.

Numerical Modeling of the Influene of Solar Activity on the Global Circulation in the Earth's Mesosphere and Lower Thermosphere

29

(a)

(b)

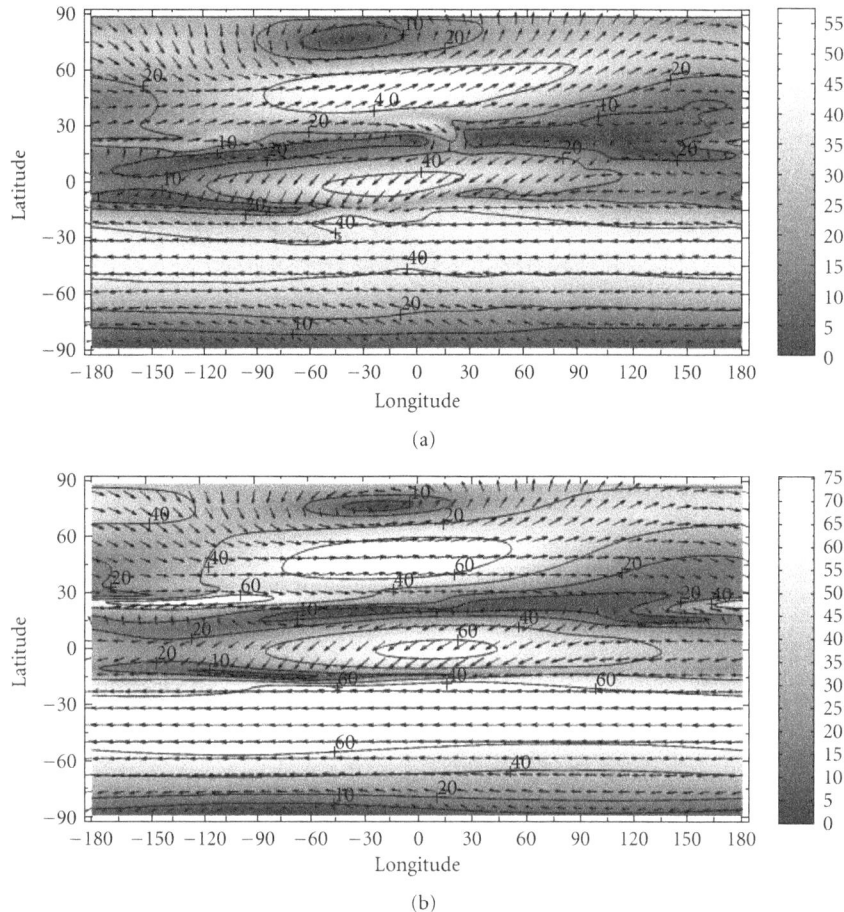

FIGURE 4: The same as in Figure 3 but at the altitude of 50 km.

3. Presentation and Discussion of Results

The utilized mathematical model of the global neutral wind system can be used for different solar cycle, seasonal, and geomagnetic conditions. For the present study, calculations were made for low-geomagnetic activity (Kp = 1) and for two different dates, namely, 16 January and 16 July, which belong to winter and summer in the northern hemisphere, respectively. To investigate the influence of solar activity on the global circulation of the atmosphere, we made calculations for conditions corresponding to two different 10.7 cm solar fluxes: moderate and high, namely, $F_{10.7} = 101$ and 230. The variations of the atmospheric parameters with time were calculated until they become stationary. The steady-state distributions of the atmospheric parameters were obtained for two considered dates on condition that inputs to the model correspond to the identical moment (10.30 UT) for each day. The temperature distributions, corresponding to this moment, were taken for each day from the NRLMSISE-00 empirical model [20]. The NRLMSISE-00 empirical atmospheric model extends from the ground to the exobase and is a major upgrade of the MSISE-90 model in the thermosphere. The model and the associated NRLMSIS database include the following data: (1) total mass density from satellite accelerometers and from orbit determination, (2) temperature from incoherent scatter radar covering 1981–1997, and (3) molecular oxygen number density, $[O_2]$, from solar ultraviolet occultation aboard the Solar Maximum Mission.

Thus, the momentum equations for the zonal, meridional, and vertical components of the neutral gas velocity acquire ultimately a form of a generalized Navier-Stokes equation for compressible gas on scales which are more than the steps of the finite-difference approximations, with the effect of the turbulence on the mean flow being taken into account by using an empirical subgrid-scale parameterization. The steps of the finite-difference approximations in the latitude and longitude directions are identical and equal to 1 degree. The height step is nonuniform and does not exceed the value of 1 km.

The simulation domain is the layer surrounding the earth globally and stretching from the ground up to the altitude of 126 km at the equator. Upper boundary conditions provide the conservation law of mass in the simulation domain. The earth's surface is supposed to coincide approximately with an oblate spheroid whose radius at the equator is more than that at the pole. More complete details of the utilized model may be found in the studies of I. V. Mingalev and V. S. Mingalev [17] and Mingalev et al. [18].

(a)

(b)

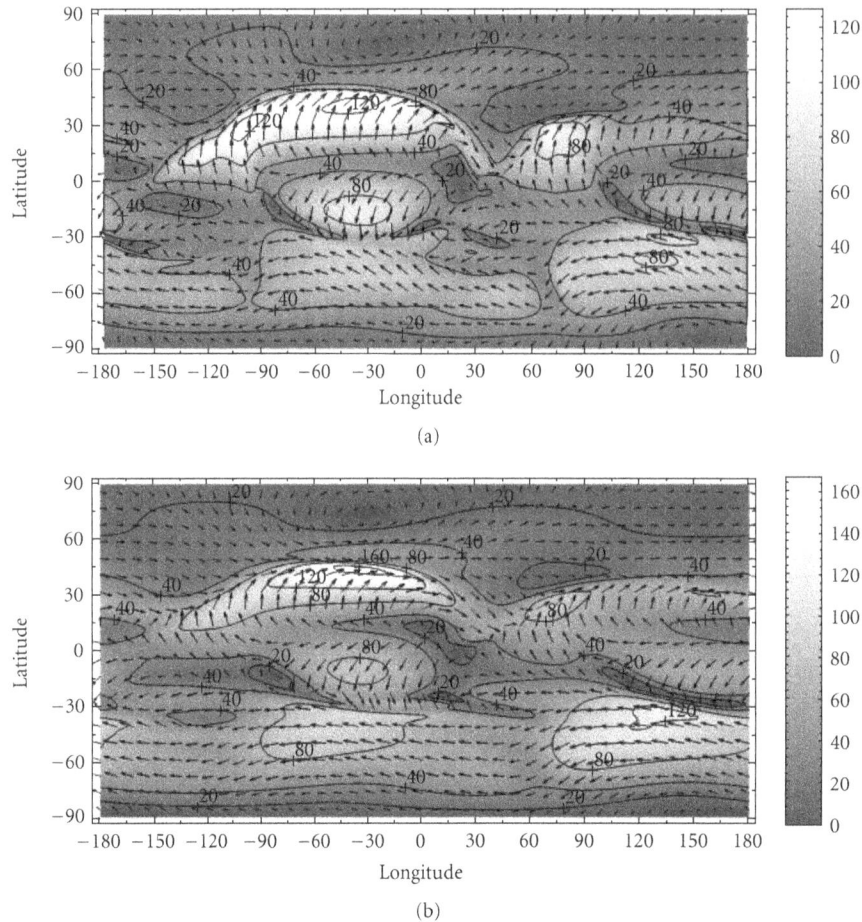

FIGURE 5: The same as in Figure 3 but at the altitude of 70 km.

The numerical solutions to system of equations, described above, were obtained on condition that the boundary conditions and inputs to the model are time independent and correspondent to 10.30 UT. Since the obtained results are different, it is convenient to present them separately.

3.1. Simulation Results for January Conditions. Let us consider simulation results, obtained for the winter period in the northern hemisphere (16 January). The atmospheric temperature is one of major input parameters of the utilized mathematical model. As was noted earlier, in the present study, this input parameter is obtained from the NRLMSISE-00 empirical model [20]. It turns out that atmospheric temperatures, calculated with the help of the NRLMSISE-00 empirical model for two distinct values of solar activity ($F_{10.7} = 101$ and 230), are very similar below approximately 100 km, while, above this altitude, they may be rather different. This fact can be easy seen from Figures 1 and 2, with the planetary distributions of the atmospheric temperature being essentially nonuniform. From Figure 1, one can see that, at the altitude of 90 km, temperatures, obtained for two considered 10.7 cm solar fluxes, are practically equal, with their differences being less than 0.001 K in whole simulation domain. From Figure 2, one can see that, at the altitude of

110 km, differences between temperatures, obtained for two considered 10.7 cm solar fluxes, can achieve a few tens of degrees at identical points of the globe. Thus, the application of the NRLMSISE-00 empirical model shows that the influence of level of solar activity on the global distribution of the atmospheric temperature ought to be absent at altitudes of the troposphere, stratosphere, and mesosphere, while this influence ought to be appreciable at altitudes of the lower thermosphere for the winter period in the northern hemisphere.

Distributions of the atmospheric parameters, calculated with the help of the mathematical model and obtained for 16 January, are shown in Figures 3–7. The results of modeling illustrate both common characteristic features and distinctions caused by different values of solar activity.

The calculated global distributions of the atmospheric parameters display the following common features. The horizontal and vertical components of the wind velocity are changeable functions of latitude and longitude at levels of the mesosphere and lower thermosphere. The horizontal domains exist where the steep gradients in the horizontal velocity field take place. The horizontal wind velocity can have various directions which may be opposite at the near points. Moreover, the horizontal domains exist in which

(a)

(b)

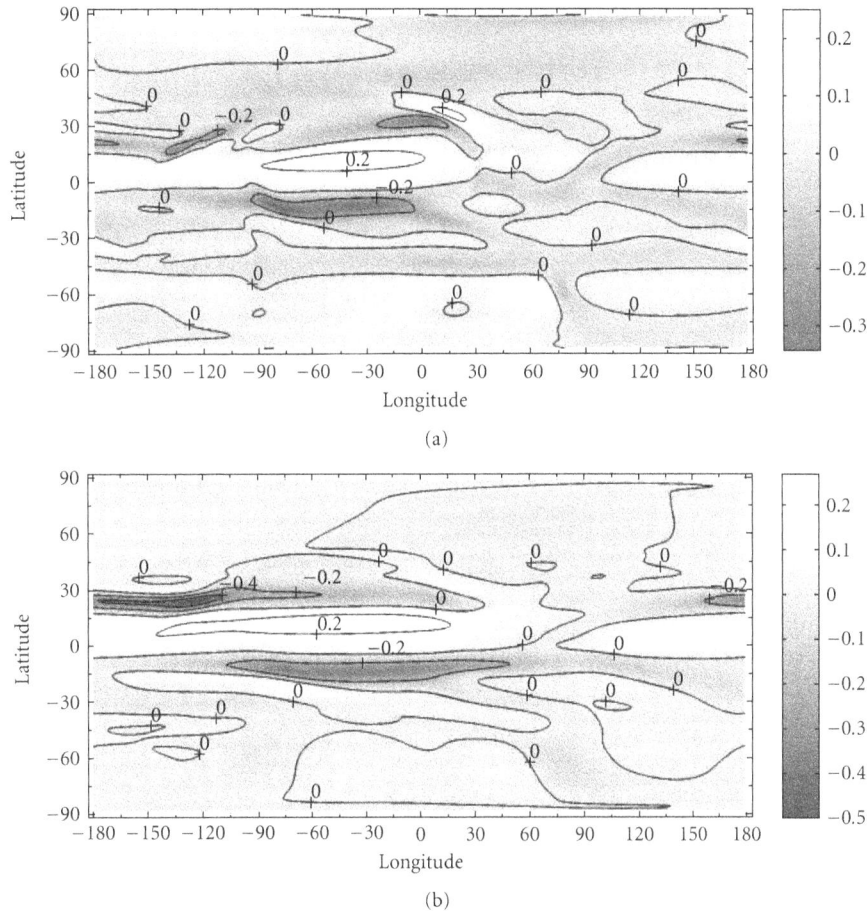

FIGURE 6: The global distributions of the simulated vertical component of the neutral wind velocity at the altitude of 50 km, obtained for 16 January and calculated for two distinct values of solar activity: $F_{10.7} = 101$ (a) and $F_{10.7} = 230$ (b). The velocities are given in m/s, with positive direction of the vertical velocity being upward.

the vertical neutral wind component has opposite directions. Maximal absolute values of the horizontal and vertical components of the wind velocity are larger at higher altitudes. At levels of the mesosphere, the horizontal wind velocity can achieve values of more than 160 m/s.

From numerous observations, it is well known that circumpolar vortices are formed at heights of the stratosphere and mesosphere in the periods close to summer and winter solstices. The circumpolar cyclone arises in the northern hemisphere under winter conditions, while the circumpolar anticyclone arises in the southern hemisphere under summer conditions.

From the results shown in Figures 3–5, we can see that, for winter period in the northern hemisphere, at levels of the stratosphere and mesosphere, the motion of the neutral gas in the northern hemisphere is primarily eastward, so a circumpolar cyclone is formed. It can be noticed that the center of the northern cyclone may be displaced from the pole. Simultaneously, the motion of the neutral gas is primarily westward in the southern hemisphere at levels of the stratosphere and mesosphere, so a circumpolar anticyclone is formed for summer period of the southern hemisphere. It can be seen that the circumpolar vortices of the northern

and southern hemispheres, simulated for moderate and high 10.7 cm solar fluxes for January conditions, correspond qualitatively to the global circulation, obtained from observations.

Let us consider simulation results, obtained for distinct values of solar activity and their distinctions. From Figure 3, one can see that, at the altitude of 30 km, the horizontal component of the wind velocity in the circumpolar cyclone of the northern hemisphere, obtained for moderate solar activity, can achieve values of less than 35 m/s, while this component, obtained for high-solar activity, exceeds a value of 40 m/s. Thus, the horizontal wind velocity in the circumpolar cyclone of the northern hemisphere, obtained for moderate solar activity, is less than that obtained for high-solar activity. Similarly, the horizontal wind velocity in the circumpolar anticyclone of the southern hemisphere, obtained for moderate-solar activity, is less than that obtained for high-solar activity.

From Figure 3, it can be seen that, at low latitudes, the horizontal wind velocity, obtained for high-solar activity, can achieve values of more than 35 m/s, while this velocity, obtained for moderate solar activity, does not exceed a value of 25 m/s. Thus, at low latitudes, the horizontal wind velocity,

(a)

(b)

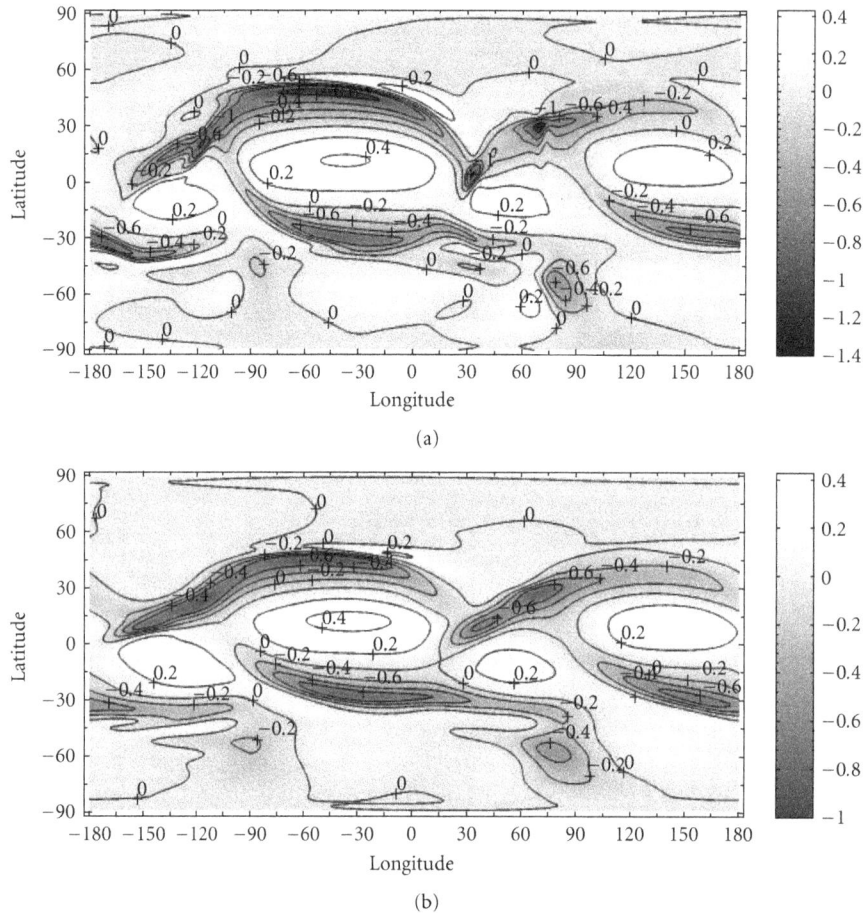

FIGURE 7: The same as in Figure 6 but at the altitude of 70 km.

obtained for high-solar activity, is more than that obtained for moderate solar activity, at the altitude of 30 km.

From Figures 4 and 5, we can see that maximal absolute values of the horizontal wind velocity, obtained for moderate-solar activity, are less than those, obtained for high-solar activity, at the altitudes of 50 and 70 km.

From the results shown in Figures 6 and 7, we can see that, for winter period in the northern hemisphere, at levels of the mesosphere, the vertical wind velocity can have opposite directions in the horizontal domains having different configurations. Maximal absolute values of the downward vertical wind component are commensurable with the maximal module of the upward vertical wind component for conditions of moderate-solar activity. On the contrary, for conditions of high-solar activity, maximal absolute values of the downward and upward vertical wind components can be rather different. At low latitudes, maximal absolute values of the vertical wind velocity, obtained for moderate solar activity, are more than those, obtained for high-solar activity, at the altitudes of the mesosphere.

Simulation results, obtained for January conditions, indicate that despite of independence of the atmospheric temperature on the 10.7 cm solar flux below approximately 100 km,

the influence of the solar activity level on the global circulation of the stratosphere and mesosphere does exist. This influence is a consequence of a relationship between large-scale circulations of the middle atmosphere and thermosphere, with the thermospheric circulation being dependent on the solar activity level, undoubtedly. The influence is conditioned by the vertical transport of air from the lower thermosphere to the mesosphere and stratosphere. This transport may be rather different under distinct solar activity conditions.

3.2. Simulation Results for July Conditions. Now, let us consider simulation results, obtained for the summer period in the northern hemisphere (16 July). The atmospheric temperatures, calculated with the help of the NRLMSISE-00 empirical model for two distinct values of solar activity ($F_{10.7} = 101$ and 230), are identical below 100 km. However, above this altitude they may be rather different at the same points of the globe. As can be seen from Figure 8, the differences between temperatures, obtained for two considered 10.7 cm solar fluxes, can achieve values of more than 30 K at the altitude of 110 km. Thus, the influence of solar activity on the global distributions of the atmospheric temperature, calculated with the help of the NRLMSISE-00 empirical

Numerical Modeling of the Influene of Solar Activity on the Global Circulation in the Earth's Mesosphere and Lower Thermosphere

33

(a)

(b)

(c)

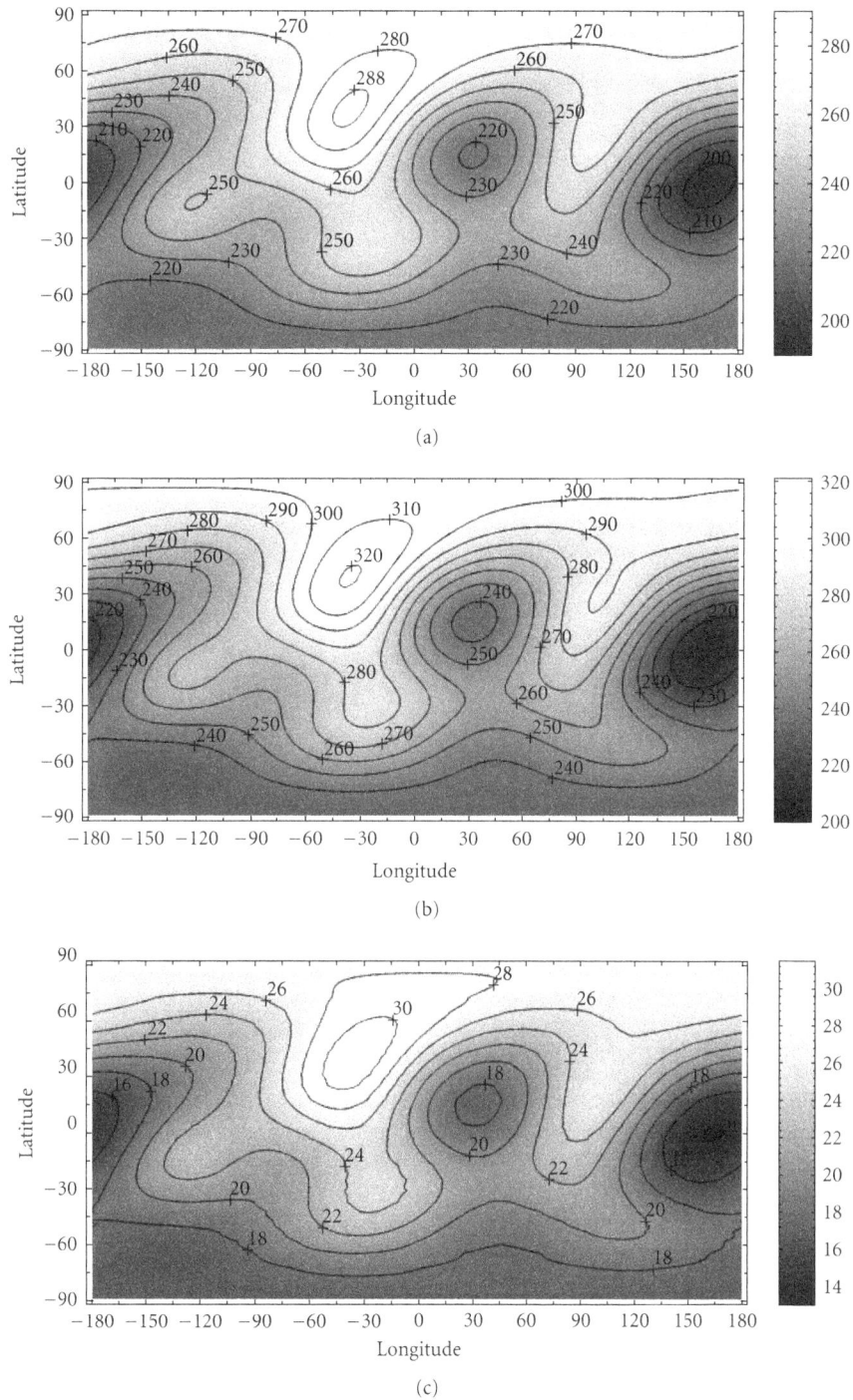

FIGURE 8: The global distributions of the atmospheric temperature (K) at 110 km altitude, obtained from the NRLMSISE-00 empirical model for 16 July, UT = 10.30 and calculated for two distinct values of solar activity: $F_{10.7} = 101$ (a) and $F_{10.7} = 230$ (b). Also the difference between the latter and the former distributions (c).

model, ought to be absent below 100 km for the summer period in the northern hemisphere.

The results of modeling, obtained for 16 July, are shown in Figures 9–13. These results illustrate common characteristic features which, in essence, are the same as in January conditions, described in the previous subsection. The

simulation results, presented in Figures 9 and 10, indicate that, for the summer period in the northern hemisphere, at levels of the stratosphere and mesosphere, the motion of the neutral gas in the northern hemisphere is primarily westward, so a circumpolar anticyclone is formed. Simultaneously, the motion of the neutral gas is primarily eastward

(a)

(b)

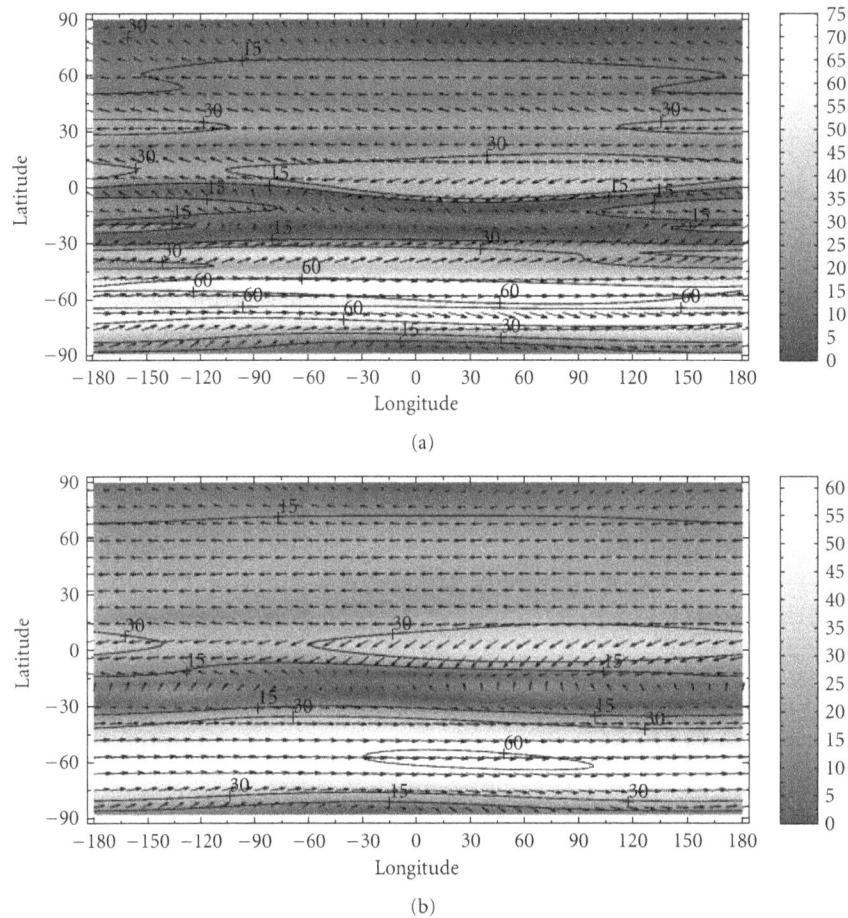

FIGURE 9: The global distributions of the vector of the simulated horizontal component of the neutral wind velocity at the altitude of 30 km, obtained for 16 July and calculated for two distinct values of solar activity: $F_{10.7} = 101$ (a) and $F_{10.7} = 230$ (b). The velocities are given in m/s.

in the southern hemisphere at levels of the stratosphere and mesosphere, so a circumpolar cyclone is formed for winter period of the southern hemisphere.

From numerous observations, it is known that a circumpolar anticyclone arises in the northern hemisphere under summer conditions, while the circumpolar cyclone arises in the southern hemisphere under winter conditions. It is easy to see that the circumpolar vortices of the northern and southern hemispheres, simulated for moderate and high 10.7 cm solar fluxes for July conditions, correspond qualitatively to the global circulation, obtained from observations.

From the results shown in Figures 9 and 10, we can see that the horizontal wind velocity in the circumpolar cyclone of the southern hemisphere, obtained for moderate-solar activity, is more than that obtained for high-solar activity. Also, from Figure 9, it can be seen that, at low latitudes, the westward horizontal wind velocity, obtained for moderate solar-activity, is more than that obtained for high-solar activity, at the altitude of 30 km. From Figure 10, it can be seen that, close to a latitude of $-30°$, a long narrow band exists in which the horizontal component of the wind velocity, calculated for high-solar activity, can achieve values of more than 110 m/s. This peculiarity is absent in the results,

calculated for moderate-solar activity. From Figure 11, we can see that maximal absolute values of the horizontal wind velocity, obtained for moderate-solar activity, are more than those, obtained for high-solar activity, at the altitude of 90 km. As it is seen from Figures 12 and 13, at levels of the mesosphere and lower thermosphere, the vertical wind velocity, calculated for the summer period in the northern hemisphere, can have opposite directions in the horizontal domains having different configurations. The horizontal domains, where the vertical wind component is downward, are similar to a long narrow band, while the horizontal domains, where the vertical component of the neutral wind is upward, are similar to a spot having a great length and large width.

From the results shown in Figure 12, we can see that maximal values of the upward vertical wind component, calculated for both moderate- and high-solar activities, are approximately equal at the altitude of 50 km. On the contrary, maximal values of the downward vertical wind component, calculated for high-solar activity, are more than those obtained for moderate-solar activity, at the altitude of 50 km. From Figure 13, we can see that maximal absolute values of the vertical wind velocity can achieve magnitudes

Numerical Modeling of the Influene of Solar Activity on the Global Circulation in the Earth's Mesosphere
and Lower Thermosphere

35

(a)

(b)

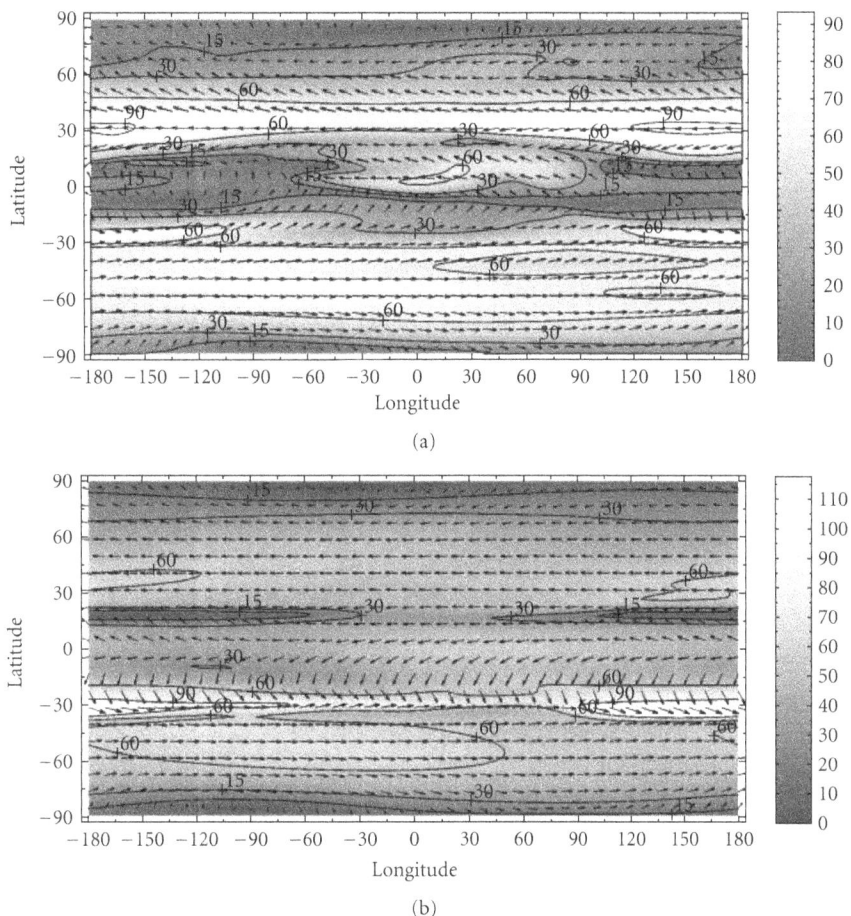

FIGURE 10: The same as in Figure 9 but at the altitude of 50 km.

of about 2 m/s at the altitude of 90 km. It is doubtful that such magnitudes may be simulated by general circulation models applying the hydrostatic approach. On the contrary, it may be expected that such magnitudes may be simulated by nonhydrostatic general circulation models, in particular by the Whole Atmosphere Community Climate Model (WACCM), utilized in the study of Smith et al. [22].

From the simulation results, obtained for the summer period in the northern hemisphere, we can see that the atmospheric temperature, calculated with the help of the NRLMSISE-00 empirical model, does not depend on the 10.7 cm solar flux below approximately 100 km. Nevertheless, below 100 km, the effect of solar activity on the global circulation of the atmosphere does exist. This effect is conditioned by a relationship between circulations of middle atmosphere and thermosphere. A variation of the 10.7 cm solar flux leads to changes of large-scale circulation of the thermosphere, in particular, the vertical wind, which can penetrate to low altitudes and affect the global circulation of the mesosphere and stratosphere.

4. Summary and Conclusions

The nonhydrostatic model of the global neutral wind system of the earth's atmosphere, developed earlier in the Polar

Geophysical Institute, was utilized to investigate how solar activity affects the formation of the large-scale global circulation of the mesosphere and lower thermosphere. The applied mathematical model differs essentially from existing global circulation models of the atmosphere. Firstly, an internal energy equation for the neutral gas is not included in the applied mathematical model. Instead, the global temperature field is assumed to be a given distribution, that is, the input parameter of the model and obtained from the NRLMSISE-00 empirical model [20]. Secondly, the vertical component of the neutral wind velocity is calculated without using the hydrostatic equation. The vertical component of the neutral wind velocity as well as the horizontal components is obtained by means of a numerical solution of a generalized Navier-Stokes equation for compressible gas, with whatever restrictions on the vertical transport of the neutral gas being absent.

The applied mathematical model was utilized for obtaining the steady-state distributions of the atmospheric parameters, using the method of establishment, for conditions, corresponding to two different dates, namely, 16 January and 16 July, for the identical moment (10.30 UT) for each considered day. Calculations were performed for low-geomagnetic activity and for conditions corresponding to two different

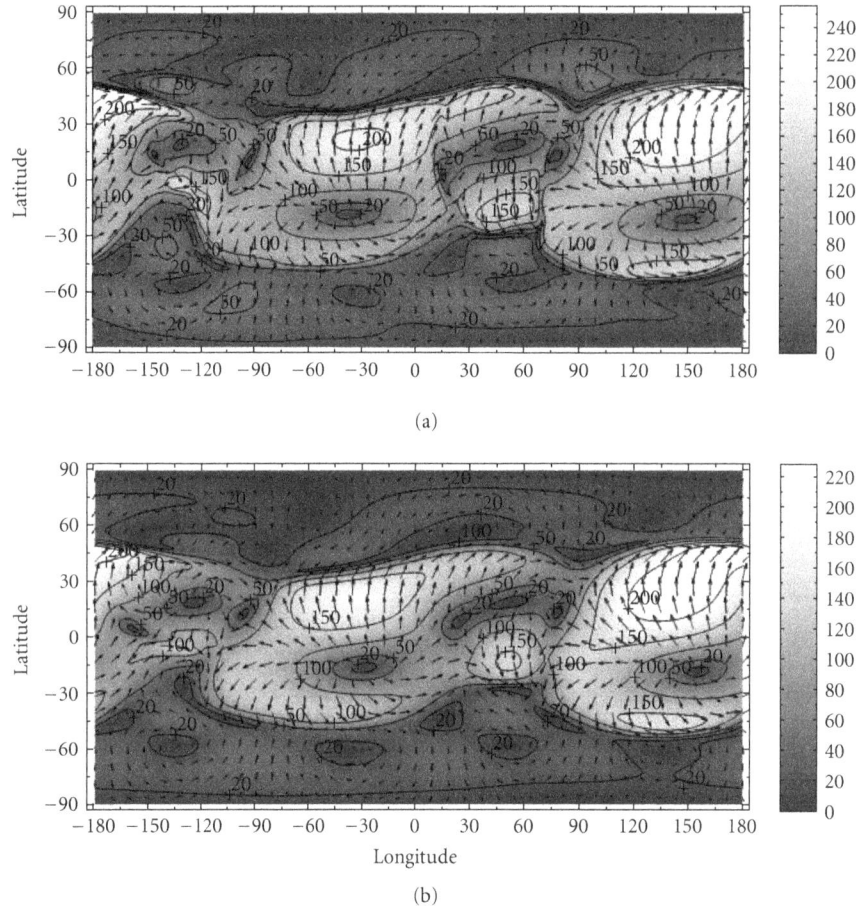

FIGURE 11: The same as in Figure 9 but at the altitude of 90 km.

10.7 cm solar fluxes: moderate and high, namely, $F_{10.7} = 101$ and 230.

The results of simulation illustrate common characteristic features. In particular, the calculated horizontal and vertical components of the wind velocity are changeable functions of latitude and longitude at levels of the mesosphere and lower thermosphere. It turned out that the calculated global distributions of the horizontal wind velocity, obtained for different dates and solar activity levels, contain large-scale circumpolar vortices of the northern and southern hemispheres. It turned out that, in the northern hemisphere, the circumpolar cyclone is formed under winter conditions and the circumpolar anticyclone is formed under summer conditions. In the southern hemisphere, the circumpolar anticyclone is formed in January and the circumpolar cyclone is formed in July. It should be emphasized that the circumpolar vortices of the northern and southern hemispheres, obtained using the mathematical model at levels of the mesosphere for January and July conditions, are consistent with existing observational data. This fact manifests the adequacy of the mathematical model utilized in the present study.

The simulation results, obtained for January conditions, indicated that solar activity ought to influence considerably on the formation of global neutral wind system in the mesosphere and lower thermosphere. In both northern and southern hemispheres, above approximately 25 km, maximal absolute values of the horizontal wind velocity, obtained for moderate-solar activity, ought to be less than those, obtained for high-solar activity. At low latitudes, maximal absolute values of the horizontal wind velocity, obtained for high solar activity, ought to be more than those, obtained for moderate solar activity, at altitudes of more than 25 km. At low latitudes, maximal absolute values of the vertical component of the wind velocity, obtained for moderate-solar activity, ought to be more than those, obtained for high-solar activity, at altitudes of the mesosphere.

The simulation results, obtained for July conditions, possess the following characteristic features. In the northern hemisphere, below approximately 40 km, maximal absolute values of the horizontal wind velocity, obtained for moderate-solar activity, ought to be less than those, obtained for high-solar activity, while, above this altitude, these values, obtained for moderate-solar activity, ought to be more than ones, obtained for high-solar activity. In the southern hemisphere, below approximately 40 km, maximal absolute values of the horizontal wind velocity, obtained for moderate-solar activity, ought to be more than those, obtained for high-solar

Numerical Modeling of the Influene of Solar Activity on the Global Circulation in the Earth's Mesosphere
and Lower Thermosphere

37

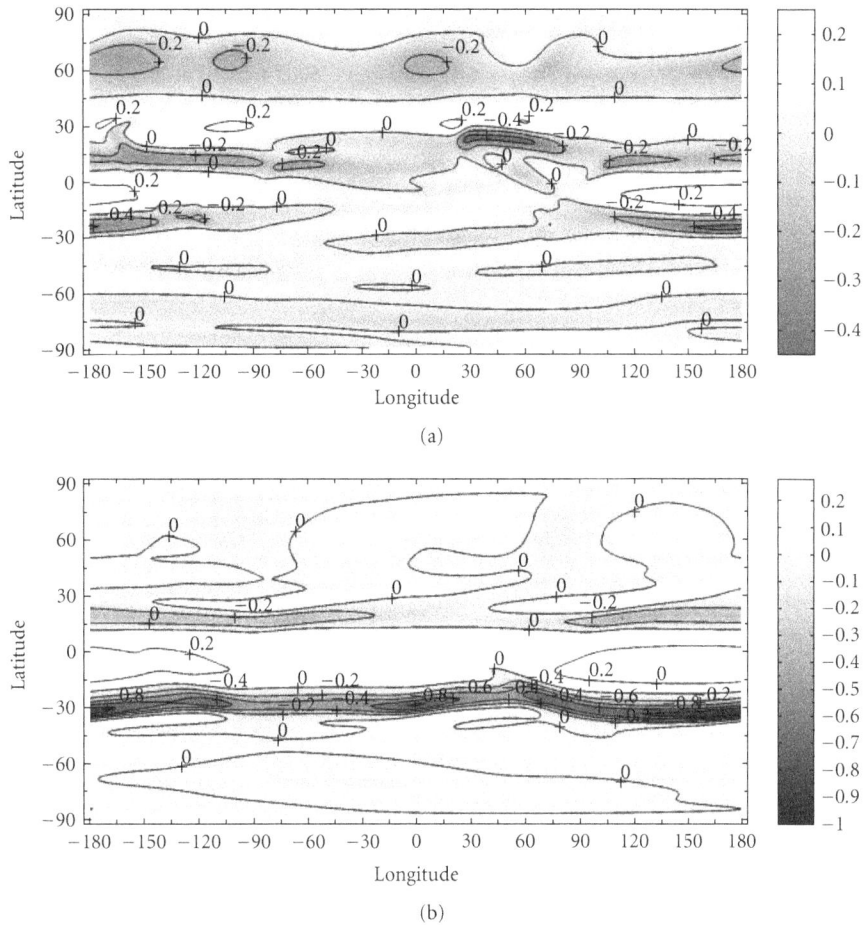

FIGURE 12: The global distributions of the simulated vertical component of the neutral wind velocity at the altitude of 50 km, obtained for 16 July and calculated for two distinct values of solar activity: $F_{10.7} = 101$ (a) and $F_{10.7} = 230$ (b). The velocities are given in m/s, with positive direction of the vertical velocity being upward.

activity, while, above this altitude, these values, obtained for moderate-solar activity, ought to be less than ones, obtained for high-solar activity. At low latitudes, below approximately 25 km, maximal absolute values of the horizontal wind velocity, obtained for moderate-solar activity, ought to be less than those, obtained for high-solar activity, while, above this altitude, these values, obtained for moderate-solar activity, ought to be more than ones, obtained for high-solar activity.

From the simulation results presented, we can see that the atmospheric temperature, calculated with the help of the NRLMSISE-00 empirical model, does not depend on the 10.7 cm solar flux below approximately 100 km. Nevertheless, the effect of solar activity on the global circulation of the atmosphere below 100 km does exist. This effect is conditioned by a relationship between global circulations of the thermosphere and middle atmosphere. In this relationship, a vertical motion of air can play a significant role. At altitudes of more than 100 km, the global distributions of the atmospheric temperatures, calculated for distinct 10.7 cm solar fluxes, are different. As a consequence, correspondent global circulations of the atmosphere at these altitudes are different, too, including the vertical wind system. Since the

vertical wind can penetrate to low altitudes, the global circulation of the mesosphere and stratosphere may be transformed. In this way, the influence of solar activity on the global circulation of the mesosphere and stratosphere is primarily realized. Incidentally, the utilized mathematical model was able to simulate this influence due to the fact that the model is nonhydrostatic.

It can be noticed that the point of view exists that the atmospheric temperature ought to depend on the 10.7 cm solar flux below 100 km (e.g., see [23, 24]). However, such dependence is not reproduced by the NRLMSISE-00 empirical model as well as by other existing empirical models of the global distributions of the atmospheric temperature. Nevertheless, the application of the NRLMSISE-00 empirical model leads to the effect of solar activity on the global circulation of the atmosphere below 100 km due to the vertical transport of air from the lower thermosphere to the mesosphere and stratosphere. It may be expected that, taking into consideration the dependence of atmospheric temperature on the 10.7 cm solar flux below 100 km, one can obtain in future a more pronounced effect of solar activity on the global circulation of the atmosphere below 100 km.

(a)

(b)

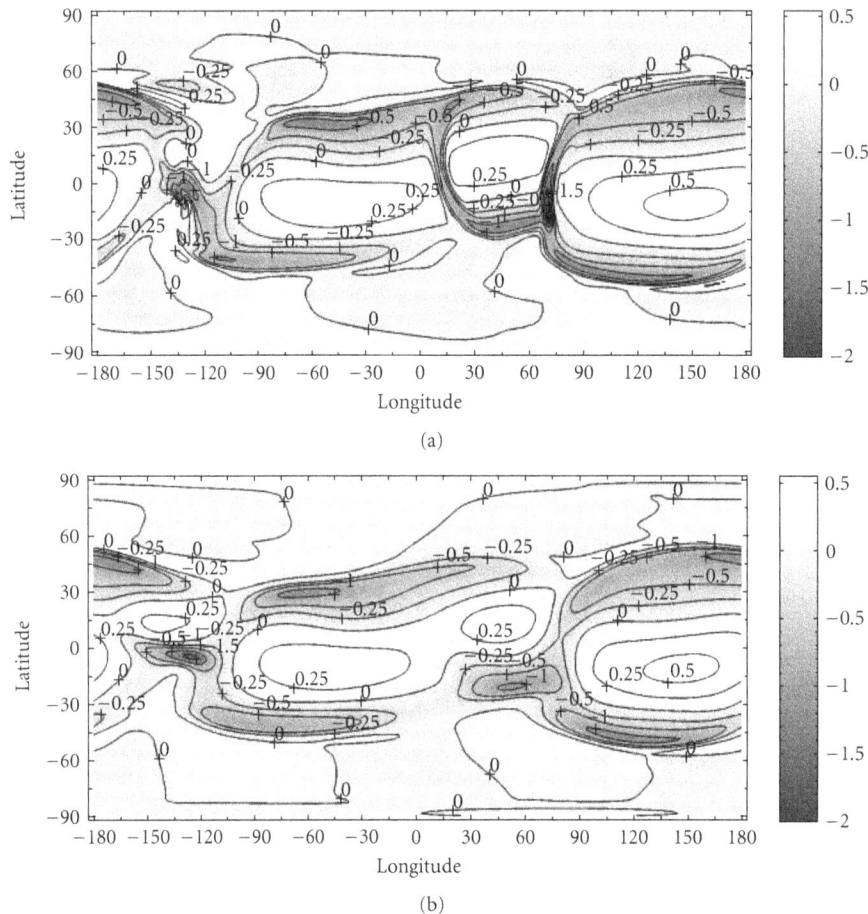

FIGURE 13: The same as in Figure 12 but at the altitude of 90 km.

Acknowledgments

This work was partly supported by Grant no. 10-01-00451 from the Russian Foundation for Basic Research. The authors would like to thank the reviewers for helpful suggestions that led to improvement in the original paper.

References

[1] S. Manabe and D. G. Hahn, "Simulation of atmospheric variability," *Monthly Weather Review*, vol. 109, no. 11, pp. 2260–2286, 1981.

[2] D. Cariolle, A. Lasserre-Bigorry, J. F. Royer, and J. F. Geleyn, "A general circulation model simulation of the springtime antarctic ozone decrease and its impact on mid-latitudes," *Journal of Geophysical Research*, vol. 95, no. 2, pp. 1883–1898, 1990.

[3] P. J. Rasch and D. L. Williamson, "The sensitivity of a general circulation model climate to the moisture transport formulation," *Journal of Geophysical Research*, vol. 96, article D7, pp. 123–137, 1991.

[4] H. F. Graf, I. Kirchner, R. Sausen, and S. Schubert, "The impact of upper-tropospheric aerosol on global atmospheric circulation," *Annales Geophysicae*, vol. 10, pp. 689–707, 1992.

[5] P. A. Stott and R. S. Harwood, "An implicit time-stepping scheme for chemical species in a Global atmospheric circulation model," *Annales Geophysicae*, vol. 11, pp. 377–388, 1993.

[6] B. Christiansen, A. Guldberg, A. W. Hansen, and L. P. Riishojgaard, "On the response of a three-dimensional general circulation model to imposed changes in the ozone distribution," *Journal of Geophysical Research D*, vol. 102, no. 11, pp. 13051–13077, 1997.

[7] V. Y. Galin, "Parametrization of radiative processes in the DNM atmospheric model," *Izvestiya, Atmospheric and Ocean Physics*, vol. 34, no. 3, pp. 339–347, 1998 (Russian).

[8] A. L. Gibelin and M. Déqué, "Anthropogenic climate change over the mediterranean region simulated by a global variable resolution model," *Climate Dynamics*, vol. 20, no. 4; pp. 327–339, 2002.

[9] M. Mendillo, H. Rishbeth, R. G. Roble, and J. Wroten, "Modelling f2-layer seasonal trends and day-to-day variability driven by coupling with the lower atmosphere," *Journal of Atmospheric and Solar-Terrestrial Physics*, vol. 64, no. 18, pp. 1911–1931, 2002.

[10] M. J. Harris, N. F. Arnold, and A. D. Aylward, "A study into the effect of the diurnal tide on the structure of the background mesosphere and thermosphere using the new coupled middle atmosphere and thermosphere (cmat) general circulation model," *Annales Geophysicae*, vol. 20, no. 2, pp. 225–235, 2002.

[11] U. Langematz, A. Claussnitzer, K. Matthes, and M. Kunze, "The climate during the maunder minimum: a simulation with the freie universität berlin climate middle atmosphere

model (fub-cmam)," *Journal of Atmospheric and Solar-Terrestrial Physics*, vol. 67, no. 1-2, pp. 55–69, 2005.

[12] S. Peteherych, G. G. Shepherd, and J. K. Walker, "Observation of vertical e-region neutral winds in two intense auroral arcs," *Planetary and Space Science*, vol. 33, no. 8, pp. 869–873, 1985.

[13] H. U. Widdel, "Vertical movements in the middle atmosphere derived from foil cloud experiments," *Journal of Atmospheric and Terrestrial Physics*, vol. 49, no. 7-8, pp. 723–741, 1987.

[14] U. P. Hoppe and T. Hansen, "Studies of vertical motions in the upper mesosphere using the EISCAT UHF radar," *Annales Geophysicae*, vol. 6, pp. 181–186, 1988.

[15] G. D. Price and F. Jacka, "The influence of geomagnetic activity on the upper mesosphere/lower thermosphere in the auroral zone. i. vertical winds," *Journal of Atmospheric and Terrestrial Physics*, vol. 53, no. 10, pp. 909–922, 1991.

[16] M. Ishii, "Relationship between thermospheric vertical wind and the location of ionospheric current in the polar region," *Advances in Polar Upper Atmosphere Research*, vol. 19, pp. 63–70, 2005.

[17] I. V. Mingalev and V. S. Mingalev, "The global circulation model of the lower and middle atmosphere of the Earth with a given temperature distribution," *Mathematical Modeling*, vol. 17, pp. 24–40, 2005 (Russian).

[18] I. V. Mingalev, V. S. Mingalev, and G. I. Mingaleva, "Numerical simulation of the global distributions of the horizontal and vertical wind in the middle atmosphere using a given neutral gas temperature field," *Journal of Atmospheric and Solar-Terrestrial Physics*, vol. 69, no. 4-5, pp. 552–568, 2007.

[19] I. V. Mingalev, O. V. Mingalev, and V. S. Mingalev, "Model simula-tion of the global circulation in the middle atmosphere for january conditions," *Advances in Geosciences*, vol. 15, pp. 11–16, 2008.

[20] J. M. Picone, A. E. Hedin, D. P. Drob, and A. C. Aikin, "Nrlmsise-00 empirical model of the atmosphere: statistical comparisons and scientific issues," *Journal of Geophysical Research A*, vol. 107, no. 12, article 1468, 16 pages, 2002.

[21] A. M. Obukhov, *Turbulence and Dynamics of Atmosphere*, Hydrometeoizdat, Leningrad, Russia, 1988.

[22] A. K. Smith, R. R. Garcia, D. R. Marsh, and J. H. Richter, "WACCM simulations of the mean circulation and trace species transport in the winter mesosphere," *Journal of Geophysical Research D*, vol. 116, no. 20, Article ID D20115, 17 pages, 2011.

[23] R. E. Cahalan, G. Wen, J. W. Harder, and P. Pilewskie, "Temperature responses to spectral solar variability on decadal time scales," *Geophysical Research Letters*, vol. 37, Article ID L07705, 5 pages, 2010.

[24] A. I. Semenov, N. N. Shefov, and V. A. Sukhodoev, "Re-analysis of the long-term hydroxyl rotational temperature trend according to measurements in Spitsbergen," in *Proceedings of the 29th Annual Seminar on Physics of Auroral Phenomena*, Polar Geophysical Institute, Apatity, Russia, 2007.

4

Reverse Time Migration: A Seismic Imaging Technique Applied to Synthetic Ultrasonic Data

Sabine Müller,[1] Ernst Niederleithinger,[1] and Thomas Bohlen[2]

[1] Division 8.2 of Non-destructive Damage Assessment and Environmental Measurement Methods,
   BAM Federal Institute for Materials Research and Testing, Unter den Eichen 87, 12205 Berlin, Germany
[2] Geophysical Institute, KIT Karlsruhe Institute of Technology, Hertzstr. 16, 76187 Karlsruhe, Germany

Correspondence should be addressed to Sabine Müller, sabine.mueller@bam.de

Academic Editor: Joerg Schleicher

Ultrasonic echo testing is a more and more frequently used technique in civil engineering to investigate concrete building elements, to measure thickness as well as to locate and characterise built-in components or inhomogeneities. Currently the Synthetic Aperture Focusing Technique (SAFT), which is closely related to Kirchhoff migration, is used in most cases for imaging. However, this method is known to have difficulties to image steeply dipping interfaces as well as lower boundaries of tubes, voids or similar objects. We have transferred a processing technique from geophysics, the Reverse Time Migration (RTM) method, to improve the imaging of complicated geometries. By using the information from wide angle reflections as well as from multiple events there are fewer limitations compared to SAFT. As a drawback the required computing power is significantly higher compared to the techniques currently used. Synthetic experiments have been performed on polyamide and concrete specimens to show the improvements compared to SAFT. We have been able to image vertical interfaces of step-like structures as well as the lower boundaries of circular objects. It has been shown that RTM is a step forward for ultrasonic testing in civil engineering.

1. Introduction

Quality assurance and damage assessment of concrete structures as buildings, bridges, and dams are major tasks in civil engineering. On one side construction drawings may be missing or are considered to be unreliable, which is a drawback for maintenance and rehabilitation measures. On the other side parts of the public infrastructure, namely, bridges, suffer from aging, increased traffic, and increasing individual truck weight. In addition nondestructive inspection methods are required to provide a reliable quality assurance for new, repaired, or rebuilt structures. Since about twenty years, ultrasonic techniques are more and more frequently used to map the interior of structures, for example, to locate layers, voids, objects, or other features. The localisation and characterisation of tendon ducts is of major importance. These ducts are used to apply external forces to the concrete via prestressed steel wires to enhance its tension resistance. After prestretching these wires, the duct is grouted with mortar in order to couple the wires and the concrete as well as

to avoid corrosion. Remaining air voids are a problem for the durability of the structure. In many cases the first step in an assessment is to locate the tendon ducts, because quite often they are not placed in accordance with the design plans.

Until 20 years ago, ultrasonic testing in civil engineering was limited to relatively simple methods as time of flight transmission measurements to assess concrete quality. Meanwhile echo measurements are a well-established nondestructive testing method in civil engineering. Recently, the introduction of new transmitters and receivers, array techniques, and data processing methods has led to the development of imaging applications in research and practice. The data acquisition geometries as well as wave phenomena are similar to the ones in seismic techniques, but on a different scale. Imaging is mainly done using SAFT (synthetic aperture focusing technique), which is closely related to Kirchhoff migration. Thereby it suffers from the same limitations, for example, problems with imaging of vertical or steeply dipping interfaces. Measurements are mainly done using a zero offset geometry. That means that receiver and

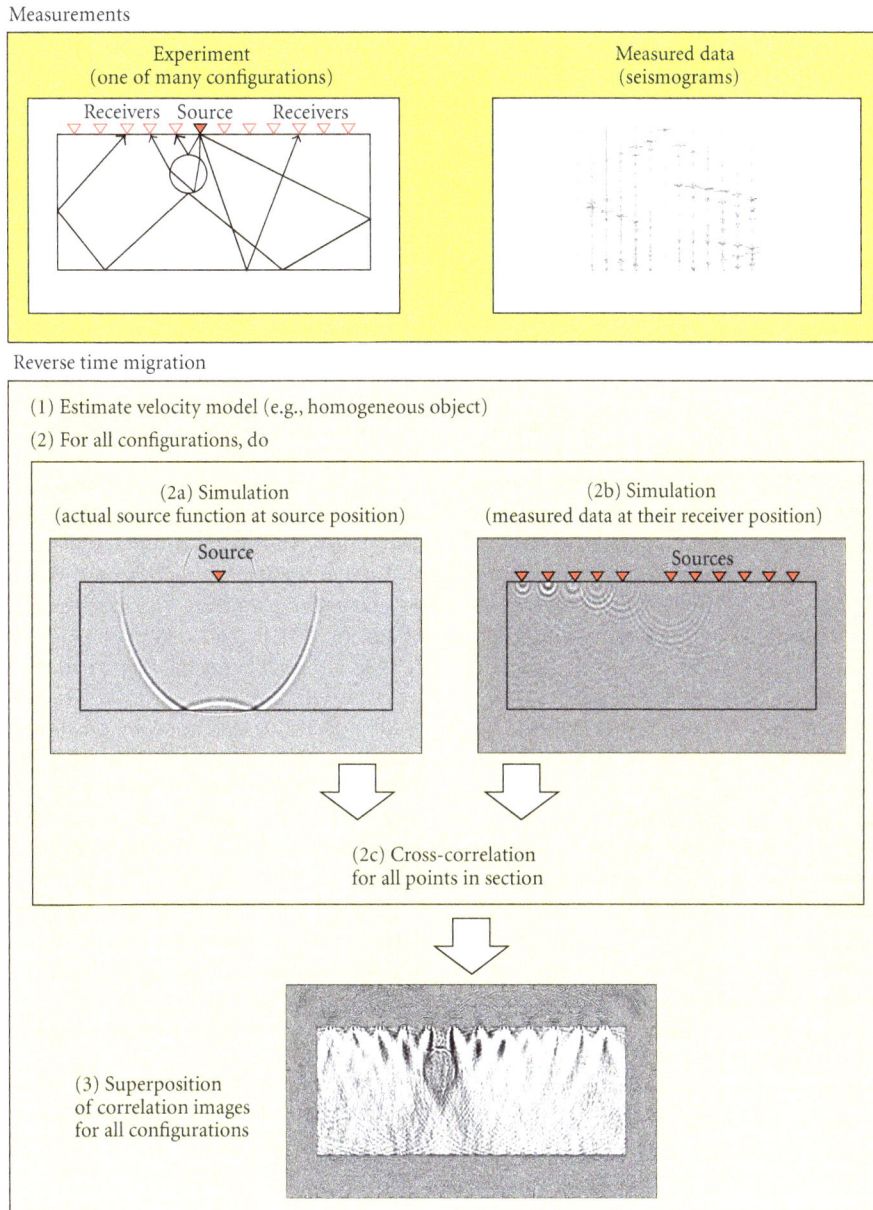

FIGURE 1: Principle of reverse time migration.

transmitter have a constant distance of some centimetres. By this reason only the top of, for example, tendon ducts can be imaged. The duct also shadows the back wall and any other structural features behind it. Details on SAFT algorithms applied to measurements on concrete have been published by Mayer et al. [1] and Schickert et al. [2]. Krause et al. [3] have compiled the state of the art on imaging grouting faults in tendon ducts. An extension of the method using phase analysis of the measured signal reflections was published by Krause et al. [4]. In exploration seismics several methods have been introduced to image complex structures including steeply dipping interfaces. A promising technique called reverse time migration (RTM) was originally introduced by Baysal et al. [5], Loewenthal and Mufti [6], and McMechan [7] in 1983. However, due to the required computing power, it was not until after 2000 that RTM became a practicable method. Some examples of recent applications of RTM in hydrocarbon exploration are shown by Farmer et al. [8]. Other authors have experimented successfully with the application of RTM in structural health monitoring of composites, for example, crack detection (Zhou et al. [9], Wang and Yuan [10]).

In this paper we show the potential of RTM for ultrasonic testing in civil engineering and the advantages compared to SAFT using synthetic models.

2. Reverse Time Migration

The basic idea of RTM is a three-step procedure of (a) forward modelling of a wave field through an appropriate

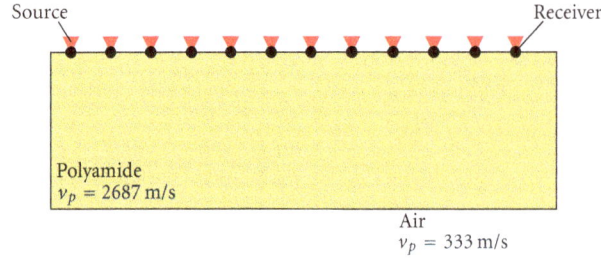

FIGURE 2: Homogeneous velocity model for migration.

velocity model as well as (b) back propagation of the measured data through the very same model and (c) superposition of both using an imaging condition. The complete wave equation is used for the simulation. There are no dip restrictions for imaging because the entire wave field including multiple reflections is used. Thus even vertical boundaries or lower edges of objects can be imaged.

The RTM algorithm used in this paper is based on a finite difference modelling code by Bohlen [11]. Currently we are using the two-dimensional acoustic wave equation. Some improvements for full waveform inversion, acoustic modelling with perfectly matched layers as boundary condition, and shot parallelisation were done by Kurzmann et al. [12]. The code can be applied to different tasks from high-frequency ultrasonic measurements to exploration seismics. It provides a set of parameters to be adapted to the specific problem. For example, arbitrary source functions can be chosen to mimic ultrasonic transducers or explosive sources. In our examples we have used a Ricker wavelet as source signal and perfectly matched layers as boundary condition.

Figure 1 shows the principle of RTM for a 2D case with coordinates x (horizontal) and z (vertical). In an experiment with several sources and receivers measurement data are acquired. The recording time must be sufficiently long to include multiple reflections. A velocity model has to be chosen for the imaging process. A homogeneous or smoothed model is used in most cases. For all shot point configurations two simulations are performed using the very same model: (a) forward simulation using an appropriate source function at the shot point to calculate the wave field $S_s(t,z,x)$ and (b) reverse modelling using the reversed measured data as a source function at the corresponding receiver points to calculate a second wave field $R_s(t,z,x)$. After the simulation an imaging condition is applied. In our case we have used cross-correlation which is normalised by the square of the source illumination strength. The correlation images of all shots are summed up to get the image $I(z,x)$ [13]:

$$I(z,x) = \sum_s \frac{\sum_t S_s(t,z,x)R_s(t,z,x)}{\sum_t S_s^2(t,z,x)}. \qquad (1)$$

3. Simulation of Typical Problems in Nondestructive Testing

We have tested the RTM with synthetic data, which we had generated with the forward modelling part of the RTM code. We created several synthetic models to study different tasks in nondestructive testing. The first models represent polyamide specimen, a homogeneous, diffraction-free material. In a second step we have simulated a concrete-like structure with multiple scatterers.

We have chosen four polyamide models to show the capabilities of the algorithm: a step, two steps, an air inclusion, and a step with an air inclusion. The models consist of 1200 by 500 grid points with a distance between grid points of $dh = 0.001$ m. The specimens have a maximum size of 1 m by 0.30 m and compressional wave velocity of $v_p = 2687$ m/s. The specimens are surrounded by an 0.10 m width air layer with a velocity of $v_p = 333$ m/s. On the top surface of the polyamide specimen we positioned the 12 sources and coincidental receivers with a distance of $x = 0.08$ m between them. If a Ricker wavelet with $f = 100$ kHz is inserted at every shot point, all 12 receivers are in record mode. The recording time is $t = 0.6$ ms with a time stepping of $dt = 0.1\,\mu$s. This way even multiple reflections are recorded.

For all following RTM reconstructions we have used a homogeneous, rectangular velocity model ($v_p = 2687$ m/s as in data generation), shown in Figure 2. This way, the algorithm has no information about the real structure, except for the outer limits and the material velocity.

3.1. Step. The first synthetic model includes a small step in the lower boundary of the polyamide specimen (Figure 3(a)). The step has a height of $z = 0.10$ m. Figure 3(b) shows the result of the RTM calculated with the homogeneous velocity model. The step is imaged in the correct position and height. The vertical border of the step is clearly visible. There are also artefacts at every multiple height of the step, which are not completely explained until now. They are probably an effect of the cross-correlation, as the velocity model used here intentionally does not include the step. So the forward wave field contains no information about the step while the back propagation (using the measured data) does. So we get a high signal in the cross-correlated image at the real step position, but also lower signals in multiple distance. The longer the recording time is, the more multiple reflections are recorded. That is why the amplitude of the artefacts increase with longer recording time. This result shows the importance to choose a proper recording time for experiments, long enough to image the entire structure, but not too long to enhance artefacts.

3.2. Two Steps. The next model consists of two steps with a height of $z = 0.06$ m for each step (Figure 4(a)). Figure 4(b) shows the migrated image generated with the homogeneous

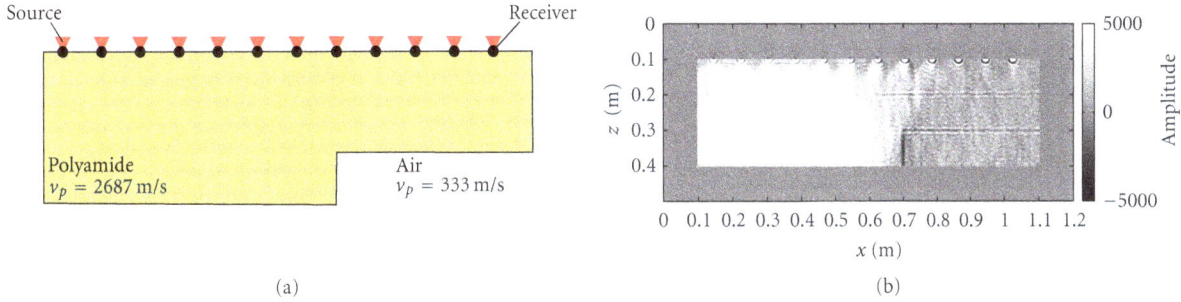

FIGURE 3: (a) Synthetic step model. (b) RTM imaging result using velocity model of Figure 2.

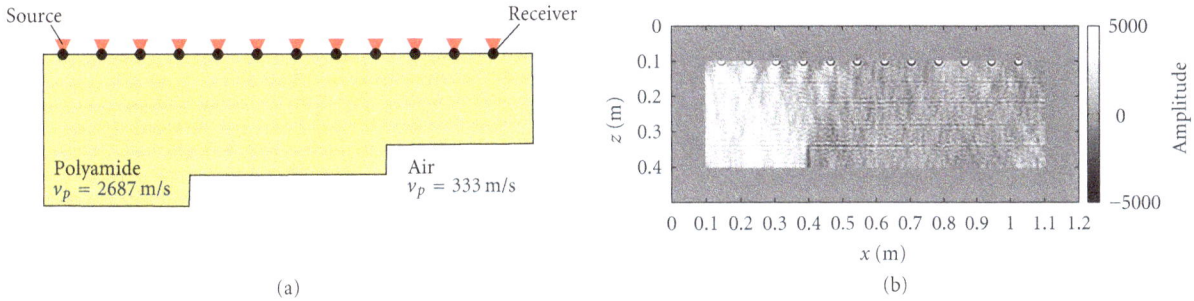

FIGURE 4: (a) Two-step model. (b) RTM imaging result using velocity model of Figure 2.

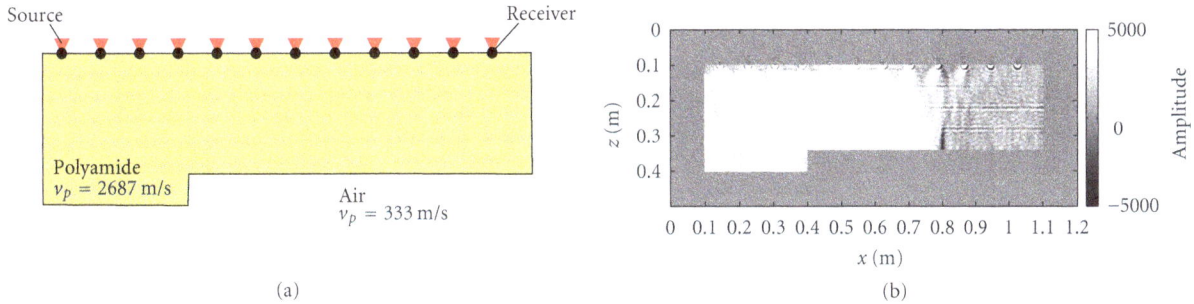

FIGURE 5: (a) Modified velocity model for migration. (b) RTM imaging result of the two steps using velocity model of (a).

rectangular model. Now only the first step is imaged at the right height and with a clear vertical border. However, the second step is missing. Only artefacts by multiple reflections are imaged. Apparently RTM fails, if the model used for reconstruction is too far from reality. Thus we have tried a kind of two-stage RTM. We used the information from the first reconstruction, the imaged step, and created a new velocity model (Figure 5(a)). Figure 5(b) shows the result of the second RTM using the improved velocity model. Now the second step is clearly visible. This example illustrates that for more complicated models a multistage RTM is required if insufficient a priori information about the structure is available.

3.3. Air Inclusion. The next model includes a circular-shaped air inclusion in polyamide. The circle has a radius of $r = 0.05$ m. The midpoint is at $x = 0.5$ m and $z = 0.2$ m (Figure 6(a)). Figure 6(b) shows the result of the RTM, again calculated using the homogeneous velocity model. The complete circular boundary of the inclusion is imaged.

This is an significant improvement compared to Kirchhoff migration or SAFT, which are capable of imaging the upper side of the inclusion only.

3.4. Step with Air Inclusion. The last polyamide model is a combination of the step model and a smaller air inclusion. The step has again a height of $z = 0.10$ m. The midpoint of the inclusion is located at $x = 0.4$ m and $z = 0.25$ m, and it has a radius of $r = 0.03$ m (Figure 7(a)). Figure 7(b) shows the migrated image. This example shows that also different defects inside a specimen can be located simultaneously. The vertical border of the step is imaged and also the boundary of the inclusion. As already seen in the first model (Section 3.1), artefacts are generated by imaging the step using a rectangular velocity model.

3.5. Scattering Model with Air Inclusion. To simulate experiments on concrete, we have designed an additional model, which includes the scattering effect of aggregates. In real concrete the size of the aggregates (e.g., gravel) is in the same

(a)

(b)

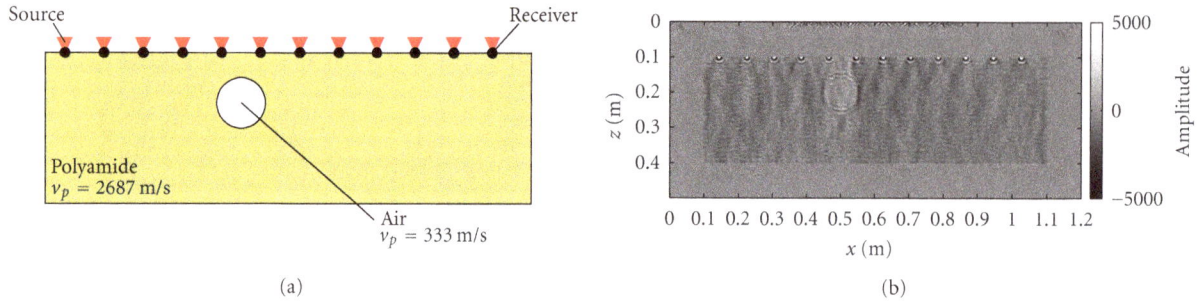

FIGURE 6: (a) Synthetic air inclusion model. (b) RTM imaging result using velocity model of Figure 2.

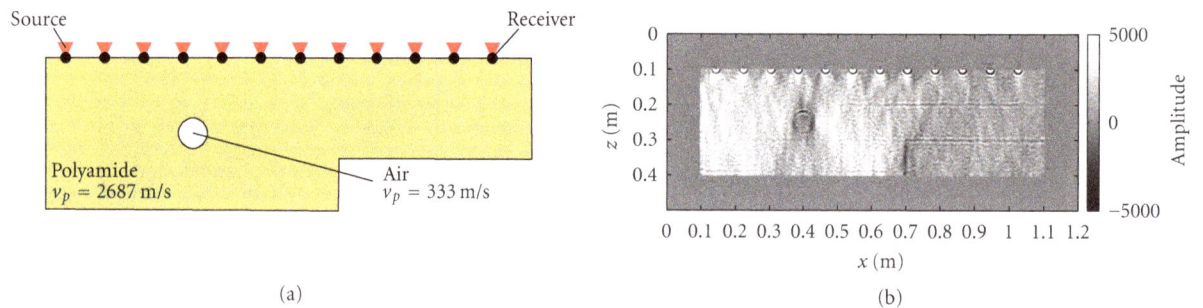

(a)

(b)

FIGURE 7: (a) Step with air inclusion model. (b) RTM imaging result using velocity model of Figure 2.

order of magnitude as the ultrasonic wavelength. We have arranged 300 circular scatterers in the synthetic specimen. Figure 8(a) shows the model. The model consists of 1200×700 grid points with a distance of $dh = 0.001$ m between them. The size of the specimen is 1 m by 0.5 m surrounded by an air layer of 0.1 m thickness. The background velocity of the specimen is $v_p = 4000$ m/s and of the surrounding air layer $v_p = 333$ m/s. The 300 randomly arranged scatterers have a velocity range from 3000 m/s to 4000 m/s and a size between 2 mm and 16 mm. The midpoint of the air inclusion is located at $x = 0.5$ m and $z = 0.3$ m, with a radius of 0.05 m.

We have performed RTM with a source/receiver distribution similar to the geometry used for the polyamide models. These experiments showed no success as the scattered wave energy blurs the image. To achieve better results we have added additional receivers to average the signal. Finally we have placed 10 sources with a distance interval of 0.1 m, and 45 receivers with a distance interval of 0.02 m. A Ricker wavelet with $f = 100$ kHz was used as source function. The signal was recorded for 0.6 ms with a time stepping $dt = 0.1\,\mu s$.

For the RTM we have chosen a homogeneous model again with the same background velocity as the real model, $v_p = 4000$ m/s, but without the scatterers and the inclusion (Figure 8(b)). Figure 8(c) shows the migrated image. The border of the inclusion is completely visible, but the amplitude is significantly lower than in the previous examples. Especially the left and right sides of the inclusion are imaged weakly. Increasing the record time did not improve the result. Indeed, using a longer record time more multiple reflections from the borders of the specimen are

included as well as scattering noise. For this reason the signal-to-noise ratio decreases, and the image of the borders of the inclusion weakens.

3.6. Comparison to SAFT. A quantitative comparison between RTM and SAFT imaging is not possible in the current stage. The decisive difference between them is the acquisition geometry. All the examples for RTM quoted above are calculated with a multioffset geometry. The available SAFT algorithms used for nondestructive testing require constant offset measurements. In addition the distance between shot points used for RTM is largely compared to SAFT. For a detailed SAFT imaging of concrete usually a distance of 0.02 m is used. For RTM we used 0.08 m. However, we have been able to show that RTM is capable to image vertical and circular inhomogeneities, which is not possible with aperture limited approaches as SAFT. An example for results acquired by SAFT and RTM on a real concrete block was published by the same authors [14].

4. Conclusion and Outlook

By using different polyamide models and a concrete-like synthetic model we have demonstrated the potential of the RTM for nondestructive testing in civil engineering. Vertical borders and lower parts of built-in elements can be imaged clearly. The image quality can thus be improved, and more faults can probably be found compared to conventional imaging techniques.

Artefacts which have been seen in RTM images have to be carefully analysed and eliminated. For this task other imaging

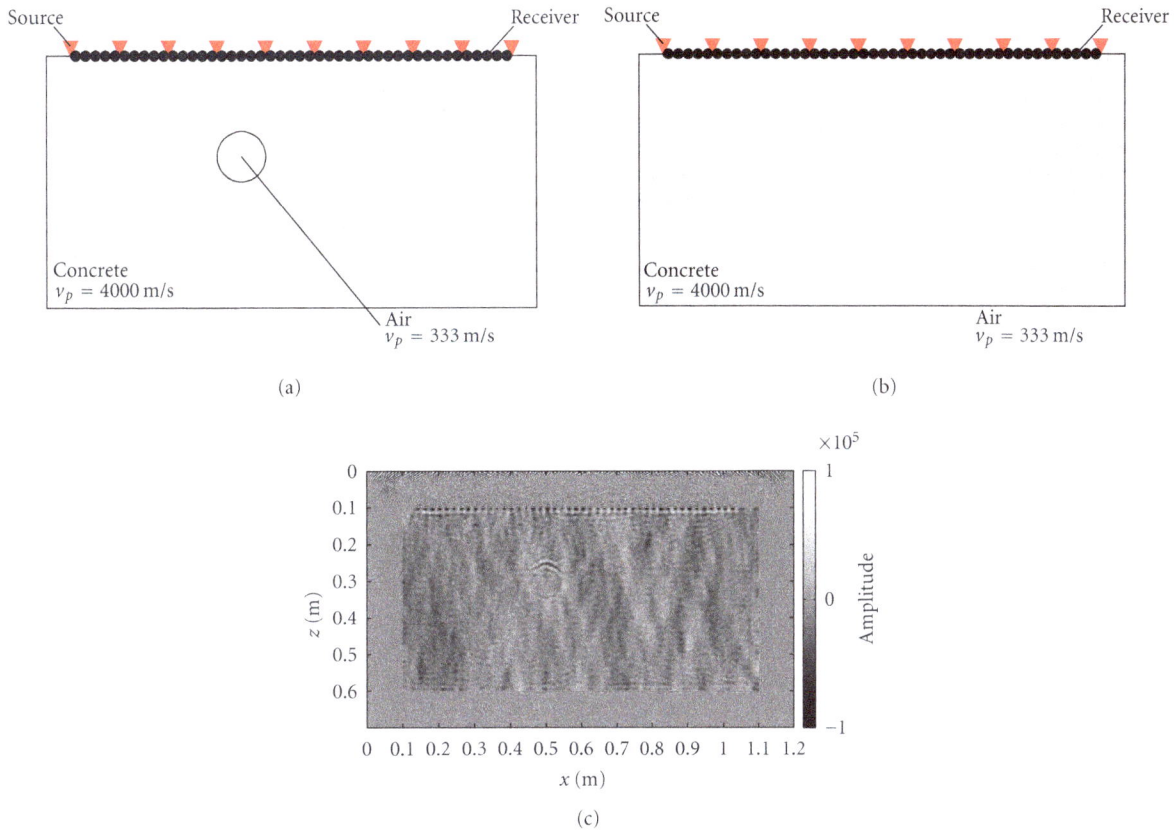

FIGURE 8: (a) Concrete-like synthetic model with an air inclusion. (b) Homogeneous velocity model for migration. (c) RTM imaging result using velocity model of (b).

conditions than cross-correlation may be used but have not been explored by us so far. In addition, improved velocity models should be chosen for the RTM to improve the resulting image. For more complicated tasks a multistage RTM can be helpful to image all features. Experiments on test specimens of polyamide and concrete under laboratory conditions are currently performed to evaluate the synthetic results. Measurement equipment as well as acquisition parameters has to be tested and optimised to obtain a good signal-to-noise ratio. The RTM code so far assumes point sources, while many sensors used in practice have a contact area with a size greater than wavelength.

Another topic to be worked on is to expand the algorithm to three dimensions and to use the full elastic wave equation. In nondestructive testing on concrete most measurements are carried out using shear waves, as smaller defects can be detected this way. To include this, the imaging conditions have to be adapted and the computing efficiency has to be improved as well. This can be done by keeping not every time step of the wave field calculation in RAM.

Acknowledgments

Wolfgang Baumann of the Zuse Institute in Berlin helped to implement the RTM code on their supercomputer. Special thanks to the author's colleagues at BAM 8.2, especially Martin Krause and Christian Köpp, who helped in the preparation of the text. The remarks of two anonymous reviewers helped to improve the paper. All this is deeply appreciated.

References

[1] K. Mayer, R. Marklein, K. J. Langenberg, and T. Kreutter, "Three-dimensional imaging system based on fourier transform synthetic aperture focusing technique," *Ultrasonics*, vol. 28, no. 4, pp. 241–255, 1990.

[2] M. Schickert, M. Krause, and W. Müller, "Ultrasonic imaging of concrete elements using reconstruction by synthetic aperture focusing technique," *Journal of Materials in Civil Engineering*, vol. 15, no. 3, pp. 235–246, 2003.

[3] M. Krause, K. Mayer, M. Friese, B. Milmann, F. Mielentz, and G. Ballier, "Progress in ultrasonic tendon duct imaging," in *Proceedings of the 7th International Symposium on Non Destructive Testing in Civil Engineering (NDTCE '09)*, Nantes, France, June 2009.

[4] M. Krause, B. Milmann, F. Mielentz et al., "Ultrasonic imaging methods for investigation of post-tensioned concrete structures: a study of interfaces at artificial grouting faults and its verification," *Journal of Nondestructive Evaluation*, vol. 27, no. 1–3, pp. 67–82, 2008.

[5] E. Baysal, D. D. Kosloff, and W. C. Sherwood, "Reverse time migration," *Geophysics*, vol. 48, no. 11, pp. 1514–1524, 1983.

[6] D. Loewenthal and I. R. Mufti, "Reversed time migration in spatial frequency domain," *Geophysics*, vol. 48, no. 5, pp. 627–635, 1983.

[7] G. A. McMechan, "Migration by extrapolation of time-dependent boundary values," *Geophysical Prospecting*, vol. 31, no. 3, pp. 413–420, 1983.

[8] P. A. Farmer, I. F. Jones, H. Zhou, R. I. Bloor, and M. C. Goodwin, "Application of reverse time migration to complex imaging problems," *First Break*, vol. 24, no. 9, pp. 65–73, 2006.

[9] L. Zhou, F. G. Yuan, and W. J. Meng, "A pre-stack migration method for damage identification in composite structures," *Smart Structures and Systems*, vol. 3, no. 4, pp. 439–454, 2007.

[10] L. Wang and F. G. Yuan, "Damage identification in a composite plate using prestack reverse-time migration technique," *Structural Health Monitoring*, vol. 4, no. 3, pp. 195–211, 2005.

[11] T. Bohlen, "Parallel 3-D viscoelastic finite difference seismic modelling," *Computers and Geosciences*, vol. 28, no. 8, pp. 887–899, 2002.

[12] A. Kurzmann, D. Koehn, A. Przebindowska, N. Nguyen, and T. Bohlen, "Acoustic full waveform tomography: performance and optimization," in *Proceedings of the 70th EAGE Conference and Technical Exhibition*, Amsterdam, The Netherlands, June 2009.

[13] B. Kaelin and A. Guitton, "Imaging condition for reverse time migration," *SEG Technical Program Expanded Abstracts*, vol. 25, no. 1, pp. 2594–2598, 2006.

[14] S. Müller, E. Niederleithinger, M. Krause, and T. Bohlen, "Reverse time migration: a seismic imaging technique applied to ultrasonic data," in *Proceedings of the SMT Conference of NDE/NDT for Highways and Bridges: Structural Materials Technology (SMT)*, New York, NY, USA, August 2010.

5

Low-Latitude Atmosphere-Ionosphere Effects Initiated by Strong Earthquakes Preparation Process

Sergey Pulinets

Space Research Institute, Russian Academy of Sciences, Moscow 117997, Russia

Correspondence should be addressed to Sergey Pulinets, pulse1549@gmail.com

Academic Editor: Yuichi Otsuka

Ionospheric and atmospheric anomalies registered around the time of strong earthquakes in low-latitude regions are reported now regularly. Majority of these reports have the character of case studies without clear physical mechanism proposed. Here we try to present the general conception of low-latitude effects using the results of the recent author's publications, including also rethinking the earlier results interpreted basing on recently established background physical mechanisms of anomalies generation. It should be underlined that only processes initiated by earthquake preparation are considered. Segregation of low-latitude regions for special consideration is connected with the important role of ionospheric equatorial anomaly in the seismoionospheric coupling and specific character of low-latitude earthquake initiated effects. Three main specific features can be marked in low-latitude ionospheric anomalies manifestation: the presence of magnetic conjugacy in majority of cases, local longitudinal asymmetry of effects observed in ionosphere in relation to the vertical projection of epicenter onto ionosphere, and equatorial anomaly reaction even on earthquakes outside equatorial anomaly (i.e., 30–40 LAT). The equality of effects morphology regardless they observed over land or over sea implies only one possible explanation that these anomalies are initiated by gaseous emanations from the Earth crust, and radon plays the major role.

1. Introduction

Many studies for the ionospheric precursors of earthquakes have been done since 1964 when great Alaska "Good Friday" ($M_w = 9.2$) earthquake took place on the 27th of March. History of these studies is described by Pulinets and Boyarchuk [1]. We can consider the starting date of systematic studies of seismo-ionospheric coupling from publication of the book *Ionospheric Precursors of the Earthquakes* [2]. Pulinets [3] proposed to consider the ionospheric variations stimulated by the earthquake preparation process as one of the constituents of day-to-day ionospheric variability. This statement was questioned in some publications [4, 5]. But after revealing the main phenomenological features of ionospheric precursors [6] and their stable and confident statistical characteristics [7–10] the discussions shifted to the direction of their physical mechanism clarification. The diversity of ionospheric effects manifestation before earthquakes (positive and negative deviations from undisturbed level, local time dependence, longitudinal effect, etc.) produced confusion in heads of physicists, and instead of unitary vision of the phenomena the diversity of the models was proposed [11]. And the most mysterious fact was the difference in precursor manifestation in high and middle latitudes and in low latitudinal regions. From the morphology of low-latitudinal effects [12, 13] it became clear that the only possible explanation could be found in the field of equatorial anomaly electrodynamics, and regardless the vertical anomalous electric field proposed in the previous models [14] some mechanism for generation of the zonal electric field should be provided. This question stood unresolved until the final conception of the lithosphere-atmosphere-ionosphere coupling appeared [15] where atmospheric effects play important role in the local modification of the global electric circuit parameters [16]. Understanding the underlying mechanism of the zonal electric field generation opened the way to complete the physical model of seismo-ionospheric effects in low-latitudes [17, 18].

The present paper will provide the short history of the low-latitude and equatorial effects of the earthquakes studies by the author and his group. Then the underlying physical mechanisms will be considered together with examples of the precursors records for the recent major earthquakes. And finally the paper will be concluded with still unresolved problems, which need further clarification.

2. Topside Sounding Results: The First Manifestation of Low-Latitude and Equatorial Effects of Earthquakes in the Ionosphere

Before the GPS era, in the absence of multisatellite constellations like COSMIC the only means that permitted to provide the global monitoring of the ionosphere variability at the level of the F2-layer maximum was the topside sounding [21]. And the very first results from IS-338 Intercosmos-19 topside sounder [12, 13, 22] demonstrated unexpected features of ionospheric anomalies before earthquakes in low-latitude regions: the longitudinal asymmetry of the effect in relation to the impending earthquake epicenter position, amplification/degradation/distortion of the equatorial anomaly shape in comparison with model solution, anomaly development at a typical local time, conjugated effects (anomalous deviations are observed not only over the impending epicenter but in magnetically conjugated point of opposite hemisphere), modification of the vertical distribution of electron concentration in topside ionosphere, and effects produced by low-mid latitude earthquakes (epicenter is 10–20 degrees poleward from the equatorial anomaly crest position) on electron concentration in equatorial anomaly. Taking into account that the majority of these effects were published earlier elsewhere, only most essential points will be illustrated lower.

In the next sections following some specific features revealed mainly with the help of topside sounding will be presented. They were published elsewhere but never were collected together. The uniqueness of these data is provided by the fact that only topside sounder is able to reveal variations of the critical frequency foF2 in global scale, and over the areas without ground based ionosondes installed.

2.1. Longitudinal Features. In Figure 1 the longitudinal variation of the critical frequency foF2 obtained with the topside sounder IS-338 at the latitude 3.2 (latitude of impending earthquake epicenter) is presented. Continuous monitoring of the longitudinal structure of the critical frequency distribution for the fixed local time demonstrated its high stability: with some variations of amplitude of critical frequency variation, its shape remained stable for the given geophysical conditions (season, solar activity, latitude, and local time). The bold line marked by number 1 is the undisturbed longitudinal distribution reflecting the so-called wave-4 longitudinal structure discovered by Intercosmos-19 satellite [23, 24]. The dash-dot line marked by number 2 shows the longitudinal distribution obtained 14 July 1980, 2 days before the M7.3 earthquake at New Guinea region (3.2N, 143.3E). The dashed line marked by number 3 was obtained on 15 July, 1 day before the main shock. Both distributions demonstrate the pronounced minimum in the vicinity of epicenter longitude, but the minimum on 14 July is shifted

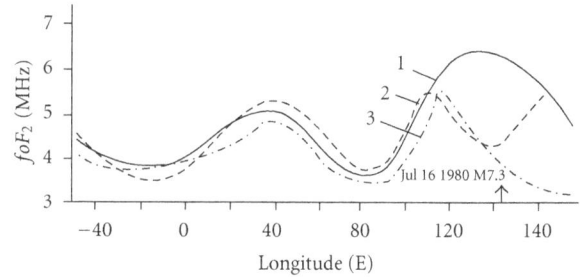

FIGURE 1: Longitudinal distributions of the critical frequency foF2 obtained by IS-338 topside sounder (Intercosmos-19 satellite) at the latitude 3N for the fixed local time (05-06 LT). Bold line (1): averaged quiet time distribution; dash-dot line (2): distribution obtained for 14 July 1980; dashed line (3): distribution obtained on 15 July 1980.

to the east from the epicenter longitude. This feature was unexplained until the mechanism of preseismic anomaly formation was clarified [16].

2.2. Modification/Distortion of Equatorial Anomaly Shape. In the left panel of Figure 2 the equatorial anomaly pre-earthquake degradation is demonstrated 2 hours before the moderate M5.1 shock in equatorial ionosphere 06.07 1979 [13]. The right panel demonstrates how the vertical distribution of the electron concentration changed. It is differential tomographic reconstruction of the latitudinal cross-section of the equatorial anomaly, one can see now the negative depletions in the crests of equatorial anomaly and high increase of electron concentration over geomagnetic equator leading to the anomaly degradation and crests disappearance. The physical model of such changes in equatorial anomaly before earthquakes is presented in [18].

Figure 3 demonstrates other cases of equatorial anomaly distortion registered by topside sounder for different earthquakes. From left to right one can see deepening of the anomaly trough between crests, formation of crests in predawn hours, and strong distortion of equatorial anomaly with trough movement to the north from geomagnetic equator. Diversity of the observed effects implies the complex physical mechanism that should be developed to explain the observed effects in the equatorial anomaly before earthquakes.

2.3. Effects in Equatorial Anomaly from Low-Midlatitude Earthquakes. Interesting feature of earthquake preparation impact on equatorial anomaly was detected during the analysis of topside sounding data around the time of Mammoth Lake seismic swarm in May 1980 [25] when during 3 days the 4 shocks with magnitude $M \geq 6$ and 4 shocks with magnitude $M \geq 5.5$ took place practically at the same point 37.6°N, −119°W. In Figure 4 one can see in the left panel the consecutive maps (2D distribution of the critical frequency scaled from topside ionograms). The top map in the left panel represents the undisturbed background level. This map was constructed from several undisturbed days in May 1980. Then, under the reference maps one can see

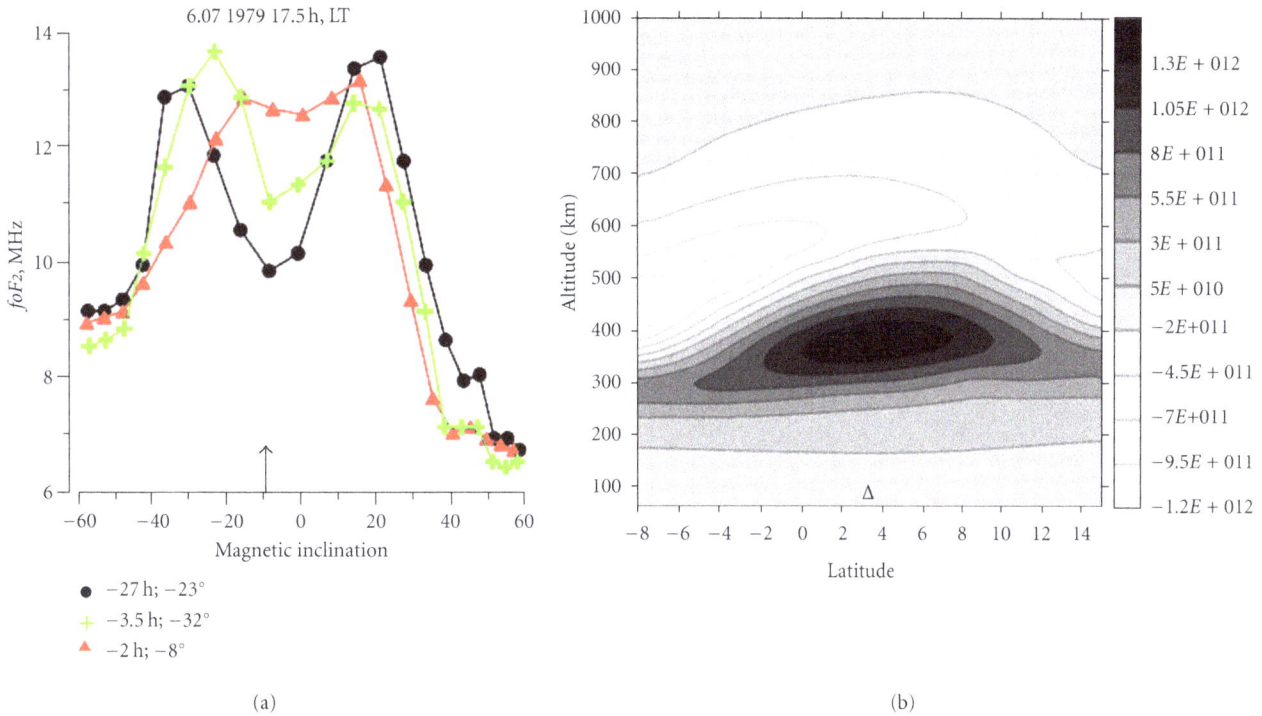

(a)

(b)

FIGURE 2: (a) consecutive passes of Intercosmos-19 satellite over equatorial anomaly. In figure legend are shown time before earthquake and longitudinal distance of the satellite orbit from the epicenter longitude sign "−" indicates satellite pass eastward and sign "+" pass westward from the epicenter longitude; (b) tomographic reconstruction of the difference of the vertical distribution of electron density between July 6 and July 5 1979 (red and black lines at the left panel).

the distributions obtained for the same local time 15-16 LT for the days 21, 22, 26, and 27 of May. In the right panel the differential maps (difference between the current day and the reference map) are shown. The main peculiarity observed on the differential maps is a formation of the positive anomaly close to the vertical projection of epicenter. But, in addition, one can easily observe the negative deviation in the northern crest of the equatorial anomaly. The absolute value of the critical frequency deviation is not too much (near 12%), but in the absolute values it is more than 1 MHz. Dynamics of the latitudinal distribution of the critical frequency from 20 to 27 of May are shown in Figure 5. These distributions demonstrate one more specific feature formation of additional density maximum at latitude 20–25°N. But still the most intriguing is the fact that the earthquake taking place far enough from the equatorial anomaly (37.6°N) produces observable effect on anomaly. The possible reason for this will be discussed next.

Some authors [26] claim that there are no effects of equatorial anomaly on low-middle latitude earthquakes in Japanese region, but it seems that it is not so. As was demonstrated by Ouzounov et al. [27] the strong increase of electron concentration before the Tohoku earthquake was observed over huge territories including the equatorial anomaly.

2.4. Conjugated Effects. For the first time the magnetically conjugated effect in the ionosphere before earthquake was detected while studying ionospheric effects before the Irpinia

earthquake M6.9 in Italy 23 November 1980 [28, 29]. Again, like what was demonstrated in Subsection 2.3, the effect was stimulated by the well-midlatitude earthquake (epicenter latitude 40.5°N). 2.5 days before the main shock the negative anomaly was detected before sunrise (03–043 LT). This anomaly is shifted from the epicenter location by 10 degrees equatorward. Accurately placing the detected anomaly on geomagnetic map we see that its center lies on the L-shell 1.3, what corresponds to the outer edge of the northern crest of equatorial anomaly. And the similar negative anomaly is detected in magnetically conjugated area of the southern hemisphere (Figure 6). Later conjugated effects were observed regularly by in situ probes onboard the DEMETER satellite [30] and using the data of GIM and COSMIC constellation which are able to produce the global distribution of the electron concentration or GPS TEC. New data confirmed the possibility of generation of the conjugated effect by the middle latitude earthquake [17] with the use of the first principle physical model of the ionosphere. The model needs the additional zonal electric field to reproduce the earthquake effect in the ionosphere, and at the present moment the crucial question is the mechanism of such field generation.

3. The Origin of the Anomalous Electric Field Generated by the Earthquake Preparation Process

Recent years are marked by increased activity in creation of the models trying to explain the changes in atmospheric

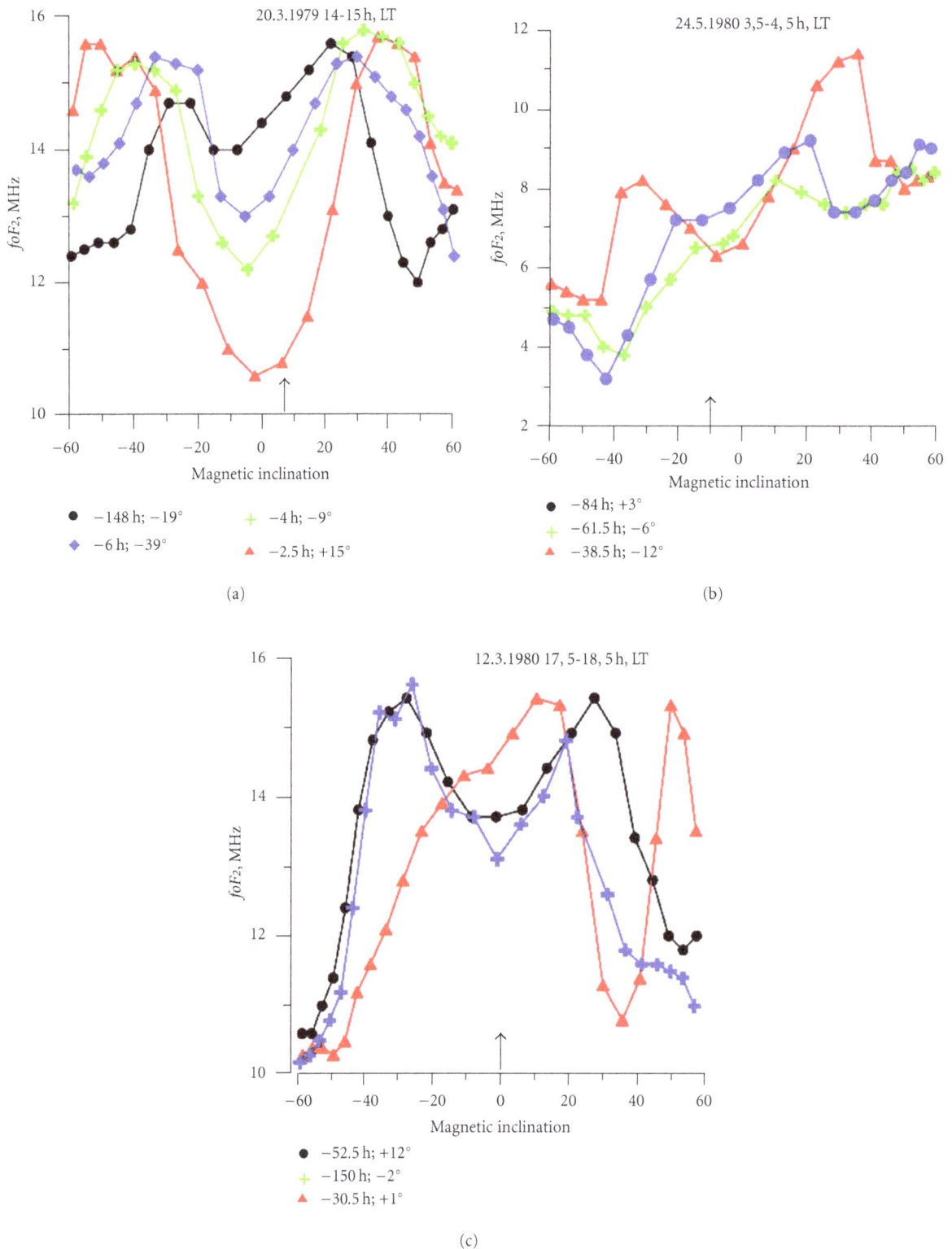

FIGURE 3: From left to right: deepening of the anomaly trough between crests, formation of crests in predawn hours, and strong distortion of equatorial anomaly with trough movement to the north from geomagnetic equator. Vertical arrows indicate the latitudinal position of the earthquake epicenter. In the figure legends the left column of numbers indicates the time (in hours) of satellite pass over the epicenter latitude in relation to the earthquake moment. Sign − indicates time before the shock and sign + time after the shock. The right column of numbers shows the longitudinal shift of the satellite orbit ground projection in relation to the epicenter longitude.

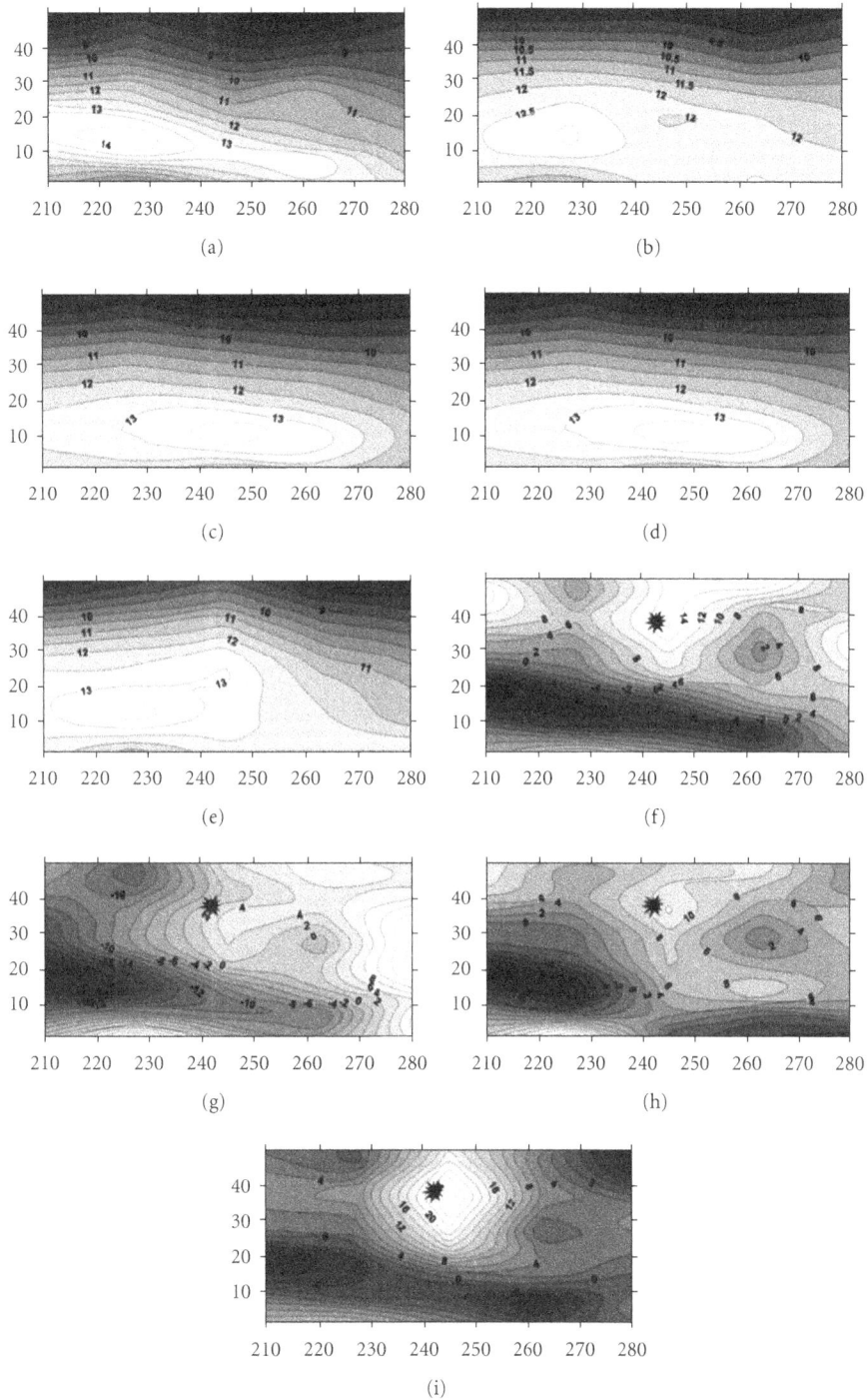

FIGURE 4: (a)–(e): absolute *(foF2* distribution) LT maps for -15-16 LT. From (a) to (e): combined map for 15 and 19 May (undisturbed conditions, background level) and 21, 22, 26, and 27 of May. (f)–(i): the maps of the critical frequency deviation (in % relative to the top panel in the left column). From (f) to (i): 21, 22, 26, and 27 May. The epicenter position is shown by an asterisk.

electricity before strong earthquakes [16, 19, 31, 32]. These theories became the background of the model calculations of the ionospheric effects [17, 18, 33, 34]. Taking into account that results of model calculations strongly depend on the initial conditions, it is worthy to consider the reality of the proposed models of atmospheric electricity modification by earthquake preparation process.

Let us start from the model proposed by Sorokin [31]. To obtain values of the anomalous electric field in the iono-sphere of order 10 mV/m (what is extremely high and com-parable with magnetic storm values) Sorokin [31] proposes the so- called "external current," which is somehow generated by the charged aerosols and ions produced by the radon ionization. It is a vertical current, so it should relate somehow

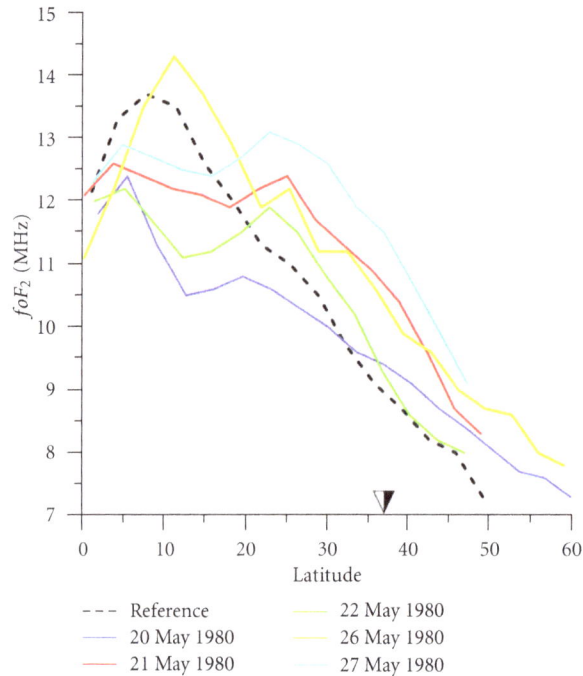

FIGURE 5: Latitudinal profiles of the critical frequency *foF2* close to the epicentral longitude (Intercosmos-19 topside sounder data). Dashed black line: 19 May, color lines: 21–27 May 1980 (see legends in the figure).

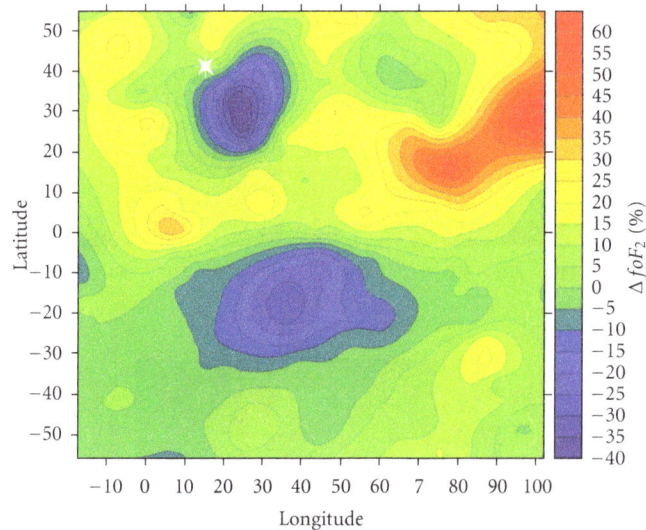

FIGURE 6: Formation of the conjugated ionospheric anomaly in the southern hemisphere before the Irpinia earthquake on 23 November 1980, M6.9, 64 hours before the main shock during predusk time (3-4 LT). White asterisk: the earthquake epicenter position.

with the natural vertical current of the global electric circuit. According to recent information [35] the vertical current density in fair weather conditions is $\sim 2 \cdot 10^{-12}$ A/m². Sorokin, in turn, proposes not the small adding to the natural current but the value $\sim 4 \cdot 10^{-6}$ A/m² at the altitude 10 km, that is, 6 orders of magnitude higher than the natural current. In addition, looking at the figures in Sorokin's paper, this current has opposite direction in relation to the fair weather electric current direction, that is, directed up. Taking into account that the vertical current should be constant through the whole bulk of atmosphere, the same value of current should

be observed not only at the altitude 10 km but at the ground surface as well (finally, the source of the current proposed by the author is underground). But we have no experimental evidences of such value of vertical current measured ever. It is absolutely unreal conception, and it a pity that some prominent theoreticians try to use this approach as initial conditions for the model calculations of the ionospheric effects.

Another mechanism was proposed by Freund [19] based on the solid state physics. Under the stress in the Earth crust the accumulation of positive hole charge carriers $h\bullet$ at

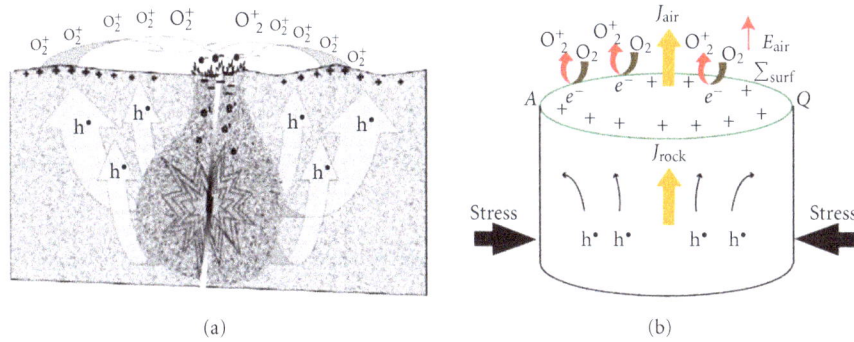

FIGURE 7: (a) conception of the positive EF field generation on the ground surface from Freund [19]; (b) modified conception of the positive EF generation on the ground surface from Koons et al. (2011).

the surface produces positive surface charges that create the vertical electric field in opposite direction to the natural one, that is, directed upwards. There exist two versions of the following development. In the first one (Figure 7(a)) somehow the ionization of air molecules happens and the flux of positive ions of O_2^+ flows into the upper layers of atmosphere. The return current is closed in corona discharge which returns positive potential into negative. What is the origin of corona discharge, and why it reverses the surface potential is not clear. In addition, we should comment that corona discharge will produce the ions of both signs, negative and positive. The second version used by Kuo et al. (Figure 7(b)) does not use the coronal discharge conception. In this version the positive ions of oxygen molecules are formed "at sharp edges of rock molecules" and go to infinity [36]. How this electric circuit is closed, not clear at all. The authors do not take into account the existence of the natural electric field at all. At the same time the natural atmospheric electric field is a fundamental feature of the global electric circuit conception, and any new electric field appearing in near ground layer of atmosphere will obviously interact with the natural one [37].

How real such conception is? No doubts that the solid physics part is absolutely correct and production of p-holes has taken place under the stress conditions. But the atmospheric part gives births to a lot of questions. Why only oxygen molecules are ionized? Atmospheric air contains a lot of gases with similar ionization potentials. So, all of them should be ionized. Positive oxygen molecules can exist in the free state in atmosphere only few nanoseconds, then they immediately enter in reactions and become hydrated. The fluxes of positive oxygen molecules in atmospheric air are completely impossible. One can find information on what happens with atmospheric air molecules after corona discharge in [20]. Figure 8 from this publication demonstrates the basic air reactions after primary ionization, from where it is clear that the fluxes of positive oxygen molecules are impossible. Actually, except the reactions shown in Figure 8 there are hundreds of other reactions forming other ions with different level of probability [38]. Every sort of these final ions will have different mobility, and to calculate the changes of atmosphere conductivity of the boundary layer is not simple task.

As concerns the surface charge, which should appear according to Freund, its propagation into the atmosphere has also serious problems. We should take into account such conception as "Coversphere" [39] The ground surface is covered by sand, clay, soil, grass, trees, and cities; we have pipelines and cables inside the near surface layer. So it is difficult to expect the pure effect like in laboratory experiments with granite or igneous rocks. And it is completely impossible to obtain such effects over the ocean surface. The statistical data of ionospheric precursory effects shown by DEMETER satellite (the most reliable source of statistical data on ionospheric precursors of earthquakes for the moment) shows that the number of ionospheric anomalies connected with earthquake is higher for the ocean, than for the land earthquakes [10]. If even to suppose that the proposed by Freund [19] mechanism works, it will be valid only for the land earthquakes, and will not work for the earthquakes which epicenter is under water surface. From the other side, there cannot be one mechanism for the land another one for the ocean earthquakes, so we should look for the other possibility.

Such possibility was proposed by Pulinets [16] (especially for the equatorial anomaly) and is connected with modification of the boundary layer conductivity by ionized action of radon emanated from the Earth crust mainly through the active tectonic faults. It should be noted in this point that according to the recent concept of radon migration in the crust, the major carrier of radon to the ground surface is the carbon dioxide and other gases such as methane, hydrogen, and helium [40]. Radon migration with gases including underwater cases is described in detail in this book. Gas discharges accompany the latest stages of earthquake preparation and bring radon to the surface even from underwater faults. Because of the novelty of this conception there no earthquake-related measurements of radon over oceans, but there are plenty of publications reporting its abundance over the ocean surface. As to the author's knowledge the first communication on the radon over the ocean surface was made by Broecker during the Symposium held in Lamont Geological Observatory (USA) in 1964 [41] This question was considered in the review paper [42] where one can find more references on

(a) Positive ion evolution

(b) Negative ion evolution

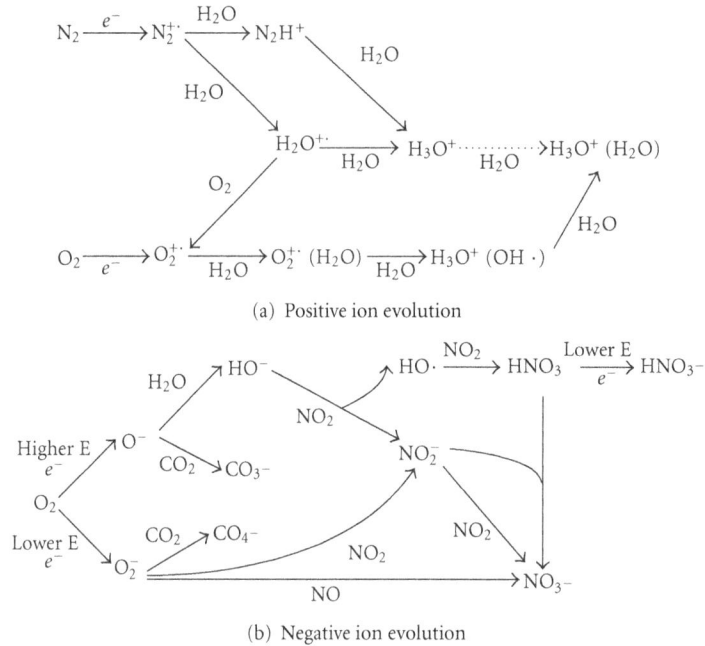

FIGURE 8: Principal ion reactions and formation the positive and negative ions after air ionization by the coronal discharge [20].

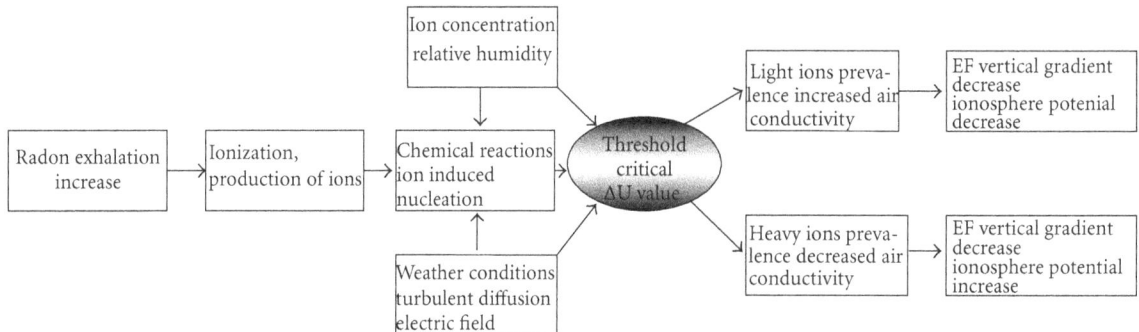

FIGURE 9: Flow chart diagram of the ionization and nucleation processes with the branching point (oval) in formation of light and heavy ions and air conductivity modification.

Broecker's publications, as more recent publications about radon measurements over the ocean surface the papers ([43] and references therein) can be proposed. Another source of radon in the near shore ocean waters (where the majority of the strong subduction earthquakes take place) is the ground water discharges carrying radon with them [44] They demonstrate the experimental evidence of the radon activity levels from underwater sources, which are sometimes much higher than for the land ones. So, the conception of air ionization with radon has no problems with both ocean and land earthquakes.

Consequences of air ionization depend on many factors such as radon activity (its concentration is different for different areas of our globe), relative air humidity (water molecules concentration determine the size to which new formed ions will grow), conditions of turbulence (the wind

will remove the ions from the areas of high ion concentration changing the chemical potential correction value), air temperature (it is also determines the chemical potential correction value). The large variety of external conditions leads to the branching point in the model shown in Figure 9 [17]. In initial stages of ionization, and also under weak ionization levels the light ions will prevail in the boundary layer of atmosphere what will lead to the general increase of the bulk conductivity of atmosphere and consecutive decrease of the ionospheric potential relative to the ground. The other opportunity rises if the ionization rate is very high, relative humidity is enough to create the large ion clusters, and calm weather conditions enable forming the large clouds of aerosol size heavy ion clusters. Both cases are demonstrated in Figure 10 (geomagnetic field is directed perpendicular to the figure plane at the bottom part of

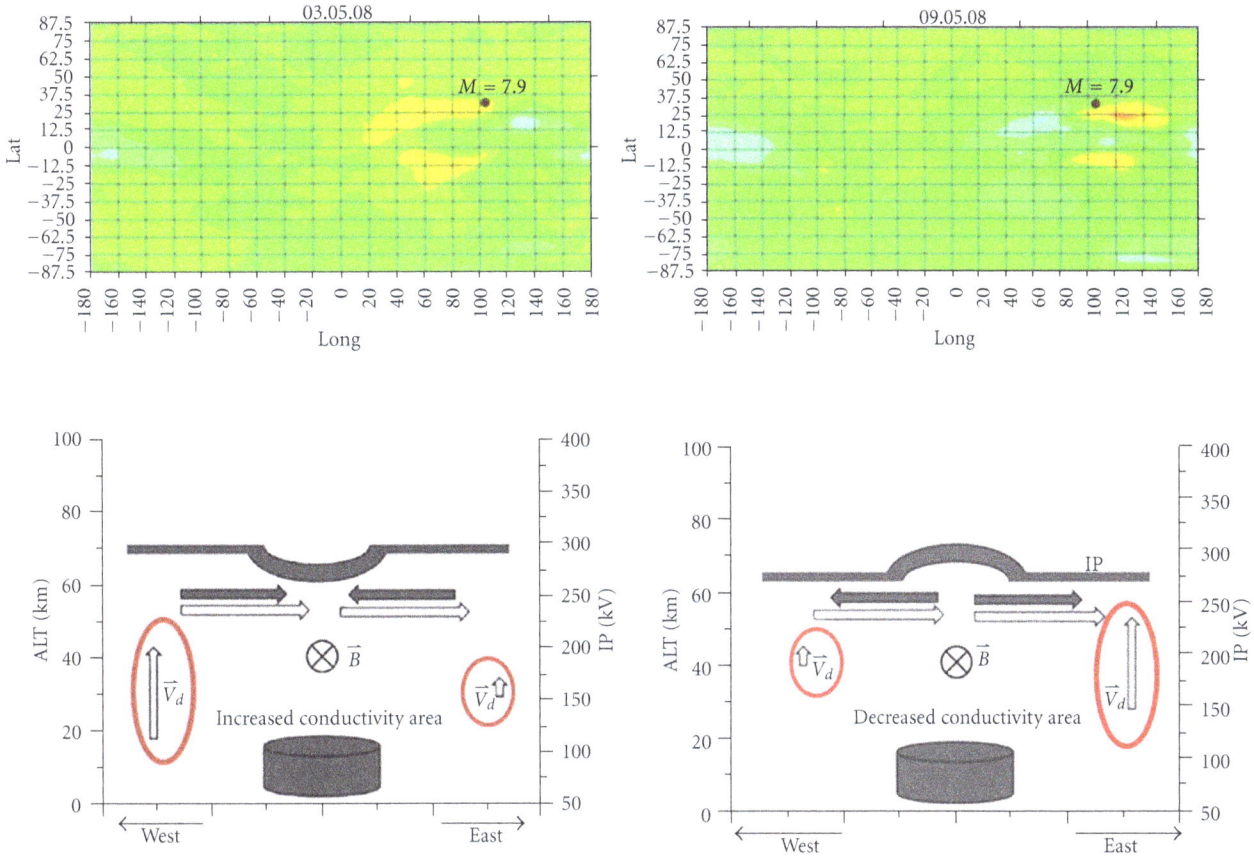

FIGURE 10: Bottom panel: schematic conception of atmosphere-ionosphere coupling through the global electric circuit: left panel: for condition of increased air conductivity, right panel: for condition of decreased air conductivity. Upper panel: the differential maps obtained from the GIM GPS TEC data for the period before the Wenchuan earthquake on 12 May 2008. Left panel: 2D distribution obtained on 3 of May 2008, right panel: 2D distribution obtained on 9 May 2008.

the figure). The case of increased air conductivity is shown in the left panel of the figure, and the opposite case in the right panel. In both cases the anomalous zonal electric field is formed from both sides of the ionospheric potential anomaly (dark horizontal arrows, the bottom part of the Figure 10). Because of the opposite direction of the anomalous electric field for cases of increased and decreased air conductivity, the anomalous electric field, added to the zonal electric field responsible for the equatorial anomaly formation (white horizontal arrows, the bottom part of the figure), will increase the vertical drift velocity to the west from the anomalous region in the case of increased air conductivity and to the east from the anomalous region in the case of decreased air conductivity (vertical arrows in ovals). Both cases were confirmed by the experimental results while studying the ionospheric effects of the Wenchuan earthquake [45] and are presented in the upper panel of the figure where the differential maps (ΔTEC) using the GIM maps as a source are presented.The left panel demonstrates the configuration registered on 3 May 2009. It corresponds to the case of increased air ionization (ions only started to form and did not grow to the low mobility level). On 9 May the so-called process of ion "ageing" took place, they grew to the aerosol size (from

1 to few microns), there mobility extremely low, and air conductivity drops, what led to development of equatorial anomaly to the east from the epicenter and its degradation to the west from it.

Such experimental confirmation for the proposed physical mechanism of atmospheric electric field generation before earthquakes [16] was obtained for many cases of the recent major seismic events in low-latitude ionosphere, and seems that it could be accepted as a plausible explanation of anomalous zonal electric field generation for the further model calculations of the effects in the ionosphere.

4. Instabilities in the Ionosphere and Plasma Bubbles Generation

The increased vertical plasma drift (in comparison with normal undisturbed conditions) provided by anomalous seismogenic zonal electric field [16] can create the increased vertical gradient scale length favorable for generation of Rayleigh-Taylor instability [46]. The possibility of plasma bubbles formation was demonstrated in modeling experiments for intensified vertical plasma drift by Kuo et al.

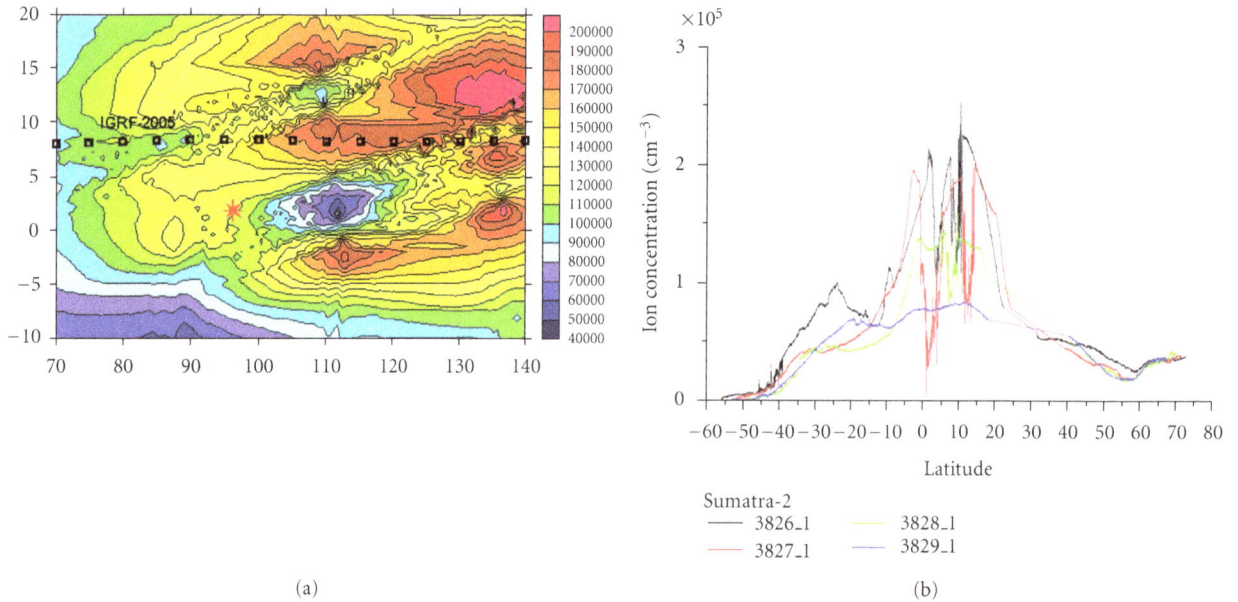

(a)   (b)

FIGURE 11: (a): 2D distribution of ion density obtained on board DEMETER satellite over the epicenter of the Sumatra 28 March 2005 M8.7 earthquake obtained for the night-time (22LT) on March 22 2009. (b): the latitudinal variations of the ion density for several night-time passes of DEMTER satellite over the area of preparation of the Sumatra 28 March 2005 M8.7 earthquake during one week before the seismic shock.

[34]. It is interesting to note that plasma bubbles were really registered before the Sumatra M8.7 earthquake on 28 March 2005 [30]. This effect was observed every day at night-time passes of DEMETER satellite (22 LT) during the period of one week before the main shock, and then disappeared. The 2D distribution of ion concentration for 22 March 2005 is shown in the Figure 11(a). As it was predicted by Pulinets [16] the anomaly is formed to the east from epicenter. One can see the formation of the crests of the equatorial anomaly and two depletions equatorward from the crests at both sides from the geomagnetic equator. The position of the geomagnetic equator (IGRF2005) is shown by black squares. One should keep in mind that DEMETER satellite orbit altitude at this time was 710 km and formation of crests at such altitude was anomalous itself, what manifested of the extremely high vertical plasma drift. Usually at these altitudes the equatorial distribution of plasma density looks like one peak over the geomagnetic equator. In Figure 11(b) the latitudinal distribution of ion density for several consecutive satellite passes over the region is shown (red one is 22 March). The shape of the distribution shows the typical plasma bubble morphology. Another check if depletions are really the plasma bubbles is the values of electron concentration and temperature which are shown at quick look plots in Figure 12. In two upper panels of Figure 12 the electron concentration and temperature are shown. The decrease of both the electron concentration and the electron temperature is characteristic only for plasma bubbles when the cold rarified plasma from the bottom ionosphere rises to the areas of the hot and dense plasma of the upper ionosphere. We can conclude that the generation of

Rayleigh-Taylor instability before the low-latitude earthquakes leading to formation of plasma bubbles is proved both experimentally and theoretically. It should be marked also that anomalous electric fields are mapped along the geomagnetic field lines into the conjugated hemisphere, what leads to the generation of anomalies from the both sides of geomagnetic equator.

One can mark in Figure 12 one more prominent feature: the strong ELF noise within the frequency band 0–250 Hz (the bottom panel of Figure 12). Such noises associated with plasma bubbles were registered by different space probes for many years [47]. The noises appearance coincides with the essential increase of the ion temperature. We can mark also that outside the plasma bubbles the electron temperature is ~2.5 times higher than the ions temperature. But inside the plasma bubbles the temperature tries to equalize, and even higher ion temperature than the electron one. Park et al. [48] observed two different types of plasma bubbles: normal equatorial plasma bubbles (EPBs) where temperature is usually lower than the ambient temperature, and bubbles with enhanced temperatures (BETs). They explain the appearance of BETs by fast poleward oxygen ion transport along magnetic flux tubes. To obtain such fluxes we should have some kind of the anomalous electric field to create the conditions for such drift. This situation needs the careful consideration of the electric field configuration in the upper ionosphere before earthquakes. Until the mechanism of ion heating is clear we can state only the presence of the strong gradient of the ion temperature, which can lead to generation of temperature gradient ion instability leading to the excitation of ELF turbulence. The other possibility is the different

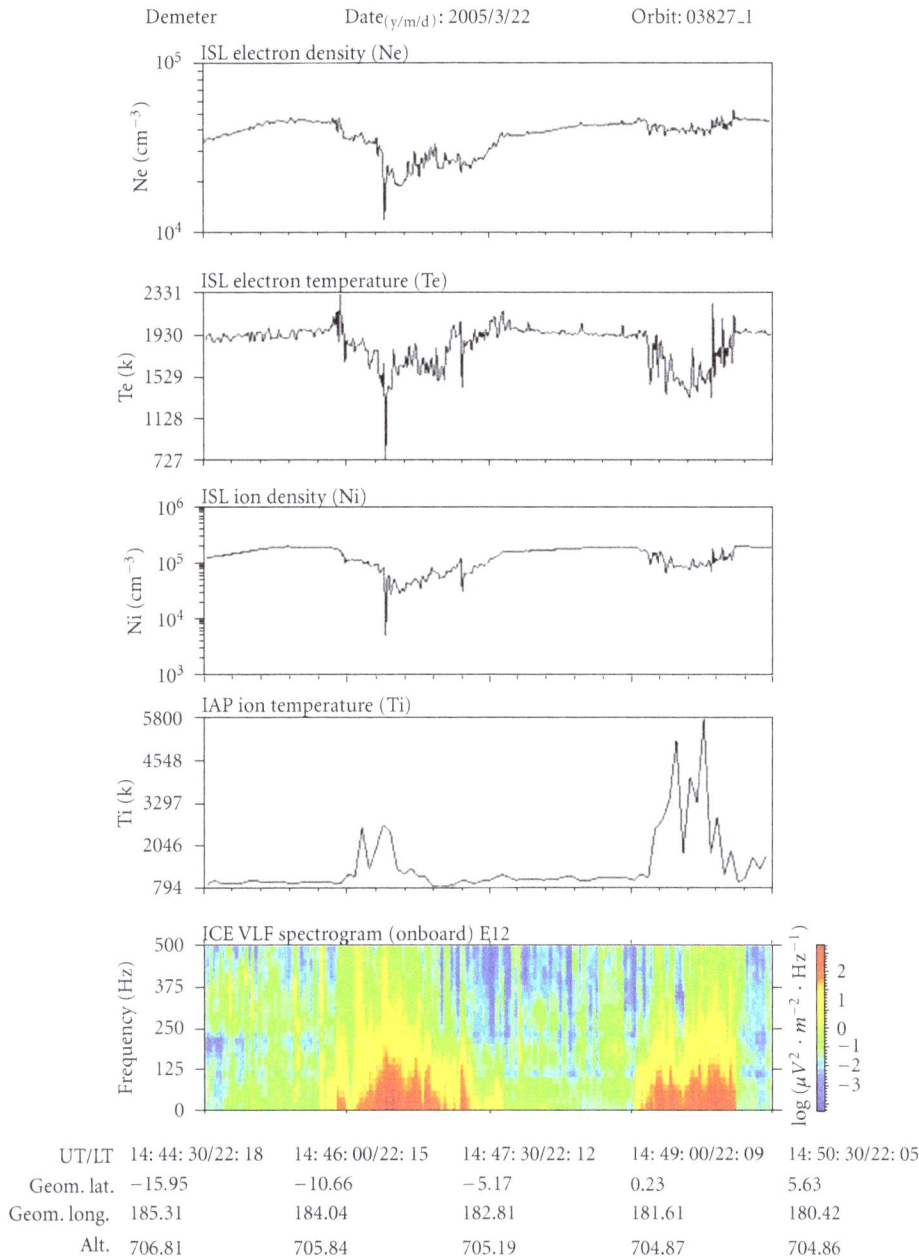

FIGURE 12: Quick look complex diagram of the DEMETER satellite data of the nigh-time pass over the area of preparation of the Sumatra 28 March 2005 M8.7 earthquake obtained on 22 March. From top to bottom: electron concentration, electron temperature, ion concentration, ion density, and dynamic spectrogram of electric component within the frequency band 0–500 Hz.

velocities of plasma inside and outside the plasma bubble reported in [49], what can lead to stimulation of two-stream instability.

5. Unusual Effects Connected with the Large Difference between the Geomagnetic and Geodetic Equator Positions

The concurrence of geomagnetic conjunction and plasma transequatorial transport due to temperature gradient (summer-winter hemisphere) can produce unusual results like what happened before the Haiti earthquake [50]. As one can see from Figure 13, the equatorial anomaly geographically almost completely lies in the southern hemisphere: the vertical line marks the position of the geodetic equator. Such configuration is probably the reason of the formation of the additional density enhancement at low-middle latitudes. According to Rishbeth [51] the seasonal asymmetry of the equatorial anomaly is formed due to the meridional wind from the summer hemisphere to the winter one. The neutral wind involves in its motion the charged component, and this

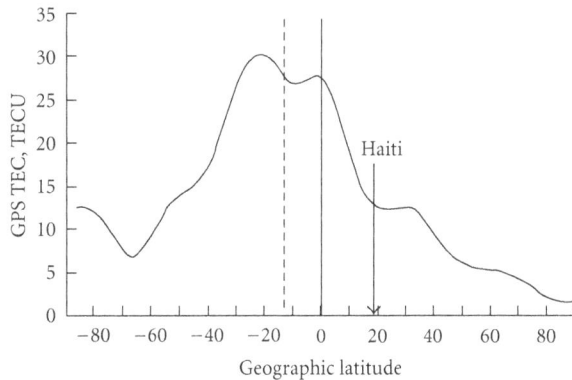

FIGURE 13: Configuration of the equatorial anomaly in relation to the geodetic equator position at the longitude of Haiti M7.9 earthquake on January 12, 2010.

leads to an increase in the concentration of the winter crest of the equatorial anomaly. But at these longitudes the winter crest of the equatorial anomaly geographically is also in the southern hemisphere. That why the wind carrying with it the denser plasma slips over the northern crest and deposits it at more high latitude region form the additional maximum at latitude near 30°N. The Haiti earthquake epicenter was situated exactly between the northern crest of the equatorial anomaly and this additional peak of ionization. The anomalous seismogenic electric field led to the complex modification of this configuration, which was expressed in the following effects:

(i) total depletion of the electron content inside the equatorial anomaly during 6 days preceding the earthquake,

(ii) local increase in the electron concentration in the ionospheric region at the epicenter latitude between the northern crest of the equatorial anomaly and additional peak of ionization at the latitude 30°N,

(iii) amplification of the additional maximum of ionization at latitude of 30°N and formation of an additional maximum at the magnetically conjugated point.

The last item needs a special comment. It demonstrates the geomagnetic feedback from the modification in the northern hemisphere to the southern one what was demonstrated in Section 2.4.

6. Conclusions

The processes taking place in the low-latitude ionosphere before strong earthquakes were described using the data of the satellite and ground based measurements, as well as the data of modeling. The earlier results obtained in 1979-1980 from topside sounding (Intercosmos-19 satellite) were confirmed by the data of the latest space probes such as DEME-TER satellite and GPS TEC data. The origin of the ionospheric anomalies is in the ionization of the boundary layer

of atmosphere and consequent changes of its electric conductivity which then leads to the changes of the ionosphere potential through the global electric circuit. It is the universal mechanism of the atmosphere-ionosphere coupling.

The deviations of electron concentration may be both positive and negative, and configuration of these variations depends on many factors. The most characteristic features are: the longitudinal asymmetry, the conjugated effect (anomalous deviations are observed not only in the vicinity of the impending earthquake epicenter vertical projection, but also in the magnetically conjugated hemisphere), the low-middle latitude earthquakes also produce the effects in equatorial anomaly, as well as in the magnetically conjugated region.

The contemporaneous ionospheric models are able adequately reproduce effects in the low-latitude ionosphere, but the initial conditions, especially the mechanism of the anomalous electric field at the ground level (or modification of atmospheric electricity in the form of air conductivity), should be established more precisely.

Acknowledgments

The research leading to these results has received funding from the European Union Sevenths Framework Program (FP7/20017-2013) under Grant Agreement no. 263502–PRE-EARTHQUAKES project: Processing Russian and European EARTH Observations for Earthquake Precursors Studies. The paper reflects only the author's views and the European Union is not liable for any use that may be made of the information contained herein.

References

[1] S. A. Pulinets and K. A. Boyarchuk, *Ionospheric Precursors of Earthquakes*, Springer, New York, NY, USA, 2004.

[2] V. A. Liperovsky, O. A. Pokhotelov, and S. A. Shalimov, *Ionospheric Precursors of the Earthquakes*, Nauka, Moscow, Russia, 1992.

[3] S. A. Pulinets, "Seismic activity as a source of the ionospheric variability," *Advances in Space Research*, vol. 22, no. 6, pp. 903–906, 1998.

[4] S. S. Kouris, P. Spalla, and B. Zolesi, "Could ionospheric variations be precursors of a seismic event? A short discussion," *Annali di Geofisica*, vol. 44, no. 2, pp. 395–402, 2001.

[5] H. Rishbeth, "Do earthquake precursors really exist?" *Eos*, vol. 88, no. 29, p. 296, 2007.

[6] S. A. Pulinets, A. D. Legen'ka, T. V. Gaivoronskaya, and V. K. Depuev, "Main phenomenological features of ionospheric precursors of strong earthquakes," *Journal of Atmospheric and Solar-Terrestrial Physics*, vol. 65, no. 16–18, pp. 1337–1347, 2003.

[7] J. Y. Liu, Y. J. Chuo, S. J. Shan et al., "Pre-earthquake ionospheric anomalies registered by continuous GPS TEC measurements," *Annales Geophysicae*, vol. 22, no. 5, pp. 1585–1593, 2004.

[8] J. Y. Liu, Y. I. Chen, Y. J. Chuo, and C. S. Chen, "A statistical investigation of preearthquake ionospheric anomaly," *Journal of Geophysical Research A*, vol. 111, no. 5, Article ID A05304, 2006.

[9] J. Y. Liu, C. H. Chen, Y. I. Chen, W. H. Yang, K. I. Oyama, and K. W. Kuo, "A statistical study of ionospheric earthquake precursors monitored by using equatorial ionization anomaly of GPS TEC in Taiwan during 2001–2007," *Journal of Asian Earth Sciences*, vol. 39, no. 1-2, pp. 76–80, 2010.

[10] M. Parrot, "Statistical analysis of Ionospheric Perturbations Observed by DEMETER in Relation with the Seismic Activity," *Earthquake Science*, vol. 24, no. 6, pp. 513–521, 2011.

[11] V. A. Liperovsky, O. A. Pokhotelov, C. V. Meister, and E. V. Liperovskaya, "Physical models of coupling in the lithosphere-atmosphere-ionosphere system before earthquakes," *Geomagnetism and Aeronomy*, vol. 48, no. 6, pp. 795–806, 2008.

[12] S. A. Pulinets, A. D. Legen'ka, and V. Kh. Depuev, "Pre-seismic effects on the equatorial anomaly," in *10th International symposium on Equatorial Aeronomy (ISEA '00)*, Antalya, Turkey, May 2000.

[13] S. A. Pulinets and A. D. Legen'ka, "Dynamics of the near-equatorial ionosphere prior to strong earthquakes," *Geomagnetism and Aeronomy*, vol. 42, no. 2, pp. 227–232, 2002.

[14] S. A. Pulinets, K. A. Boyarchuk, V. V. Hegai, V. P. Kim, and A. M. Lomonosov, "Quasielectrostatic model of atmosphere-thermosphere-ionosphere coupling," *Advances in Space Research*, vol. 26, no. 8, pp. 1209–1218, 2000.

[15] S. Pulinets and D. Ouzounov, "Lithosphere-Atmosphere-Ionosphere Coupling (LAIC) model—an unified concept for earthquake precursors validation," *Journal of Asian Earth Sciences*, vol. 41, no. 4-5, pp. 371–382, 2011.

[16] S. A. Pulinets, "Physical mechanism of the vertical electric field generation over active tectonic faults," *Advances in Space Research*, vol. 44, no. 6, pp. 767–773, 2009.

[17] M. V. Klimenko, V. V. Klimenko, I. E. Zakharenkova, S. A. Pulinets, B. Zhao, and M. N. Tsidilina, "Formation mechanism of great positive TEC disturbances prior to Wenchuan earthquake on May 12, 2008," *Advances in Space Research*, vol. 48, no. 3, pp. 488–499, 2011.

[18] M. V. Klimenko, V. V. Klimenko, I. E. Zakharenkova, and S. A. Pulinets, "Variations of equatorial electrojet as possible seismo-ionospheric precursor at the occurrence of TEC anomalies before strong earthquake," *Advances in Space Research*, vol. 49, pp. 509–517, 2012.

[19] F. Freund, "Toward a unified solid state theory for pre-earthquake signals," *Acta Geophysica*, vol. 58, no. 5, pp. 719–766, 2010.

[20] K. Sekimoto and M. Takayama, "Fundamental processes of corona discharge, surface analysis of traces stained with discharge on brass plate in negative corona," *Journal of the Institute of Electrostatics Japan*, vol. 33, no. 1, pp. 38–42, 2009.

[21] S. A. Pulinets and R. F. Benson, *Radio-Frequency Sounders in Space, Review of Radio Science*, W. Ross Stone, Ed., Oxford University Press, 1999.

[22] S. A. Pulinets, A. D. Legen'ka, A. T. Karpachev et al., "The earthquakes prediction possibility on the base of topside sounding data," IZMIRAN preprint N 34a(981), 25 p., 1991.

[23] N. P. Benkova, M. G. Deminov, A. T. Karpachev et al., "Longitude features shown by topside sounder data and their importance in ionospheric mapping," *Advances in Space Research*, vol. 10, no. 8, pp. 57–66, 1990.

[24] G. F. Deminova, "Wavelike structure of longitudinal variations of the nighttime equatorial anomaly," *Geomagnetism and Aeronomy*, vol. 35, no. 4, pp. 576–579, 1996.

[25] S. A. Pulinets and V. H. Depuev, "Ionospheric variability around the time of Mammoth Lakes seismic swarm of May 1980 in California," *Proceedings of the IRI Task Force Activity 2003, ICTP Publishing, IC/IR/2004/1*, pp. 85–96, 2004.

[26] S. Kon, M. Nishihashi, and K. Hattori, "Ionospheric anomalies possibly associated with M≥6.0 earthquakes in the Japan area during 1998–2010: case studies and statistical study," *Journal of Asian Earth Sciences*, vol. 41, no. 4-5, pp. 410–420, 2011.

[27] D. Ouzounov, S. Pulinets, A. Romanov et al., "Atmosphere-Ionosphere Response to the M9 Tohoku Earthquake Revealed by Joined Satellite and Ground Observations: Preliminary Results," *Earthquake Science*, vol. 24, pp. 557–564, 2011.

[28] S. A. Pulinets and A. D. Legen'ka, "First simultaneous observations of the topside density variations and VLF emissions before the Irpinia earthquake, November, 23, 1980 in magnetically conjugated regions," in *Proceedings of the International Workshop on Seismo Electromagnetics*, pp. 56–59, Chofu, Japan, 1997.

[29] S. A. Pulinets, P. Biagi, V. Tramutoli, A. D. Legen'ka, and V. K. Depuev, "Irpinia earthquake 23 November 1980—lesson from nature reviled by joint data analysis," *Annals of Geophysics*, vol. 50, no. 1, pp. 61–78, 2007.

[30] S. Pulinets, D. Ouzounov, and M. Parrot, "Conjugated near-equatorial effects registered by DEMETER satellite before Sumatra earthquake M8.7 of March 28, 2005," in *DEMETER Workshop*, Toulouse, France, June 2006.

[31] V. M. Sorokin, "Plasma and electromagnetic effects in the ionosphere related to the dynamics of charged aerosols in the lower atmosphere," *Russian Journal of Physical Chemistry B*, vol. 1, no. 2, pp. 138–170, 2007.

[32] F. T. Freund, I. G. Kulahci, G. Cyr et al., "Air ionization at rock surfaces and pre-earthquake signals," *Journal of Atmospheric and Solar-Terrestrial Physics*, vol. 71, no. 17-18, pp. 1824–1834, 2009.

[33] A. A. Namgaladze, M. V. Klimenko, V. V. Klimenko, and I. E. Zakharenkova, "Physical mechanism and mathematical modeling of earthquake ionospheric precursors registered in total electron content," *Geomagnetism and Aeronomy*, vol. 49, no. 2, pp. 252–262, 2009.

[34] C. L. Kuo, J. D. Huba, G. Joyce, and L. C. Lee, "Ionosphere plasma bubbles and density variations induced by pre-earthquake rock currents and associated surface charges," *Journal of Geophysical Research*, vol. 116, Article ID A10317, 2011.

[35] M. J. Rycroft and R. G. Harrison, "Electromagnetic atmosphere-plasma coupling: the global atmospheric electric circuit," *Space Science Reviews*. In press.

[36] C. L. Kuo, J. D. Huba, G. Joyce, and L. C. Lee, "Ionosphere plasma bubbles and density variations induced by pre-earthquake rock currents and associated surface charges," *Journal of Geophysical Research A*, vol. 116, no. 10, Article ID A10317, 2011.

[37] W. A. Hoppel, "Theory of the electrode effect," *Journal of Atmospheric and Terrestrial Physics*, vol. 29, no. 6, pp. 709–721, 1967.

[38] K. A. Boyarchuk, A. V. Karelin, and R. V. Shirokov, *Basic Models of the Ionized Atmosphere Kinetics*, FGUP "NPP VNIIEM", Moscow, Russia, 2006.

[39] K. Qin, L. Wu, and G. Quo, "Multi-parameters thermal anomalies before New Zealand Ms7.0 earthquake," in *Geoscience and Remote Sensing Symposium (IGARSS '11)*, pp. 2496–2499, 2011.

[40] L. F. Khilyuk, G. V. Chillingar, J. O. Robertson Jr., and B. Endres, *Gas Migration. Events Preceding Earthquakes*, Gulf Publishing Company, Houston, Tex, USA, 2000.

[41] W. S. Broecker, "An application of natural radon to problems in ocean circulation," in *Symposium on Diffusion in Oceans and Fresh Waters*, pp. 116–145, Palisades, NY, USA, 1964.

[42] G. Schumann, "Radon isotopes and daughters in the atmosphere," *Archiv für Meteorologie, Geophysik und Bioklimatologie Serie A*, vol. 21, no. 2-3, pp. 149–170, 1972.

[43] H. Kawabata, H. Narita, K. Harada, S. Tsunogai, and M. Kusakabe, "Air-sea gas transfer velocity in stormy winter estimated from radon deficiency," *Journal of Oceanography*, vol. 59, no. 5, pp. 651–661, 2003.

[44] N. Jacob, D. S. Suresh Babu, and K. Shivanna, "Radon as an indicator of submarine groundwater discharge in coastal regions," *Current Science*, vol. 97, no. 9, pp. 1313–1320, 2009.

[45] S. A. Pulinets, V. G. Bondur, M. N. Tsidilina, and M. V. Gaponova, "Verification of the concept of seismoionospheric coupling under quiet heliogeomagnetic conditions, using the Wenchuan (China) earthquake of May 12, 2008, as an example," *Geomagnetism and Aeronomy*, vol. 50, no. 2, pp. 231–242, 2010.

[46] R. W. Schunk and A. F. Nagy, *Ionospheres: Physics, Plasma Physics and Chemistry*, Cambridge University Press, Cambridge, UK, 2000.

[47] H. C. Koons, J. L. Roeder, and P. Rodriguez, "Plasma waves observed inside plasma bubbles in the equatorial F region," AEROSPACE REPORT NO. TR-98(8570)-1, The Aerospace Corporation Publishing, El Segundo, Calif, USA, pp. 4577–4583, 1988.

[48] J. Park, K. W. Min, V. P. Kim et al., "Equatorial plasma bubbles with enhanced ion and electron temperatures," *Journal of Geophysical Research A*, vol. 113, no. 9, Article ID A09318, 2008.

[49] C. S. Huang, O. De La Beaujardiere, R. F. Pfaff et al., "Zonal drift of plasma particles inside equatorial plasma bubbles and its relation to the zonal drift of the bubble structure," *Journal of Geophysical Research A*, vol. 115, no. 7, Article ID A07316, 2010.

[50] S. A. Pulinets and K. G. Tsybulya, "Unique variations of the total electron content in the preparation period of Haitian earthquake (M7.9) on January 12, 2010," *Geomagnetism and Aeronomy*, vol. 50, no. 5, pp. 686–689, 2010.

[51] H. Rishbeth, "Dynamics of the equatorial F-region," *Journal of Atmospheric and Terrestrial Physics*, vol. 39, no. 9-10, pp. 1159–1168, 1977.

6

Accounting for Antenna in Half-Space Fresnel Coefficient Estimation

A. D'Alterio and R. Solimene

Dipartimento di Ingegneria dell'Informazione, Seconda Università di Napoli, Via Roma 29, 81031 Aversa, Italy

Correspondence should be addressed to A. D'Alterio, antonietta.dalterio@unina2.it

Academic Editor: Francesco Soldovieri

The problem of retrieving the Fresnel reflection coefficients of a half-space medium starting from measurements collected under a reflection mode multistatic configuration is dealt with. According to our previous results, reflection coefficient estimation is cast as the inversion of linear operator. However, here, we take a step ahead towards more realistic scenarios as the role of antennas (both transmitting and receiving) is embodied in the estimation procedure. Numerical results are presented to show the effectiveness of the method for different types of half-space media.

1. Introduction

Subsurface imaging problem is relevant in several applicative contexts that range from geophysical to civil engineering applications [1].

In this framework, regardless of the imaging algorithm one may want to use, the knowledge of soil parameters is necessary in order to obtain properly focused images and free from artifacts proliferation [2].

By contrast, in realistic scenarios, such parameters are generally unknown or, at best, known with some degree of uncertainty. Therefore, a soil parameter estimation procedure must be run before imaging.

Many procedures for estimating soil parameters are widespread in the literature. Reflectometry methods are very common [3]. There are methods which rely on fitting the moveout of hyperbolic diffraction pattern or measure travel time to a scatterer buried at known depth [4]. Other methods exploit different offset data and perform velocity or amplitude analysis to gather soil properties [5, 6]. Iterative imaging instead identifies soil parameters as those which return the more focalized reconstruction of a cooperative target [7]. Finally, further methods first retrieve reflection coefficient and then infer the soil properties by minimizing

a nonlinear cost function by using optimization procedure [8].

Most of the methods quoted above require far zone approximation, so that asymptotic ray approximation works and generally assume soil as a homogeneous (at least transversally) half space. Moreover, time domain data, equivalently multifrequency data, are employed. In particular, this requires dealing with a non-linear inversions when the reflection coefficient is used to infer soil properties [8]. As well known, non-linear inversion are generally computationally demanding and can suffer from reliability problems due to the occurrence of false solutions. In these cases, one can take advantage from a priori information about the soil which allows to reduce the number searched for unknowns. However, this entails that soil dispersive law must be a priori known.

Determining soil electromagnetic parameters is a mandatory step to achieve soil analysis in order to assess, for example, water content and in general material composition. However, this is not necessarily hold for imaging. Indeed, what is really needed to obtain properly focused image of buried scatterers is the relevant Green's function. Under half-space homogenous assumption for soil (which is an assumption commonly used, as can be argued from previous references), Green's function can be directly determined from the half-space Fresnel reflection coefficients [9]. Of course, from the Fresnel coefficients soil properties can be inferred as well, but as said above this step is not required for imaging. However, it is remarked that following this new perspective requires to retrieve the reflection coefficients not only as a function of the time frequency but also as a function of the spatial frequencies (i.e., for different angles of incidence).

Recently, in [10] a new estimation method, which achieves such a task, was proposed. There, proximal GPR measurements, collected at different positions, were exploited to develop an estimation method based on multiview (multi-offset) information. Therefore, far-zone approximation were no longer necessary. More precisely, the problem is cast as the inversion of a linear integral operator linking the reflected field and the Fresnel reflection coefficients. When such a problem is solved, an estimation of the reflection coefficient is available for different angle of incidence but for a fixed-time frequency. Therefore, the procedure must be run for each adopted frequency. However, this inconvenient is not too critical, as to image shallowly buried scatterers the number of necessary frequency can be not so high [11, 12] and is traded off by the advantage that soil dispersive laws are not required.

The method presented in [10] was first developed and numerically validated by considering two-dimensional scalar cases. Then, the procedure was extended to three-dimensional cases where reflection coefficient is indeed a matrix. Therefore, in that cases the problem was cast as the inversion of a linear integral operator where the relevant unknowns (i.e., the Fresnel coefficients) appear embedded within a dyad. As shown in [13], depending on antennas' polarization, this gives rise to different strategies for achieving the solution.

However, in that contribution, reflected field measurements were considered as data, instead, in practical cases, what is actually measured is the voltage signal and not the field. Hence, the estimation scheme must account also for the role of the receiving antenna. This is just the further step towards a more realistic scenario done in this paper.

The plan of the paper is the following. In Section 2, we describe the adopted measurement configuration and introduce the problem mathematical formulation. In Section 3, the procedure for retrieving the Fresnel reflection coefficients is introduced, whereas in Section 4 numerical results are reported for different types of soil. Conclusions follow.

2. Geometry and Problem Formulation

The scattering problem depicted in Figure 1 is of concern.

The scenario consists of a two-layered medium separated by a planar interface at $z = 0$. The upper layer is assumed to be the free-space with ε_0 and μ_0 being its dielectric permittivity and magnetic permeability, respectively. The

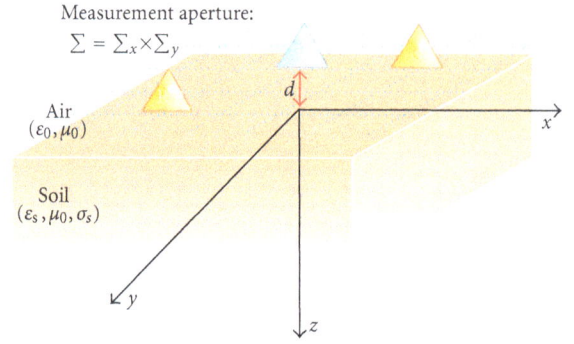

FIGURE 1: Geometry of the problem.

lower half-space is representative of the soil which is assumed nonmagnetic (i.e., its magnetic permeability is equal to the one of free space) and homogeneous with a relative dielectric permittivity and electric conductivity, denoted as ε_s and σ_S, respectively.

Accordingly, in the frequency domain the dielectric permittivity is actually a complex function which depends on the angular frequency ω and thus the relative equivalent permittivity is given by $\varepsilon_{eq}(\omega) = \varepsilon_S(\omega) - j\sigma_S/\omega$.

The transmitting antenna is located in the upper half-space at height d from the air/soil interface in $P = (x_s, y_s, -d)$. It is assumed that it is far enough so that the effect of the soil on the antenna radiation properties is negligible or a proper gating procedure, as suggested in [14], is employed before the estimation takes place.

We consider a multistatic measurement configuration where the reflected field is collected over a set of different positions taken over the observation domain $\Sigma = \Sigma_x \times \Sigma_y$ which is parallel to the air/soil interface and located at the same height as the source. $\Sigma_x = [-X_O, X_O]$ and $\Sigma_y = [-Y_O, Y_O]$ are the extent of the observation domain along the x- and y-axes, respectively.

As discussed in the Introduction, we aim at retrieving Fresnel reflection coefficients from reflected field measurements. Moreover, this must be achieved for all the relevant spatial spectral harmonics as these are necessary for Green's function determination for a subsequent imaging stage.

Accordingly, it is natural to express the link between the reflected field, $\underline{E}_r(x_O, y_O, k_0)$, and the reflection coefficients, in terms of plane-wave spectrum notation, that is,

$$\underline{E}_r(x_O, y_O, k_0)$$

$$= \frac{1}{(2\pi)^2} \iint_D \underline{\underline{\Gamma}}(k_x, k_y, k_0) \underline{f}_S(k_x, k_y, k_0)$$

$$\times \exp[-jk_x(x_O - x_S)] \exp[-jk_y(y_O - y_S)]$$

$$\times \exp(-j2k_z d) \, dk_x dk_y,$$

$$(1)$$

where $\underline{\underline{\Gamma}}(k_x, k_y, k_0)$ is the dyadic reflection coefficient at the air/soil interface [13]:

$$
\underline{\underline{\Gamma}}\left(k_k, k_y, k_0\right)
$$

$$
= \begin{bmatrix} \left(\Gamma_{\text{TE}} \dfrac{k_y^2}{k_t^2} + \Gamma_{\text{TM}} \dfrac{k_x^2}{k_t^2}\right) & \left(\Gamma_{\text{TM}} \dfrac{k_x k_y}{k_t^2} - \Gamma_{\text{TE}} \dfrac{k_x k_y}{k_t^2}\right) & 0 \\ \left(\Gamma_{\text{TM}} \dfrac{k_x k_y}{k_t^2} - \Gamma_{\text{TE}} \dfrac{k_x k_y}{k_t^2}\right) & \left(\Gamma_{\text{TE}} \dfrac{k_x^2}{k_t^2} + \Gamma_{\text{TM}} \dfrac{k_y^2}{k_t^2}\right) & 0 \\ 0 & 0 & -\Gamma_{\text{TM}} \end{bmatrix}
$$

$$(2)$$

Γ_{TE} and Γ_{TM} are the Fresnel reflection coefficients, $\underline{f}_S = (f_x, f_y, f_z)$ is the source plane-wave spectrum, k_x, k_y, k_z are the wavenumber components corresponding to x, y, z, respectively, and $k_t^2 = k_x^2 + k_y^2$. Finally, $k_0 = \omega\sqrt{\varepsilon_0\mu_0}$ is the free-space wavenumber.

Note that in (1) the spectral integration interval D corresponds to the minimum square enclosing the circle of radius k_0 which accounts for the so-called "visible domain."

Equation (1) represents the link to be inverted in order to retrieve reflection coefficients. In particular, it should be solved for Γ_{TE} and Γ_{TM} which appear embodied within $\underline{\underline{\Gamma}}$.

A way to achieve such a task has been described in [13]. That paper reports promising results but has to be meant as a proof of principle. Indeed, the inversion of (1) relies on field measurements whereas what can be actually measured is antenna output voltage. In other words, the role of receiving antenna has to be accounted for and (1) properly modified.

To this end, here, a simple truncated waveguide is considered as transmitting and receiving antenna. In particular, the source point $x_s, y_s, -d$ and the observation point $x_O, y_O, -d$ correspond to the centre of the antenna apertures. For convenience, we also introduce a local reference frame x', y', z' (see Figure 2) to address points within the antenna aperture for each antenna position.

By neglecting finite aperture effects and by assuming that the guide is matched, then the field at the antenna terminal section $z' = 0$ (for each receiving antenna positions) is just the incoming field reflected by the soil $\underline{E}_r(x', y', 0)$.

Hence, we can write the "voltage" at $z' = 0$ associated to the waveguide fundamental mode as [15]

$$
V_{10}^{\text{TE}}(0) = \iint_S \underline{E}_r(x', y', 0) \times \underline{h}_{10}^{\text{TE}} \cdot \hat{i}_{z'}\, dS \tag{3}
$$

In particular, if the working frequency is properly chosen and L sufficiently greater than the wavelength (see Figure 2), this is the only relevant contribution at the reference plane in $z' = -L$ as high-order modes are in cutoff.

Accordingly, the voltage at the reference section writes as

$$
V = V_{10}^{\text{TE}}(0)\exp(-ik_z'L), \tag{4}
$$

where $k_z' = \sqrt{k_0^2 - (\pi/a)^2}$ being the z-component of the waveguide wavevector.

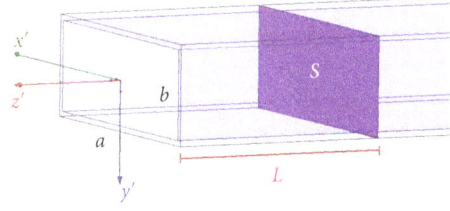

FIGURE 2: Truncated waveguide antenna.

Eventually, the relationship upon which the estimation procedure is based is obtained by inserting (1) in (4); that is

$$
\begin{aligned}
&V(x_O, y_O, k_0) \\
&= \exp(-ik_z'L) \\
&\quad \times \iint_S \frac{1}{(2\pi)^2} \iint_D \underline{\underline{\Gamma}}\left(k_x, k_y, k_0\right) \\
&\qquad \times \underline{f}_S\left(k_x, k_y, k_0\right) \times \underline{h}_{10}^{\text{TE}} \\
&\qquad \times \exp\left[-jk_x(x' + x_O - x_S)\right] \\
&\qquad \times \exp\left[-jk_y(y' + y_O - y_S)\right] \\
&\qquad \times \exp(-j2k_z d)\, dk_x\, dk_y\, dS.
\end{aligned} \tag{5}
$$

3. Estimation Procedure

In this section we introduce the procedure to estimate Γ_{TE} and Γ_{TM} (embedded in the coefficients of $\underline{\underline{\Gamma}}$) starting from voltage measurements taken at a grid of points over Σ.

To this end, different inversion strategies can be employed to invert (5) and de-embed Γ_{TE} and Γ_{TM} from $\underline{\underline{\Gamma}}$ [13]. Basically, by changing the polarization of the transmitting and/or the receiving antennas, a couple of linear integral equations are obtained whose solution gives two different parameters of the reflection matrix (2). Hence, from these two independent equations, the Fresnel coefficients are obtained.

Here, we adopt the strategy depicted in Figure 3. In particular, the two needed voltage measurements (for each observation point) are obtained by considering first the transmitting and receiving antennas polarized along the x-axis. Accordingly, (5) particularizes as

$$
\begin{aligned}
&V_1(x_O, y_O, k_0) \\
&= \frac{1}{(2\pi)^2}\sqrt{\frac{2}{ab}} \iint_D \Gamma_{11}\left(k_x, k_y, k_0\right) f_x^2\left(k_x, k_y, k_0\right) \\
&\quad \times \exp\left[-jk_x(x_O - x_S)\right] \\
&\quad \times \exp\left[-jk_y(y_O - y_S)\right]\exp(-j2k_z d) \\
&\quad \times \exp(-jk_z'L)\, dk_x dk_y,
\end{aligned} \tag{6}
$$

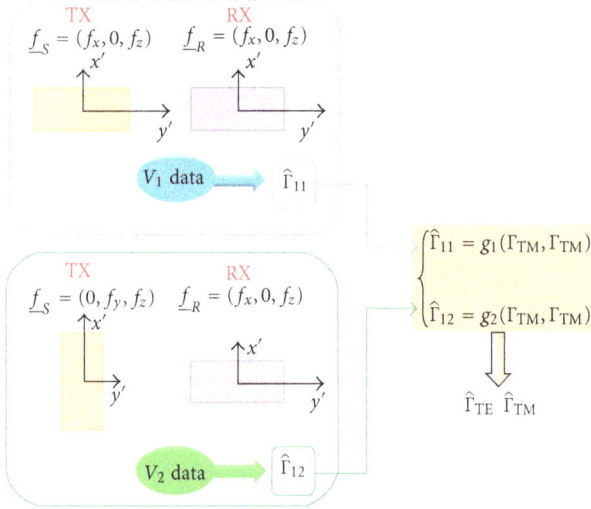

FIGURE 3: Illustrating the overall estimation procedure.

where we considered that $\underline{f}_S = \underline{f}_R = (f_x, 0, f_z)$, Γ_{11} is the (1,1) entry of $\underline{\underline{\Gamma}}$, a and b are the guide transverse dimensions (with $a > b$), and $\underline{k}_t = (k_x, k_y, 0)$.

The second equation is obtained by rotating the transmitting antenna of 90 degrees so that it results to be polarized along the y-axis; that is, $\underline{f}_S = (0, f_x, f_z)$. In this case (5) becomes

$$V_2(x_O, y_O, k_0)$$

$$= \frac{1}{(2\pi)^2} \sqrt{\frac{2}{ab}} \iint_D \Gamma_{12}(k_x, k_y, k_0) f_x^2(k_x, k_y, k_0)$$

$$\times \exp[-jk_x(x_O - x_S)]$$

$$\times \exp[-jk_y(y_O - y_S)]$$

$$\times \exp(-j2k_z d)$$

$$\times \exp(-jk'_z L) dk_x dk_y,$$

(7)

where now the term Γ_{12} of $\underline{\underline{\Gamma}}$ is involved.

We have now a pair of integral equations whose inversion returns an estimation for $\Gamma_{11}(\cdot)$ and $\Gamma_{21}(\cdot)$, that is, $\hat{\Gamma}_{11}(\cdot)$ and $\hat{\Gamma}_{21}(\cdot)$, from which, according to (2), the Fresnel reflection coefficients can be obtained via algebraic relations as

$$\hat{\Gamma}_{TE}(k_x, k_y, k_0) = \frac{\Gamma_1(k_x, k_y, k_0) - k_x^2 \Gamma_2(k_x, k_y, k_0)}{k_t^2},$$

$$\hat{\Gamma}_{TM}(k_x, k_y, k_0) = \frac{\Gamma_1(k_x, k_y, k_0) + k_y^2 \Gamma_2(k_x, k_y, k_0)}{k_t^2},$$

(8)

where $\Gamma_1(\cdot) = k_t^2 \hat{\Gamma}_{11}(\cdot)$, $\Gamma_2(\cdot) = (k_t^2/k_x k_y)\hat{\Gamma}_{21}(\cdot)$, $\hat{\Gamma}_{TE}(\cdot)$ and $\hat{\Gamma}_{TM}(\cdot)$ are the estimated Fresnel reflection coefficients.

In this plan, the inversion of the two integral equations (6) and (7) is crucial in order to get a reasonable reflection

coefficient estimations. In particular, since such operators are compact, the corresponding inversion is an ill-posed inverse problem [16]. Accordingly, a suitable inversion scheme has to be adopted to establish a proper compromise between accuracy and stability against uncertainties. To cope with this point, inversions are achieved by exploiting the truncated singular value decomposition (SVD) method [16]. Details concerning the singular value threshold N_T, the number of required measurements and unknown representation (to obtain discrete counterparts of (6) and (7)), and noise propagation can be found in [13]. Here, we report only the relevant equations. On denoting as $\{u_n, v_n, \sigma_n\}_{n=0}^{\infty}$ the singular system of one of the involved integral operators, where u_n, v_n, and σ_n are the nth left singular function, the nth right singular function, and the nth singular value, respectively, the corresponding solutions can be expressed as

$$\hat{\Gamma}_{11}(k_x, k_y, k_0) = \sum_{n=0}^{N_T} \frac{\langle V_1, v_n \rangle}{\sigma_n} u_n,$$

$$\hat{\Gamma}_{21}(k_x, k_y, k_0) = \sum_{n=0}^{N_T} \frac{\langle V_2, v_n \rangle}{\sigma_n} u_n.$$

(9)

We end this section by remarking that (8) allows to retrieve the Fresnel coefficient as a function of the spectral angular variables and for a fixed frequency. When more frequency are needed, as usually occurs during imaging, the procedure outlined above must be repeated for each adopted frequency.

4. Numerical Results

In this section, we report some numerical reconstructions obtained by exploiting synthetic data with the aim to assess the performance achievable by the estimation procedure. The source is located at $\underline{r}_S = (0, 0, -d)$, with $d = 0.1$ m, whereas receiving antenna span a grid of $N_x \times N_y = 20 \times 20$ points taken uniformly over the observation domain $\Sigma = [-X_O, X_O] \times [-Y_O, Y_O] = [-1, 1] \times [-1, 1]$ m^2 and a frequency $f = 1.5$ GHz is adopted.

The waveguide transverse dimensions are $a = 0.16$ m, $b = 0.08$ m whereas the reference plane is fixed at $h = 0.1$ m from the aperture.

We consider three kinds of soil addressed for convenience, in the following, as soil 1, soil 2, and soil 3. More in detail, soil 1 is characterized by $\varepsilon_{eq} = 4\varepsilon_0$, hence describes a nondissipative as well as nondispersive medium; soil 2 accounts for Ohmic losses, hence $\varepsilon_{eq} = 4\varepsilon_0 - j(0.1/\omega)$; finally soil 3 considers both Ohmic losses and dielectric hysteretic behaviours so that $\varepsilon_{eq}(\omega)/\varepsilon_0 = \varepsilon_\infty - j(\sigma_S/\omega\varepsilon_0) + ((\varepsilon_S - \varepsilon_\infty)/(1 + j\omega\tau_e))$ with $\varepsilon_\infty = 4$, $\varepsilon_S = 16$, $\sigma_S = 0.1$ S/m, and $\tau_e = 64$ ns. Moreover, in order to test the procedures for data corrupted by noise, some examples are obtained by adding, to both voltage data, a complex zero mean white Gaussian noise n so as to have a SNR $= 20 \log(\|V_{1,2}\|/\|n\|) = 30$ dB.

The reconstructions corresponding to the three types of soil are reported in Figures 4 and 5, Figures 6 and 7, Figures 8 and 9, respectively. In those figures, for comparison purposes, beside being noiseless and noisy estimations also actual Fresnel reflection coefficients are shown.

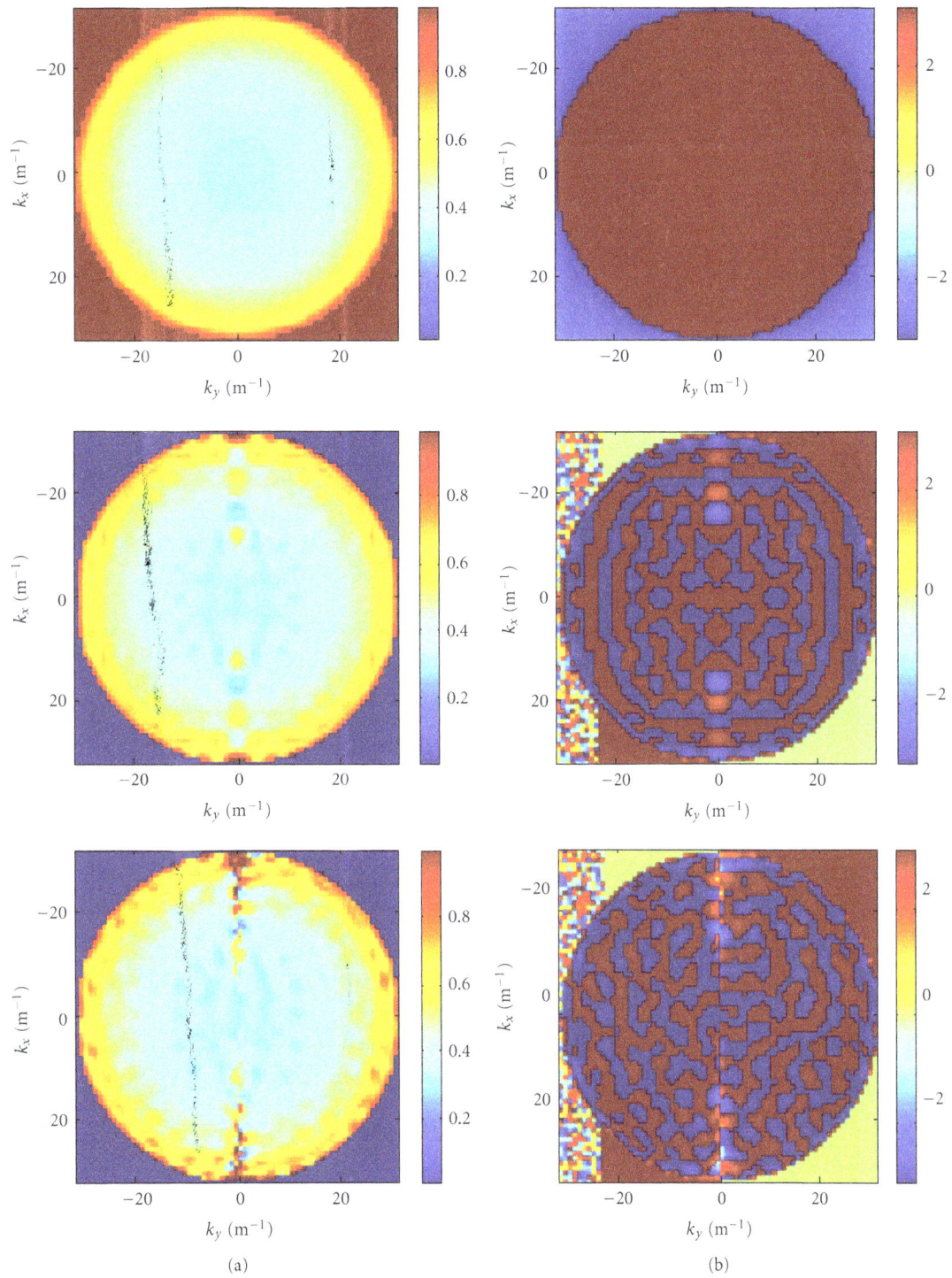

FIGURE 4: Soil 1. Amplitude (a) and phase (b) of Γ_{TE}. Top line: actual values; middle line: noiseless estimation; bottom line: noisy estimation.

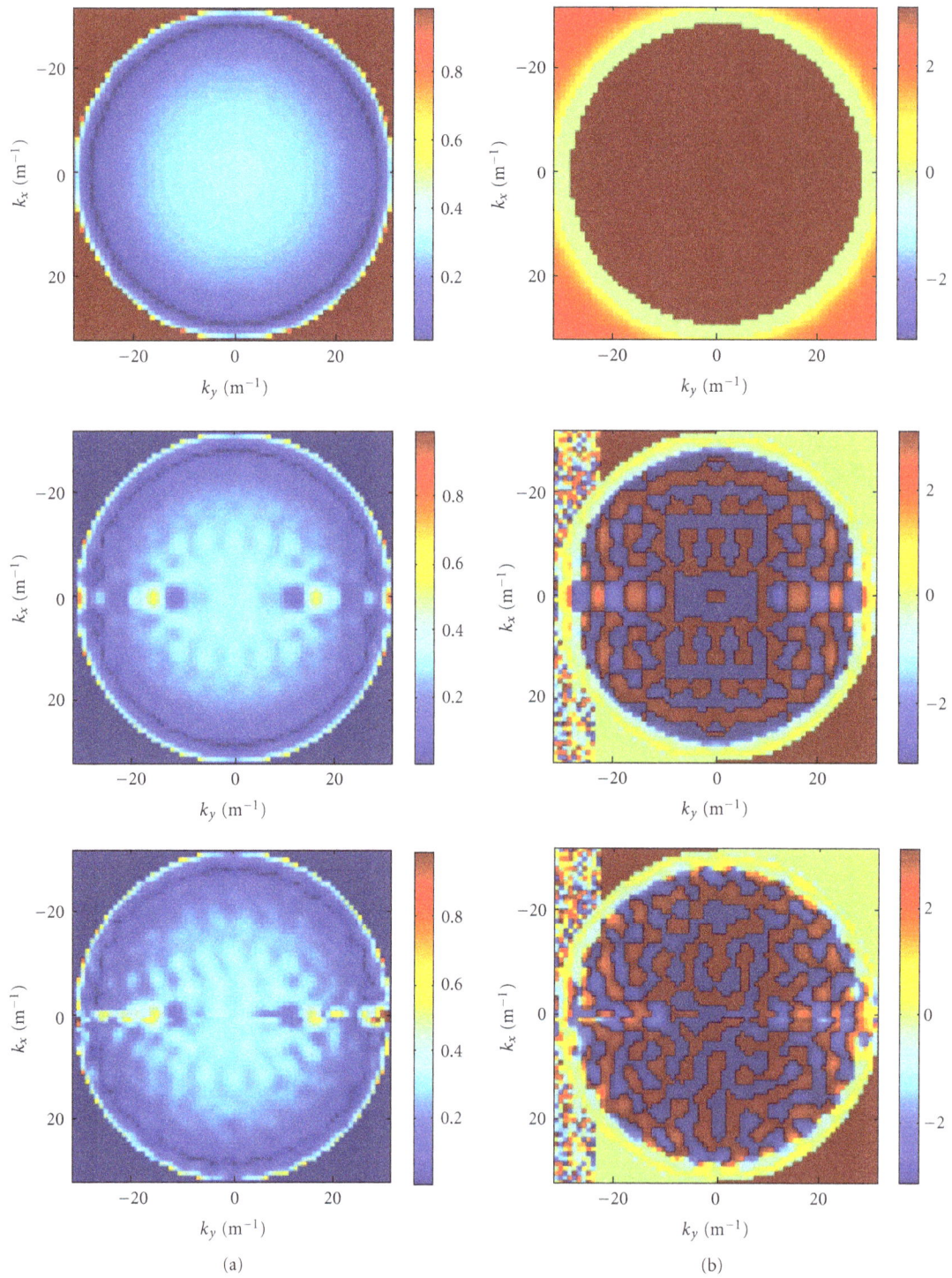

FIGURE 5: Soil 1. Amplitude (a) and phase (b) of Γ_{TM}. Top line: actual values; middle line: noiseless estimation; bottom line: noisy estimation.

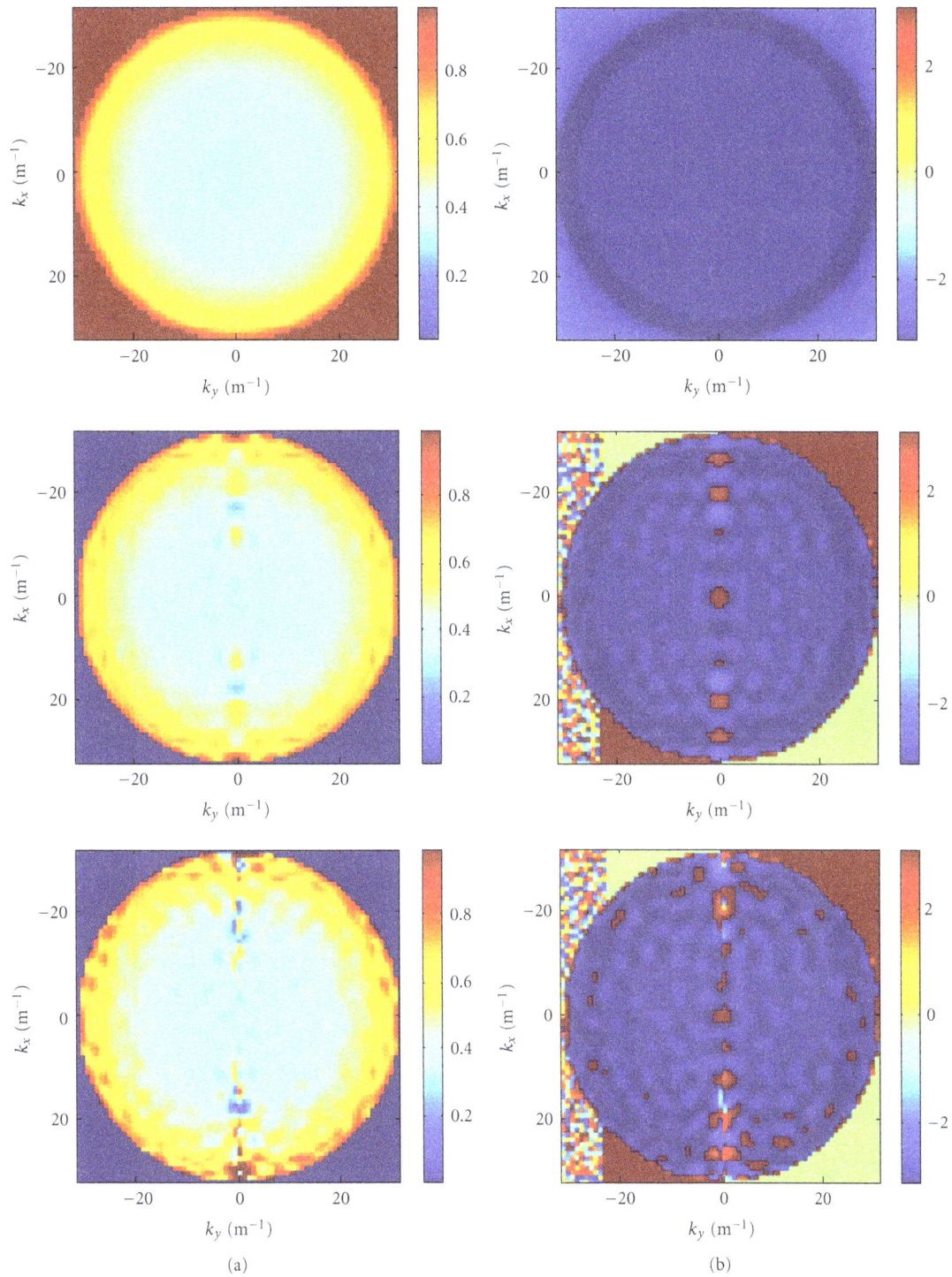

FIGURE 6: Soil 2. Amplitude (a) and phase (b) of Γ_{TE}. Top line: actual values; middle line: noiseless estimation; bottom line: noisy estimation.

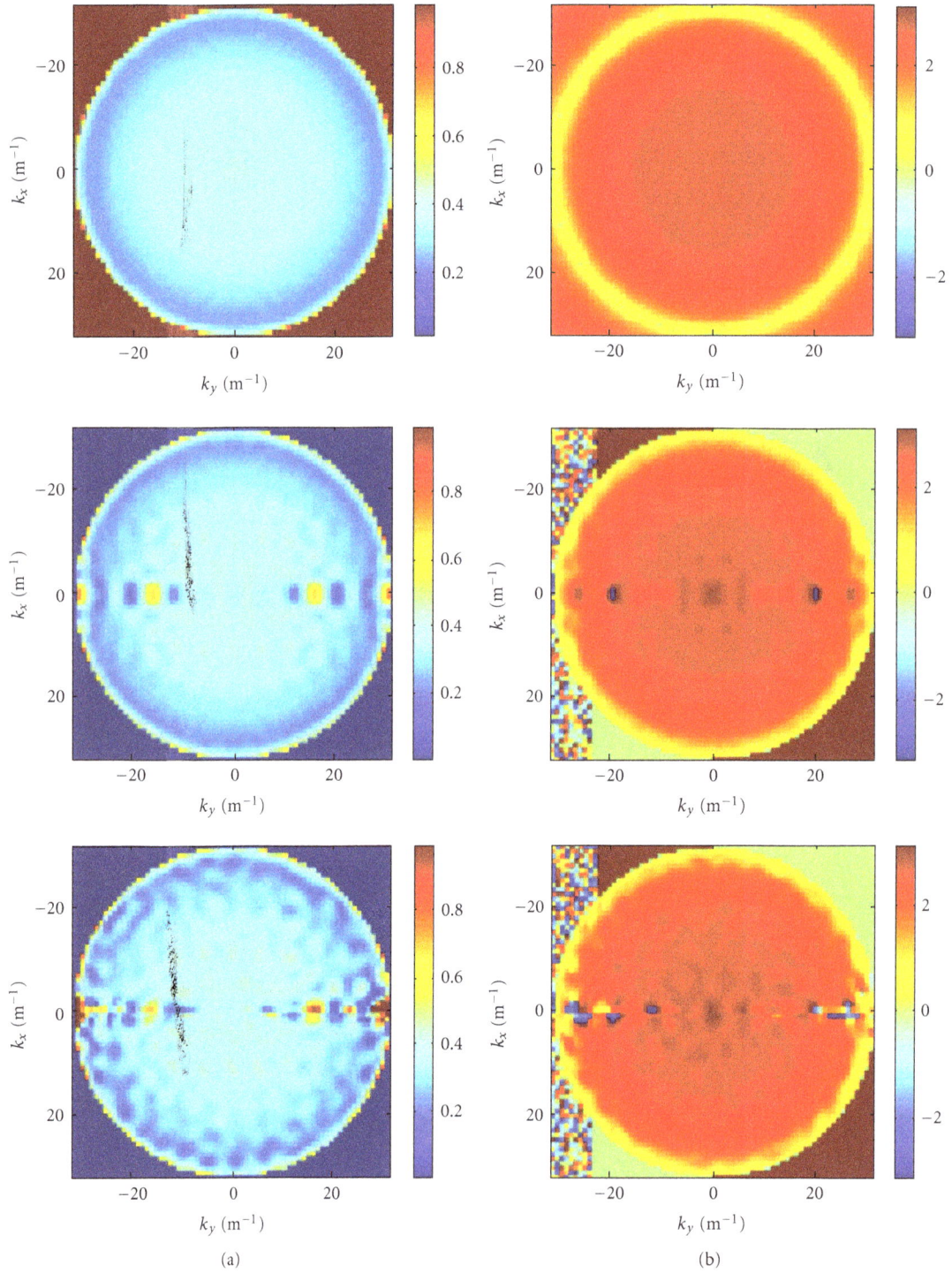

FIGURE 7: Soil 2. Amplitude (a) and phase (b) of Γ_{TM}. Top line: actual values; middle line: noiseless estimation; bottom line: noisy estimation.

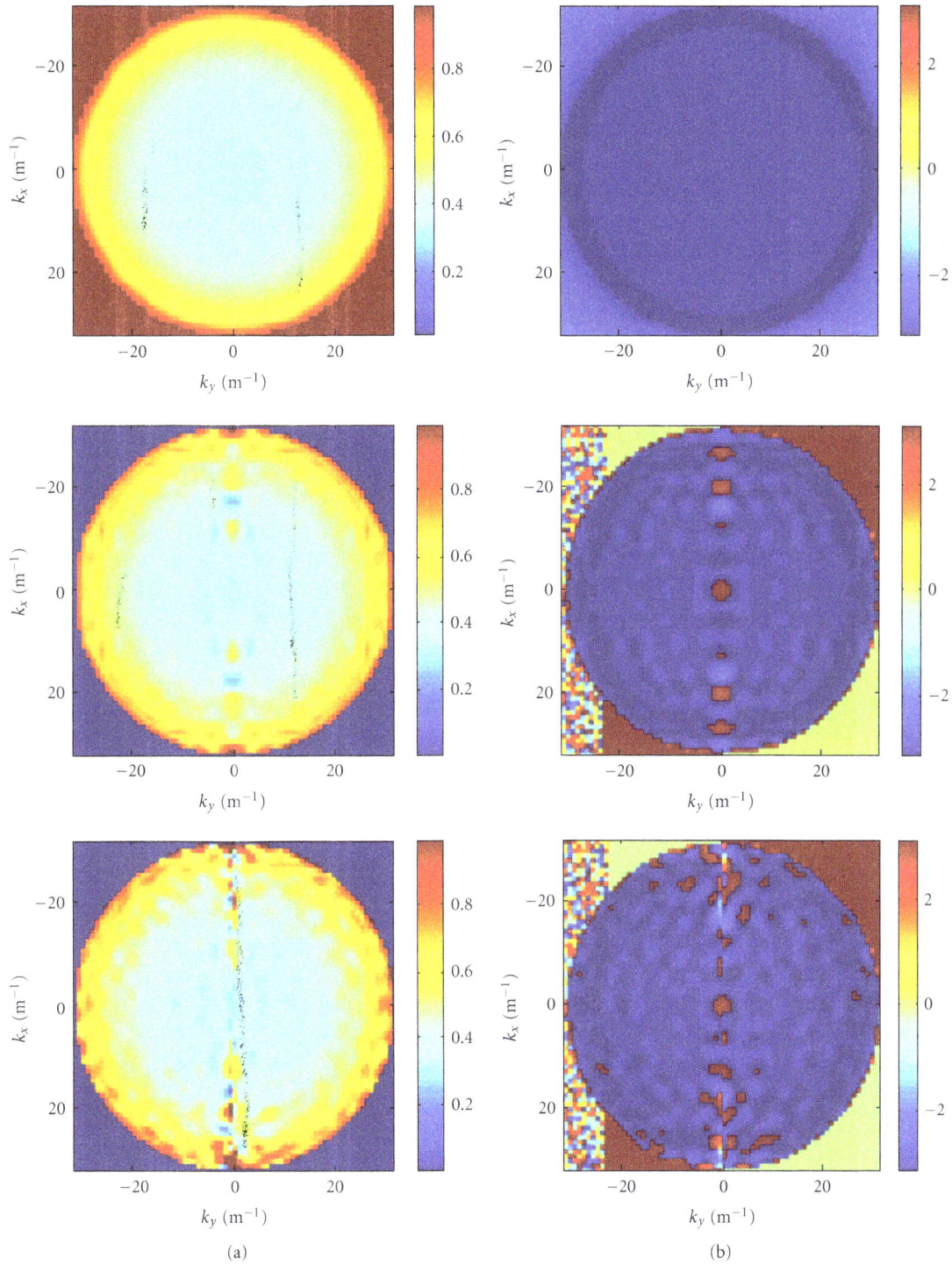

FIGURE 8: Soil 3. Amplitude (a) and phase (b) of Γ_{TE}. Top line: actual values; middle line: noiseless estimation; bottom line: noisy estimation.

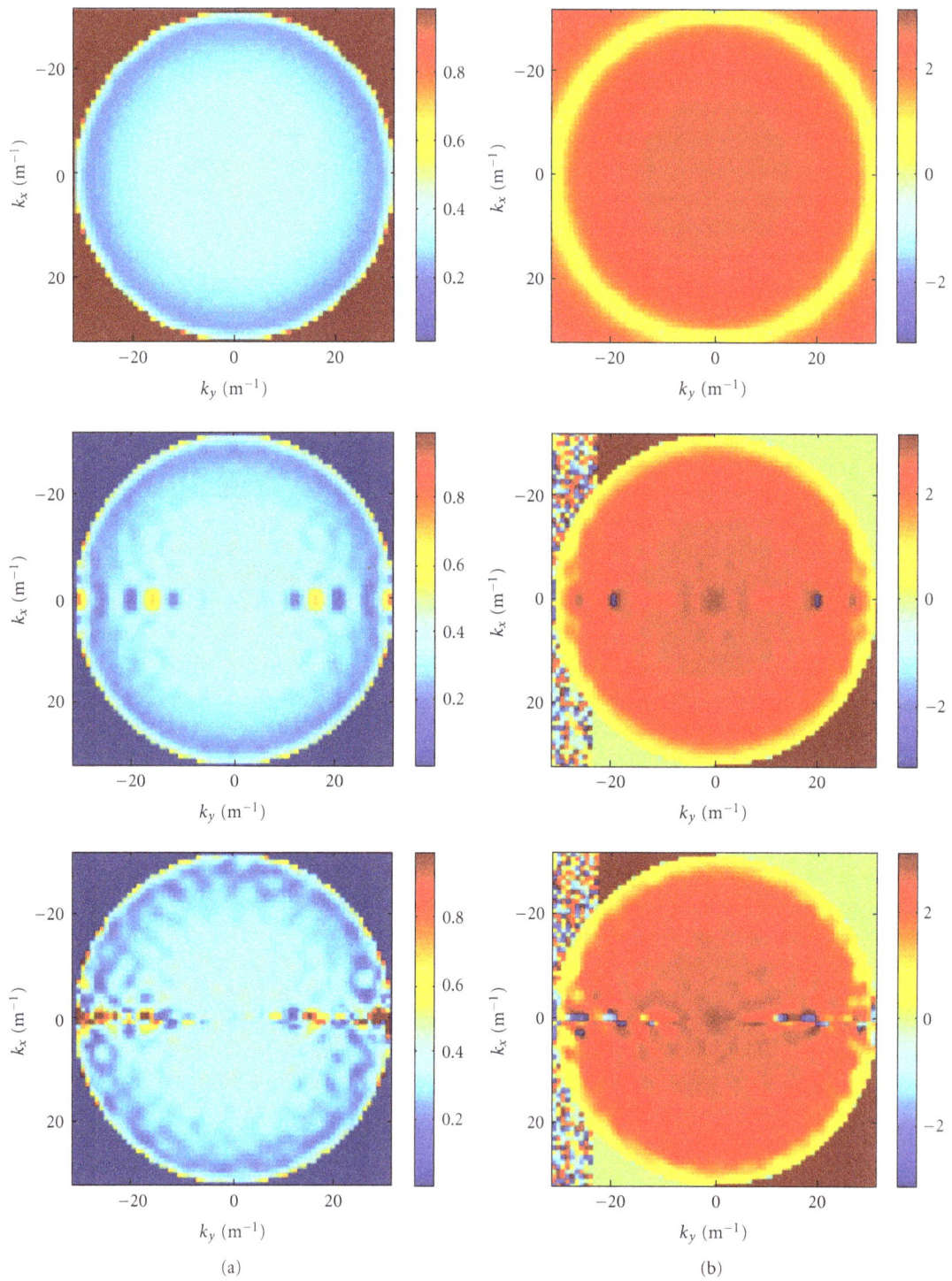

FIGURE 9: Soil 3. Amplitude (a) and phase (b) of Γ_{TM}. Top line: actual values; middle line: noiseless estimation; bottom line: noisy estimation.

By comparing actual and estimated coefficients, it can be noted that in all the cases estimation fails outside the visible domain whereas inside that domain some oscillations appear. This could be expected as they result from regularization, we applied through (9) to invert (6) and (7), which entails retrieving a filtered version of the unknown. By contrast, reconstructions appear rather "stable" against noise. More in detail, estimations fit well actual reflection coefficients expect along the lines at $k_y = 0$ for Γ_{TE} and $k_x = 0$ for Γ_{TM}. This is due to the "numerical" singularities that (8) introduce while passing from the retrieved scattering parameters $\hat{\Gamma}_{11}(\cdot)$ and $\hat{\Gamma}_{21}(\cdot)$ to the Fresnel coefficients. However, this drawback can be circumvented by getting rid estimation along the "singular" lines and replacing such values by some interpolation scheme.

5. Conclusions

In this contribution, we extended our previous work concerning the Fresnel reflection coefficients estimation from reflected field data [13] by including also the role of the receiving antenna.

It is shown that, even in the case of homogeneous soil, the corresponding mathematical problem is rather interesting as it amounts to solving a vector linear integral equation where the unknowns, the Fresnel coefficients, appear embodied within a dyadic term. To solve such a problem a two-step procedure is used. First, a pair of half-space scattering parameters are determined by inverting two integral equations by means of the TSVD scheme. Then, Fresnel reflection coefficients are easily estimated by algebraic equations.

The method allows to retrieve the Fresnel coefficients as function of the angle of incidence (spatial harmonics); therefore it can be used to build up Green's function which is what is really necessary to obtain focalized subsurface images. The estimation procedure must be run for each adopted frequency. By contrast soil dispersive laws are not required to be a priori known.

Numerical results are used to show the effectiveness of estimation proposed procedure for different and typical kinds of soil. Except for certain numerical singularities, estimation's fits well the actual reflection coefficient within the visible domain, which in turn generally is prominent contribution with respect to the evanescent waves.

References

[1] D. Daniels, *Ground Penetrating Radar*, IEE Press, London, UK, 2nd edition, 2004.

[2] F. Soldovieri and R. Solimene, "Ground penetrating radar subsurface imaging of buried objects," in *Radar Technology*, G. Kouemou, Ed., In-Tech, 2010.

[3] D. A. Robinson, "Measurement of the solid dielectric permittivity of clay minerals and granular samples using a time-domain reflectometry immersion method," *Vadose Zone Journal*, vol. 3, no. 2, pp. 705–713, 2004.

[4] A. P. Annan, *Ground Penetrating Radar Workshop Notes*, Sensors & Software, Mississauga, Canada, 2001.

[5] X. Zeng, G. A. McMechan, and T. Xu, "Synthesis of amplitude-versus-offset variations in ground-penetrating radar data," *Geophysics*, vol. 65, no. 1, pp. 113–125, 2000.

[6] J. Deparis and S. Garambois, "On the use of dispersive APVO GPR curves for thin-bed properties estimation: theory and application to fracture characterization," *Geophysics*, vol. 74, no. 1, pp. J1–J12, 2009.

[7] F. Soldovieri, G. Prisco, and R. Persico, "A strategy for the determination of the dielectric permittivity of a lossy soil exploiting GPR surface measurements and a cooperative target," *Journal of Applied Geophysics*, vol. 67, no. 4, pp. 288–295, 2009.

[8] S. Lambot, E. C. Slob, I. van den Bosch, B. Stockbroeckx, B. Scheers, and M. Vanclooster, "Estimating soil electric properties from monostatic ground-penetrating radar signal inversion in the frequency domain," *Water Resources Research*, vol. 40, no. 4, pp. W042051–W0420512, 2004.

[9] P. Meincke and T. B. Hansen, "Plane-wave characterization of antennas close to a planar interface," *IEEE Transactions on Geoscience and Remote Sensing*, vol. 42, no. 6, pp. 1222–1232, 2004.

[10] R. Solimene, F. Soldovieri, and A. D'Alterio, "Determination of the fresnel reflection coefficient of a half-space for medium estimation purposes," *Progress in Electromagnetics Research B*, no. 27, pp. 61–82, 2011.

[11] R. Persico, F. Soldovieri, and G. Leone, "A microwave tomographic imaging approach for multibistatic configuration: the choice of the frequency step," *IEEE Transactions on Instrumentation and Measurement*, vol. 55, no. 6, pp. 1926–1934, 2006.

[12] R. Solimene, F. Ahmad, and F. Soldovieri, "A novel CS-TSVD strategy to perform data reduction in linear inverse scattering problems," *IEEE Geoscience and Remote Sensing Letters*, vol. 9, no. 5, pp. 881–885, 2012.

[13] R. Solimene, A. D'Alterio, and F. Soldovieri, "Determining Fresnel reflection coefficients in 3D half-space geometry by GPR multistatic data," *Near Surface Geophysics*, vol. 9, no. 3, pp. 265–275, 2011.

[14] R. Solimene, A. D'Alterio, and F. Soldovieri, "Half-space estimation by time gating based strategy," in *Proceedings of the 13th International Conference on Ground Penetrating Radar (GPR '10)*, pp. 683–687, Lecce, Italy, June 2010.

[15] R. E. Collin, *Antennas and Radiowave Propagation*, McGraw-Hill, New York, NY, USA, 1985.

[16] M. Bertero and P. Boccacci, *Introduction to Inverse Problems in Imaging*, IOP, Bristol, UK, 1998.

7

Experimental Studies on the Changes in Resistivity and Its Anisotropy Using Electrical Resistivity Tomography

Tao Zhu, Jian-Guo Zhou, and Jin-Qi Hao

Institute of Geophysics, China Earthquake Administration, Beijing 100081, China

Correspondence should be addressed to Tao Zhu, zxl_tao@yahoo.com.cn

Academic Editor: Sabatino Piscitelli

Three measuring lines were arranged on one of free planes of magnetite cuboid samples. Apparent resistivity data were acquired by MIR-2007 resistivity meter when samples were under uniaxial compression of servocontrol YAW-5000F loadingmachine in laboratory. Then we constructed the residual resistivity images using electrical resistivity tomography (ERT) and plotted the diagrams of apparent resistivity anisotropy coefficient (ARAC) λ^* and the included angle α between the major axis of apparent resistivity anisotropy ellipse and the axis of load with pressure and effective depth. Our results show that with increasing pressure, resistivity and the decreased (D region) and increased (I region) resistivity regions have complex behaviors, but when pressure is higher than a certain value, the average resistivity decrease and the area of D region expand gradually in all time with the increase of pressure, which may be significant to the monitoring and prediction of earthquake, volcanic activities, and large-scale geologic motions. The effects of pressure on λ^* and α are not very outstanding for dry magnetite samples.

1. Introduction

The striking changes in electrical resistivity and its anisotropy were widely observed before earthquake, mine earthquake, volcanic activities, and geologic motions [1–12]. In order to address the change characteristics of resistivity and its anisotropy, many investigators conducted a plenty of experiments on rock samples under compression and during fracture, shear, and frictional sliding in lab [13–25] and theoretic studies [26–30]. Experimental results reported by Brace et al. [13] and Brace and Orange [14–16] in lab showed that striking changes in electrical resistivity were observed for a variety of igneous and sedimentary crystalline rocks when water-saturated crystalline rocks were stressed to fracture. Resistivity decreased by as much as an order of magnitude. Most of the change occurred at stresses above half the fracture stress. The electrical resistivity of hematite decreased gradually with increasing shock pressure to 440 kb and decreased discontinuously to less than 1 Ω cm from several tens of Ω cm at pressure of 440–520 kb [18], and that of a saturated granite during frictional sliding showed clear variations premonitory to the occurrence of stick-slip: resistivity decreased by a few percent with increasing shear stress; the

minimum coincided with a sudden release of shear stress of a few bars, accompanied by an immediate return of resistivity [17]. Morrow and Brace [19] and Yamazaki [31, 32] reported that the electrical resistance of tuffs changed very largely, particularly for small strains and could be applied to earthquake prediction [1]. Studies on the changes in resistivity anisotropy were conducted by Chen et al. [33] firstly in lab. They used the water-saturated crystalline rock samples and arranged three measuring lines parallel to, perpendicular to, and intersecting at the angle of 45° with maximum principle stress axis on one of its free planes, and studied the relations between apparent resistivity anisotropy and pressure. Considering the effects of groundwater, An et al. studied the changes in resistivity and its anisotropy with pressure in the presence of feed water [34]. In order to recognize the expansion of fissure with stress, Chen et al. [22–25] studied the water-saturated granite samples and man-made samples under uniaxial compression, triaxial compression with low confining pressure, and frictional sliding. This previous work played very important role in recognizing and understanding the change behaviors of resistivity and its anisotropy with pressure, but they had to analyze the curves of resistivity and its anisotropy with pressure because of only a few electrodes

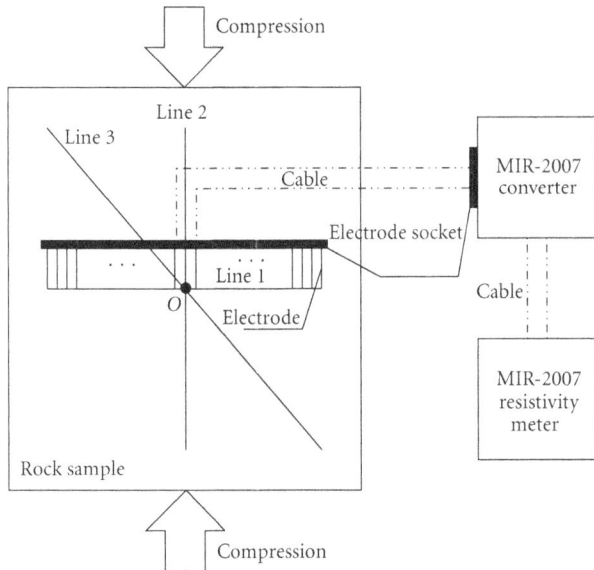

FIGURE 1: Sketch of test system.

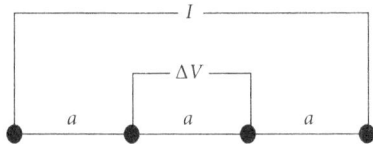

FIGURE 2: Sketch of Wenner-α array. I indicates the intensity of current between two current electrodes A and B. ΔV indicates the potential difference between two potential electrodes M and N.

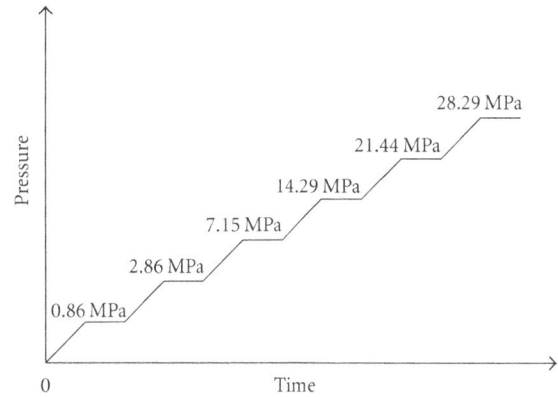

(a) Sample YN1

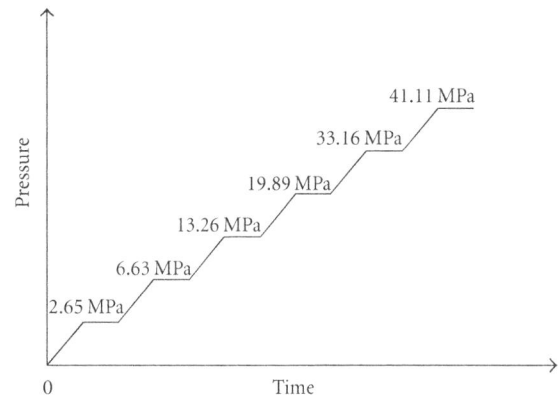

(b) Sample YN2

FIGURE 3: The loading curves of samples.

designed on one measuring line. In order to study the changes in resistivity and its anisotropy with pressure, we use MIR-2007 resistivity meter to acquire apparent resistivity data of dry magnetite samples in lab. In resistivity measurements, 37–120 electrodes and 12–30 "n" factors [35] and Wenner-α array are used, and 210–1205 apparent resistivity measurements are acquired. We construct the resistivity images at each pressure value using ERT and plot the diagrams of λ^* and α at each effective depth. Then the changes in resistivity and its image and anisotropy with pressure and depth are studied. We find that when the pressure is higher than a certain pressure, average resistivity will decrease, and the area of D region will expand gradually, and most of λ^* and α change slightly with the increase of pressure.

2. Experiments

2.1. Samples. Two cuboid magnetite samples used in this study were collected in Yinan, Shandong province, China. Physical properties and geometric dimensions of specimens were listed in Table 1.

They mainly consist of magnetite (larger than 60%), which leads to their good conductivity; so it is possible to conduct resistivity measurements easily. Their compression planes are the plane of 31.8×22.0 cm for sample YN1 and of 29.0×26.0 cm for sample YN2 in dimension.

2.2. Experimental System. The experimental system used is shown in Figure 1. The uniaxial servocontrol YAW-5000F loading machine (http://shijin.shuoyi.com/shtml/shijin/product/57fd78bf28f0244d.html) manufactured by Jinnan Shijin Group Co. Ltd. was used for axial load. Three measuring lines are arranged on one of free planes. Line 1 and 2 are perpendicular and parallel to the axis of load, and Line 3 intersects this axis at the angle of 45° (Figure 1). Three measuring lines all intersect at the point O which almost locates the center of the free plane. Copper wires spot welded on the free planes are used as electrodes of resistivity measurements. The parameters of measuring lines of two samples are listed in Table 2. MIR-2007 resistivity meter used in this experiment was specially designed for resistivity measurement of rock sample in lab by Beijing Geological Instrument Factory (http://bjdzyqc.shuoyi.com/shtml/bjdzyqc/index.htm). The input impedance of the meter is about 4104 MΩ, and its accuracy of potential measurement is about 0.15%. Wenner-α array (Figure 2) was used in data acquisition.

2.3. Loading Curves and Data Acquisition. The samples were compressed by uniaxial load (Figure 1). Their loading curves are shown in Figure 3. Resistivity measurements were conducted along measuring lines one by one when pressure increased to 0, 0.86 MPa, 2.86 MPa, 7.15 MPa, 14.29 MPa, 21.44 MPa, and 28.29 MPa at uniform rate of 1.43×10^{-2} MPa/s for sample YN1 and to 0, 2.65 MPa, 6.63 MPa,

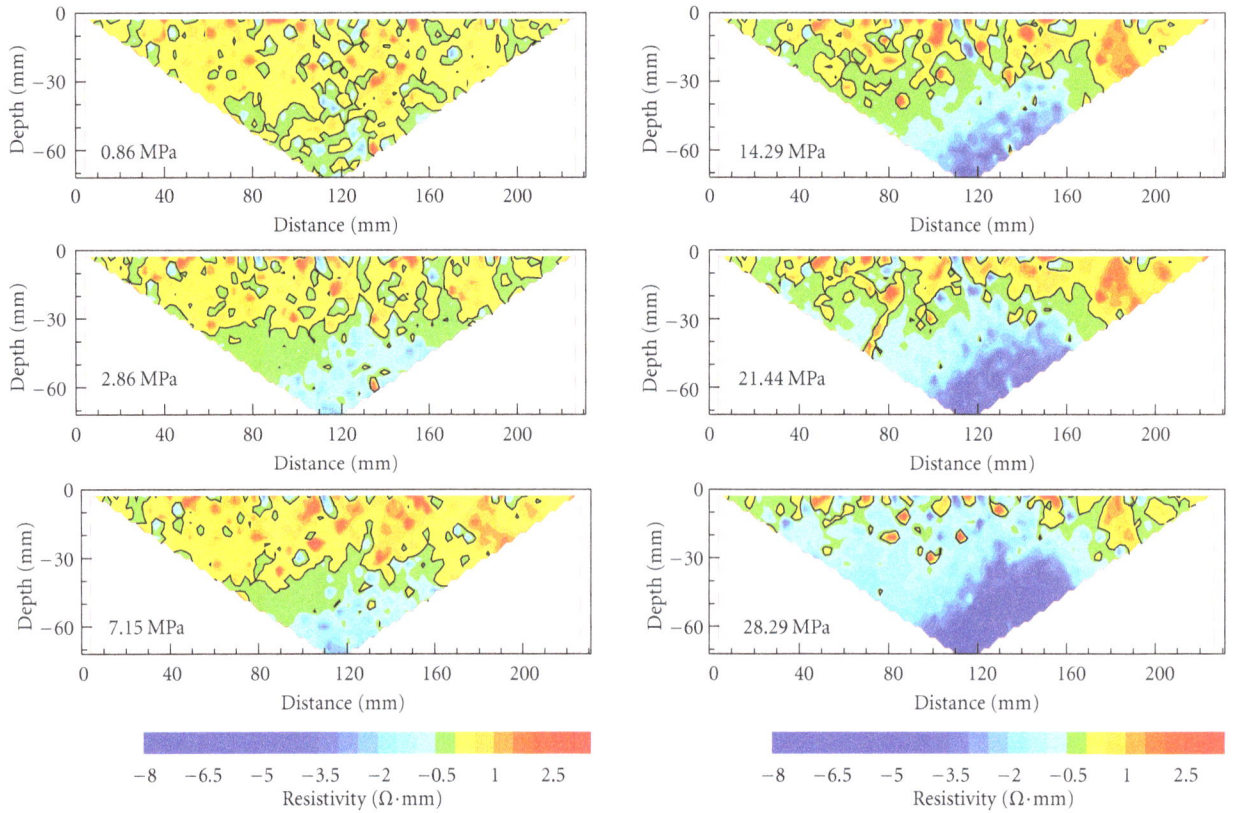

(a)

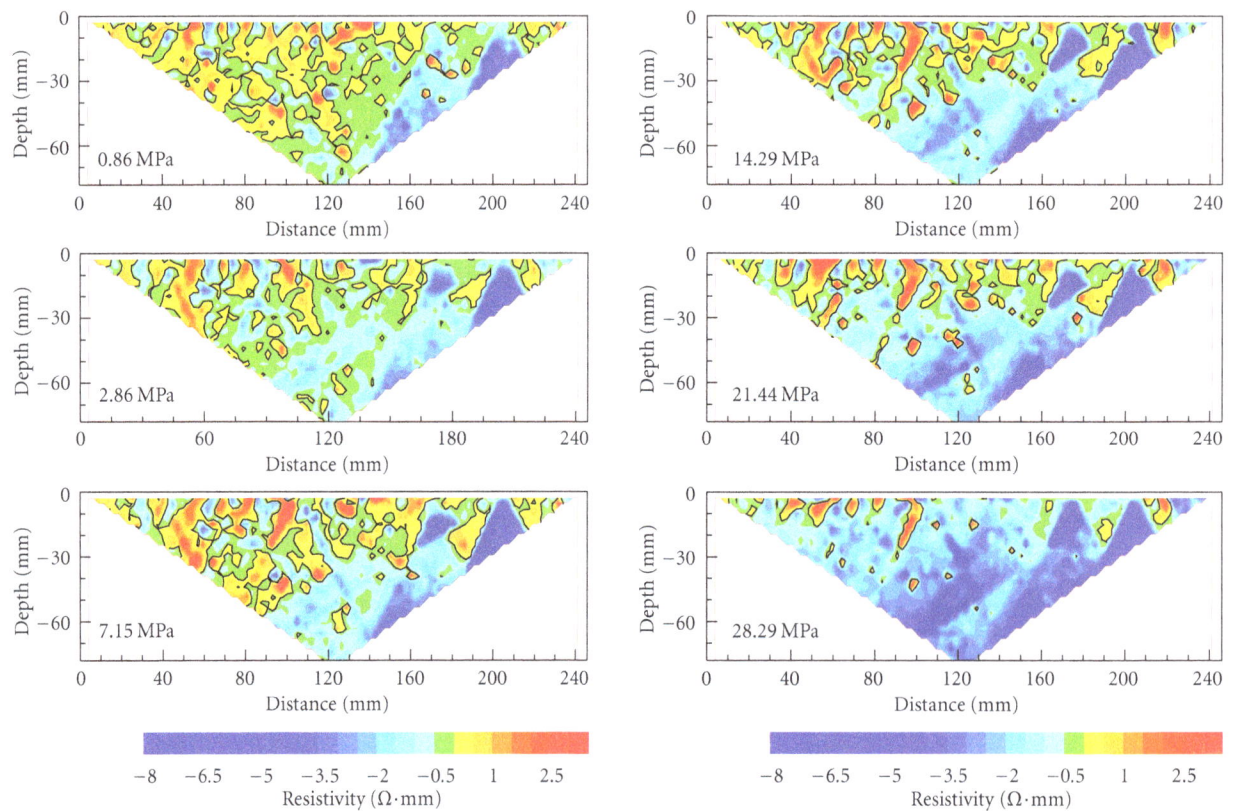

(b)

FIGURE 4: Continued.

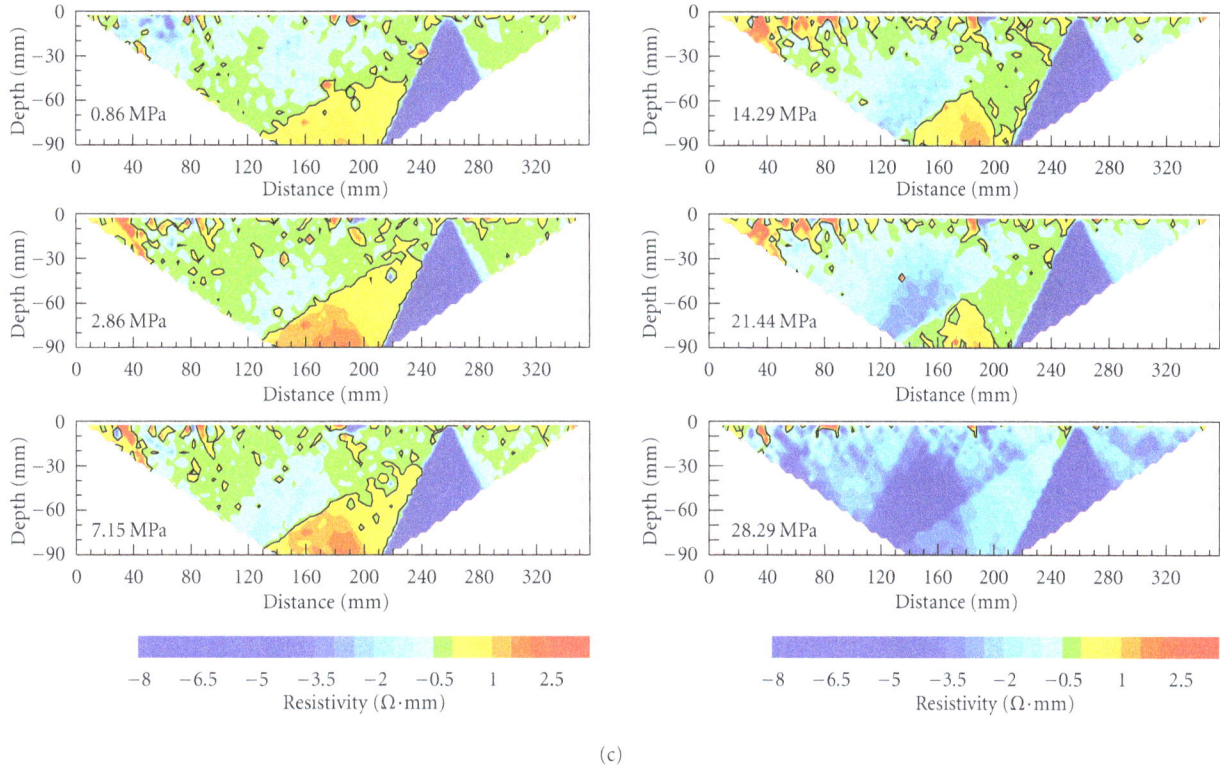

(c)

FIGURE 4: The residual resistivity images of sample YN1. (a) Line 1, (b) Line 2, and (c) Line 3.

TABLE 1: Physical properties and geometric dimensions of specimens.

Sample no.	Density (g/cm³)	Dimension (cm)	Modal analysis
YN1	5.04	34.0 × 31.8 × 22.0	64 Magnetite, 29 Quartz, 7 Feldspar
YN2	5.07	34.0 × 29.0 × 26.0	68 Magnetite, 27 Quartz, 5 Feldspar

TABLE 2: Parameters of measuring lines.

ML	NE	SE	NF
	Sample YN1		
Line 1	78	3 mm	24
Line 2	83	3 mm	26
Line 3	120	3 mm	30
	Sample YN2		
Line 1	37	5 mm	12
Line 2	52	5 mm	17
Line 3	66	5 mm	21

Note: ML, NE, SE, and NF indicate measuring line, number of electrode, spacing of electrode, and "n" factors, respectively.

13.26 MPa, 19.89 MPa, 33.16 MPa, and 41.11 MPa at 2.65×10^{-2} MPa/s for sample YN2 and remained constant at each value. It took about 3 hours for sample YN1 and 1.5 hours for sample YN2 for resistivity measurement at each pressure value. In the meantime, time-lapse strain was recorded. To our regret, the strain curves remained straight lines throughout the experiment. We thought that the pressure

was not enough high to lead to the measurable deformation recorded by our strain gauge.

3. Results and Analysis

3.1. Residual Apparent Resistivity Image (RARI). Residual apparent resistivity values were calculated via

$$\rho_{NR} = \rho_N - \rho_0, \tag{1}$$

where ρ_{NR} and ρ_N represent the residual and measured apparent resistivity value at the pressure value of N MPa, and ρ_0 indicates the apparent resistivity measurement at zero pressure. N takes 0.86 MPa, 2.86 MPa, 7.15 MPa, 14.29 MPa, 21.44 MPa, and 28.29 MPa for sample YN1 and 2.65 MPa, 6.63 MPa, 13.26 MPa, 19.89 MPa, 33.16 MPa, and 41.11 MPa for sample YN2.

As are shown in Figures 4 and 5, the RARIs along three measuring lines have different change behaviors with pressure when the pressure is smaller than 7.15 MPa for sample YN1 (Figure 4) and 6.63 MPa for sample YN2 (Figure 5). Figure 4 indicates that the RARIs could be divided simply into D (blue to green zone) and I (yellow to red zone) region.

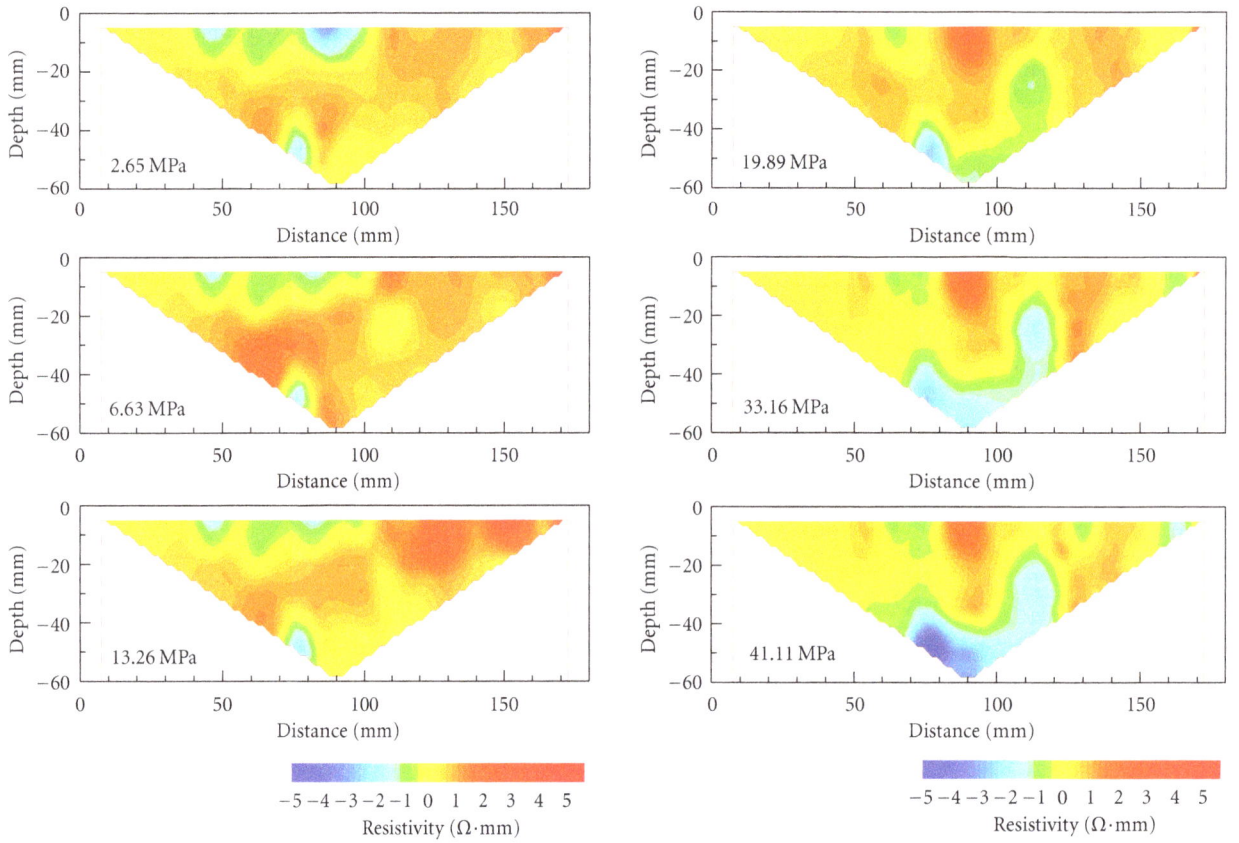

(a)

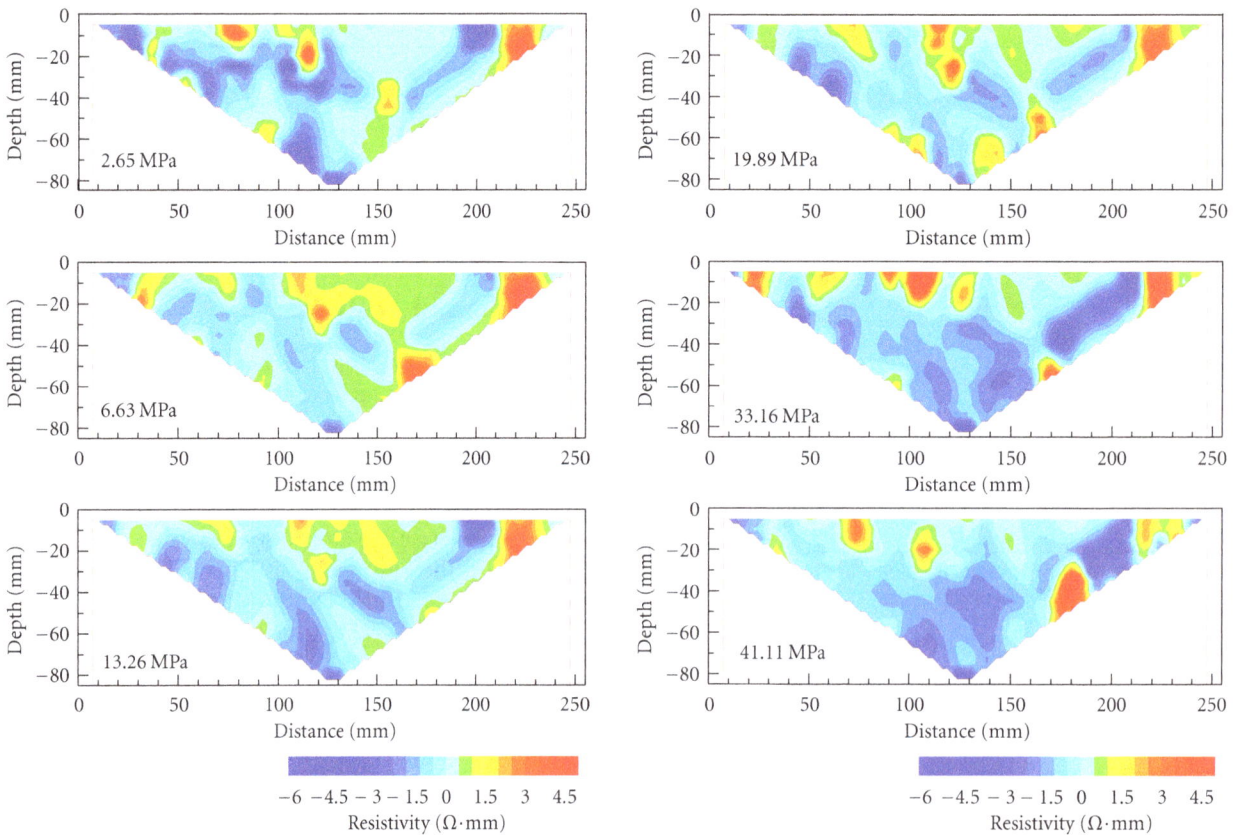

(b)

FIGURE 5: Continued.

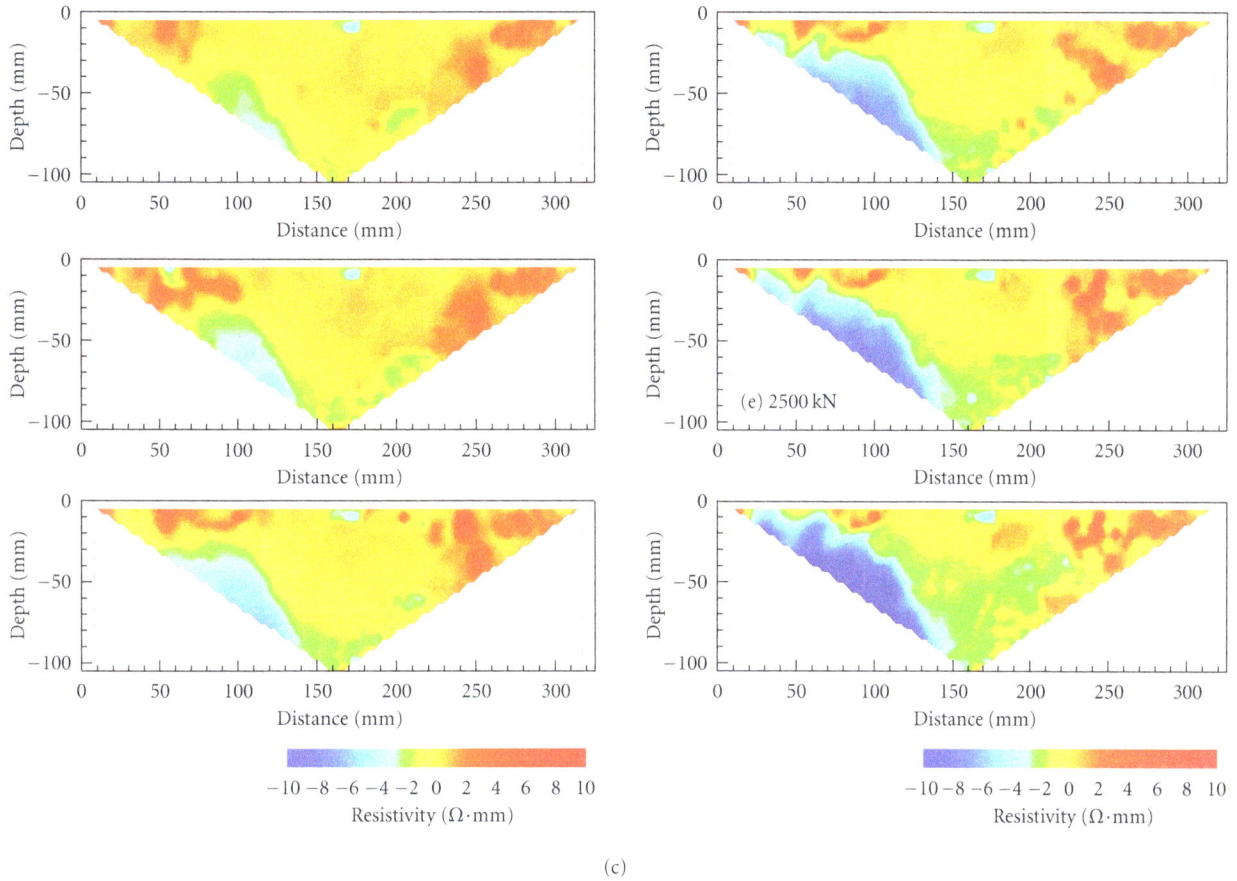

(c)

FIGURE 5: The residual resistivity images of sample YN2. (a) Line 1, (b) Line 2, and (c) Line 3.

With the increase of pressure, the average electrical resistivities of D region in Figure 4(a) increase (i.e., the average decreased magnitude becomes smaller) (compared the RARIs at 0.86 MPa and 2.86 MPa) firstly and then decrease (i.e., the average decreased magnitude becomes larger), (compared the RARIs at 2.86 MPa and 7.15 MPa), while those of I region in Figure 4(a) decrease firstly and then increase. The area of D region expands strikingly firstly and then almost remains constant, while that of I region reduces outstandingly firstly and then almost keeps constant. The electrical resistivities of D region in Figure 4(b) decrease in all time, and those of I region decrease firstly and then increase. The area of D region expands firstly and then reduces. The electrical resistivities of D and I region in Figure 4(c) both increase firstly and then decrease. The area of D region reduces firstly and then almost remains constant while that of I region increases firstly and then almost remains constant. With the increase of pressure from 2.65 MPa to 6.63 MPa, the electrical resistivities of D and I regions in Figures 5(a) and 5(b) all increase slightly. The area of D region expands slightly, and that of I region reduces slightly in Figure 5(a) while it is going in the opposite direction in Figure 5(b). The electrical resistivities of D and I region in Figure 5(c) both decrease slightly. The area of D region expands slightly while that of I region reduces slightly. When the pressure is larger than 7.15 MPa

for sample YN1 (Figure 4) and 6.63 MPa for sample YN2 (Figure 5), the electrical resistivities of D regions in Figures 4 and 5 all decrease gradually while those of I regions probably have complex change behaviors with the increase of pressure. For instance, they decrease firstly (As shown in the RARIs at 6.63 MPa and 13.26 MPa in Figure 5(a)), then increase (As shown in the RARIs at 13.26 MPa and 19.89 MPa in Figure 5(a)) and decrease gradually, at last (As shown in the RARIs at 19.89 MPa, 33.16 MPa, and 41.11 MPa in Figure 5(a)). However, the areas of D regions in Figures 4 and 5 all expand gradually and those of I regions are in opposite direction.

Our results show, as the previous investigators [14–19] reported, that electrical resistivities of rocks have complex behaviors with the increase of pressure. However, due to only two electrodes used in (volume) resistivity measurements, insufficient resistivity measurements could be used to construct resistivity image, and only the curves of resistivity versus pressure could be used to study the changes in resistivity with pressure in these previous studies. In this paper, 37–120 electrodes and 12–30 "n" factors are used, and then 210–2205 resistivity measurements are acquired, which allow us to construct resistivity images easily we have the chance to study the changes in resistivity and its image with pressure in laboratory. Our results indicate that when pressure increases to a certain value, with the increase of pressure, the electrical

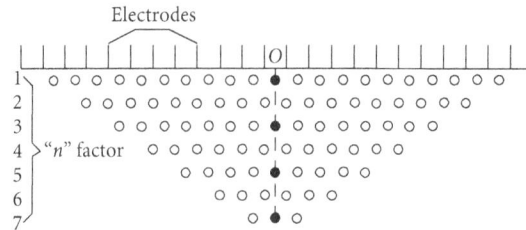

FIGURE 6: The illustration of extracting apparent resistivity set for Wenner-α array. In the studies on apparent resistivity anisotropy at point O, we need the apparent resistivity data arranged along the vertical dash line. According to the characteristics of Wenner-α array, the apparent resistivity measurements could be extracted every one "n" factor. That is, the apparent resistivity measurements just corresponding to odd "n" factors and on the vertical dash line will be extracted out. Black solid circles represent the extracted data.

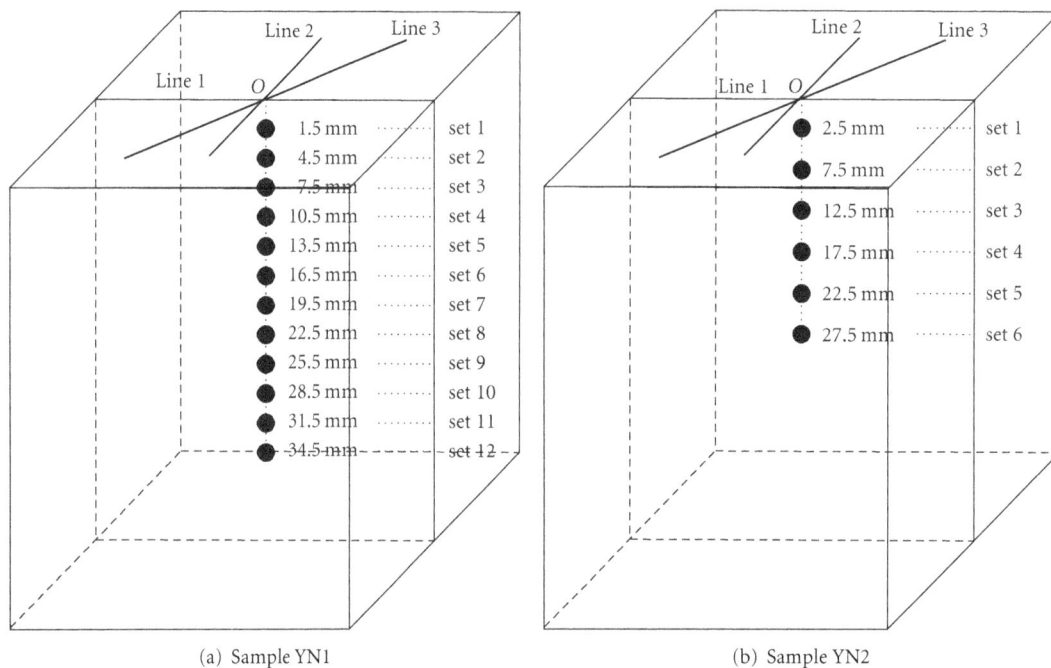

FIGURE 7: The diagram of apparent resistivity sets corresponding to effective depths. Because the effective depth of Wenner-α array was evaluated to be about 0.5 time of electrode spacing [36, 37], the 12 apparent resistivity sets were corresponding to the effective depths of about 1.5 mm, 4.5 mm, 7.5 mm, 10.5 mm, 13.5 mm, 16.5 mm, 19.5 mm, 22.5 mm, 25.5 mm, 28.5 mm, 31.5 mm, and 34.5 mm for sample YN1, and 6 ones were corresponding to the effective depths of about 2.5 mm, 7.5 mm, 12.5 mm, 17.5 mm, 22.5 mm, and 27.5 mm for sample YN2 according to Figure 6.

resistivities will decrease, and the area of D region will expand gradually in all time, which may be significant to the monitoring and prediction of earthquake, volcanic activities, and large-scale geologic motions.

3.2. Apparent Resistivity Anisotropy

3.2.1. Data Extraction. Studies on resistivity anisotropy require that resistivity measurements are conducted along at least two measuring lines arranged in different directions and that they must intersect at a point. For example, Mao et al. [7, 8] used the resistivity data measured along E-W and N-S trending measuring lines in studying the changes in the degree of earth resistivity anisotropy during the course of earthquake preparation. Chen et al. [22–25] designed

four measuring lines in studying the changes of apparent resistivity anisotropy in lab. In this paper, we arranged three measuring lines intersecting at point O (Figure 1), which allowed us to study the changes in apparent resistivity anisotropy at point O with pressure and depth. According to the used minimum "n" factor and the characteristics of Wenner-α array in resistivity measurements, we could extract 12 apparent resistivity sets (Figure 6 illustrates how to extract the apparent resistivity set.) which are corresponding to 12 effective depths (Figure 7(a)) for sample YN1 and 6 apparent resistivity sets which are corresponding to 6 effective depths (Figure 7(b)) for sample YN2. It must be noted that one apparent resistivity set contains three apparent resistivity values.

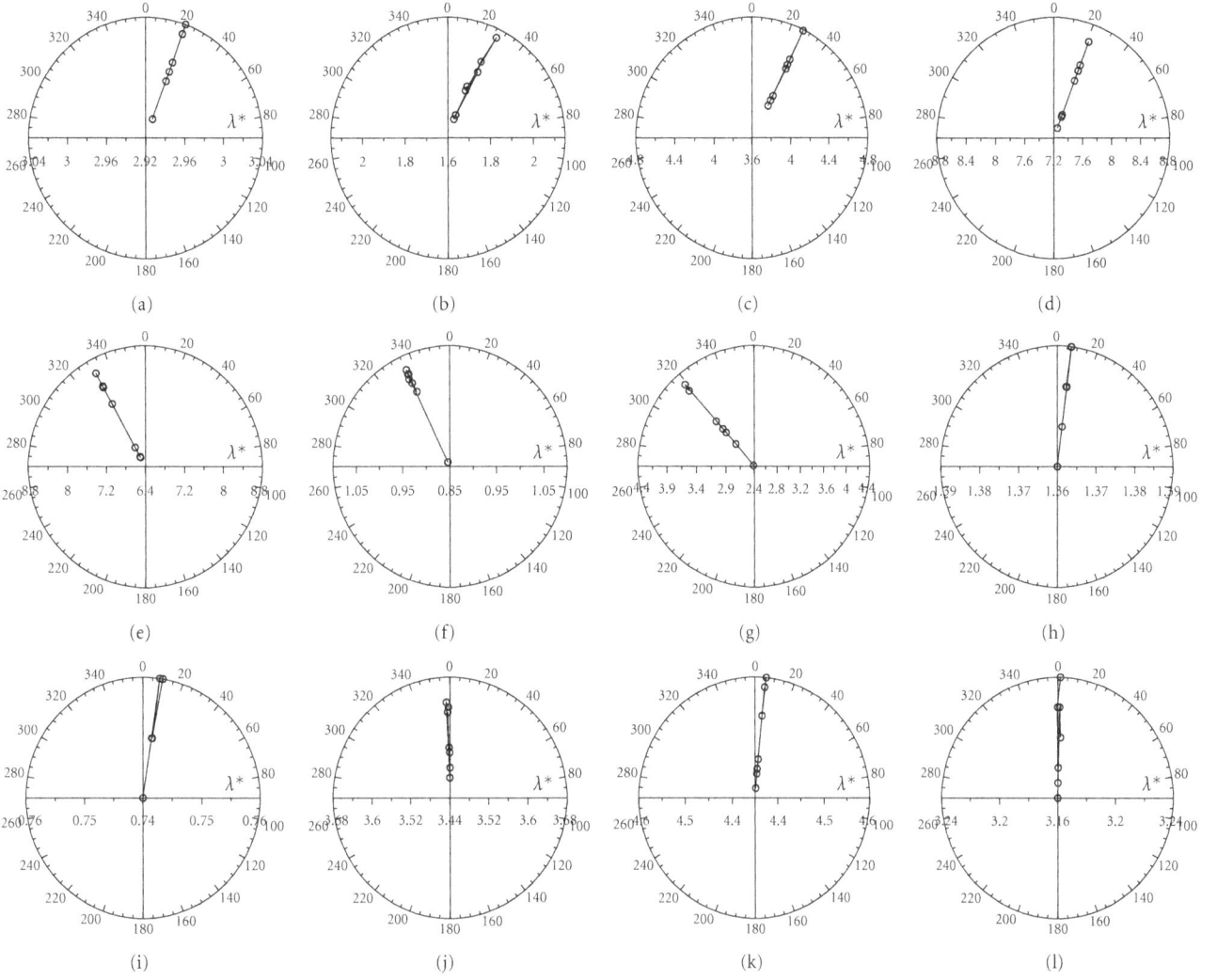

FIGURE 8: The diagrams of λ^* and α with pressure for sample YN1. (a)–(l) represent the diagrams at the effective depths of 1.5 mm, 4.5 mm, 7.5 mm, 10.5 mm, 13.5 mm, 16.5 mm, 19.5 mm, 22.5 mm, 25.5 mm, 28.5 mm, 31.5 mm, and 34.5 mm.

3.2.2. Apparent Resistivity Anisotropy Coefficient (ARAC).

Chen et al. [25, 33] presented the formula to calculate ARAC λ^*

$$\lambda^{*2} = \frac{2\sqrt{m^2 + Q^2}}{l - \sqrt{m^2 + Q^2}},$$

$$l = \rho_{s1}^{-2} + \rho_{s2}^{-2},$$

$$m = \rho_{s1}^{-2} - \rho_{s2}^{-2}, \qquad (2)$$

$$Q = 2\rho_{s3}^{-2} - l,$$

where ρ_{s1}, ρ_{s2}, and ρ_{s3} represent the apparent resistivities measured along three measuring lines perpendicular to, parallel to, and intersecting at the angle of 45° with the axis of load.

Seen from the curves of λ^* and α with pressure and depth for sample YN1 (Figure 8), the absolute differences between the minimum and the maximum λ^* at 10.5 mm (Figure 8(d)), 13.5 mm (Figure 8(e)), and 19.5 mm

(Figure 8(g)) effective depths are about 1.6, 2.0, and 2.0, while those at the rest depths are all smaller than 0.5. The maximum change of α, only about 7°, occurs at 4.5 mm depth. For sample YN2, the absolute difference between the minimum and the maximum λ^* at 17.5 mm (Figure 9(d)) is about 4.0, while those at the rest depths are all smaller than 1.0. The maximum change of α, about 80°, occurs at 22.5 mm depth, but those at the rest depths are all smaller than 5°. These results indicate that the effects of pressure on λ^* and α are not very striking for dry magnetite samples.

4. Discussions and Conclusions

In the previous studies, the volume resistivity changes with pressure were actually studied [14–19]. It was impossible to construct resistivity image because of insufficient data; the curves of resistivity versus pressure were analyzed. In the present paper, resistivity measurements are conducted in the case of designed 37–120 electrodes and 12–30 "n" factors, and 210–2205 resistivity values acquired are sufficient to

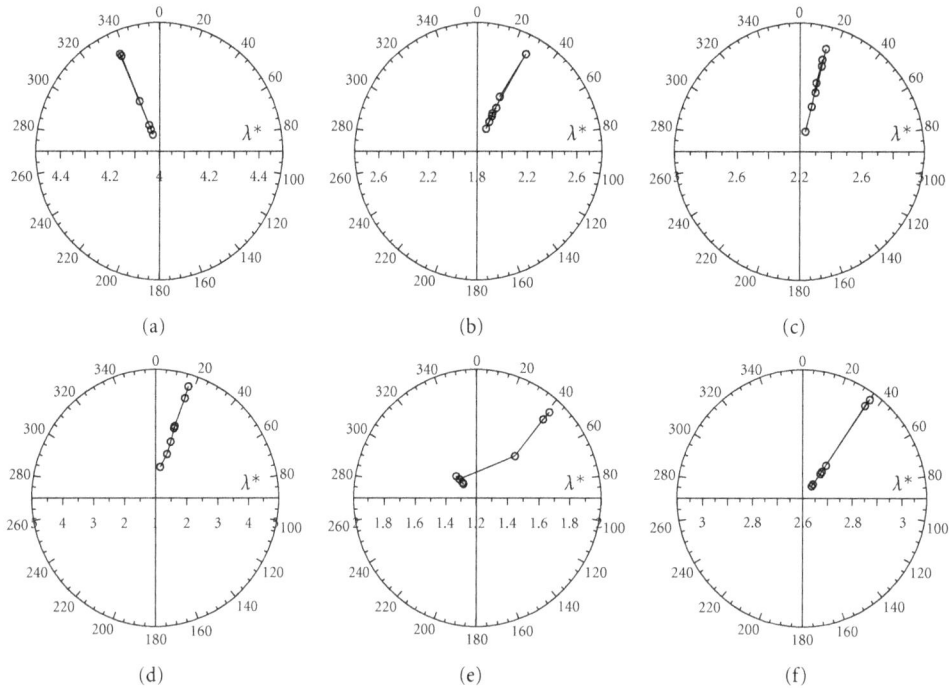

FIGURE 9: The diagrams of λ^* and α with pressure for sample YN2. (a)–(f) represent the diagrams at the effective depths of 2.5 mm, 7.5 mm, 12.5 mm, 17.5 mm, 22.5 mm, and 27.5 mm effective depths.

allow us to use ERT to construct resistivity images which may contain much aplenty information than curves. Thus we can study the changes in not only resistivity but also in the areas of D and I region with pressure. Our results indicate that resistivity and the areas of D and I region have complex behaviors with increasing pressure when pressure is lower than a certain value such as 7.15 MPa for sample YN1 and 6.63 MPa for sample YN2 in this paper, but when pressure is higher than this value, their behaviors become very simple with increasing pressure. For instance, when pressure is higher than 7.15 MPa for sample YN1 and 6.63 MPa for sample YN2, the average resistivity of D region will decrease, and its area will expand gradually in all time with the increase of pressure. These changes were ever observed in situ by Feng et al. [9] prior to Tangshan ML 5.0 and 4.4, which probably indicate that our experimental results may be helpful and significant to the monitoring and prediction of earthquake, even for volcanic activities and large-scale geologic motions.

In the previous studies, the changes in apparent resistivity anisotropy at one or two depths with pressure were studied [22–25, 33, 34] because of at most eight electrodes and two "n" factors used in resistivity measurements, which lead to the difficulty in the studies on its changes with depth. In this paper, as described above, at least 78 electrodes and 24 "n" factors for sample YN1 and 37 electrodes and 12 "n" factors for sample YN2 (Table 2) and Wenner-α array were used in resistivity measurements, which allow us to calculate λ^* and α at 6 or 12 effective depths (Figure 7) and give us the chance to study their changes with pressure at more depths. Our results indicate that λ^* and α do not change outstandingly with the increase of pressure at a certain depth. However,

the striking changes in apparent resistivity anisotropy were observed before earthquake [7, 8, 38]. We thought that we did not use water-saturated samples while the underground rocks are often in saturation because of the presence of rich groundwater. Therefore, water-saturated samples would be considered in our future study.

Acknowledgments

The authors thank two anonymous reviewers and the editor Sabatino Piscitelli for their constructive suggestions for this paper. This research was supported by the Special Project for the Fundamental R&D of the Institute (DQJB07B09).

References

[1] Y. Yamazaki, "Electrical conductivity of strained rocks, 5, residual strains associated with large earthquakes as observed by a resistivity variometer," *Bulletin of the Earthquake Research Institute, University of Tokyo*, vol. 47, p. 99, 1969.

[2] D. Fitterman and T. Madden, "Resistivity observations during creep events at melendy Ranch, California," *Journal of Geophysical Research*, vol. 82, no. 33, pp. 5401–5408, 1977.

[3] W. Stopinski and R. Teisseyre, "Precursory rock resistivity variations related to mining tremors," *Acta Geophysica Polonica*, vol. 30, no. 4, pp. 293–320, 1982.

[4] D. B. Jackson, J. Kauahikaua, and C. J. Zablocki, "Resistivity monitoring of an active volcano using the controlled- source electromagnetic technique: Kilauea, Hawaii," *Journal of Geophysical Research*, vol. 90, no. 14, pp. 12–555, 1985.

[5] J. R. Kayal and B. Banerjee, "Anomalous behaviour of precursor resistivity in Shillong area, Northeast India," *Geophysical Journal*, vol. 94, no. 1, pp. 97–103, 1988.

[6] T. Yukutake, T. Yoshino, H. Utada, H. Watanabe, Y. Hamano, and T. Shimomura, "Changes in the electrical resistivity of the central cone, Miharayama, of Oshima Volcano observed by a direct current method," *Journal of Geomagnetism & Geoelectricity*, vol. 42, no. 3, pp. 151–169, 1990.

[7] T. Mao, T. Wang, J. Yao, J. Lu, and H. Zhang, "The variations of the degree of ground resistivity anisotropy during the tangshan earthquake," *Acta Seismologica Sinica*, vol. 8, no. 4, pp. 621–627, 1995.

[8] T. Mao, G. Xu, S. Fan, M. Zhao, and J. Sun, "Dynamic evolution patterns of the degree of ground resistivity anisotropy and the seismogenic process," *Acta Seismologica Sinica*, vol. 12, no. 2, pp. 199–206, 1999.

[9] R. Feng, J. Hao, and J. Zhou, "Resistivity tomography in earthquake monitoring," *Chinese Journal of Geophysics*, vol. 44, no. 6, pp. 819–830, 2001.

[10] Z. Feng, J. Yang, W. Mei, J. Geng, X. Wang, and Y. Liu, "Preliminary study on characteristics of seismic precursor of earth resistivity anisotropy degree in east China," *Acta Seismologica Sinica*, vol. 17, no. 2, pp. 244–250, 2004.

[11] H. Utada, "Interpretation of time changes in the apparent resistivity observed prior to the 1986 eruption of Izu-Oshima volcano, Japan," *Journal of Volcanology and Geothermal Research*, vol. 126, no. 1-2, pp. 97–107, 2003.

[12] J. An, P. Zhou, M. Ma, F. Chen, Y. Dong, and P. Zhong, "Experiments on exploring and monitoring landslip-mass using geoelectric resistivity observations," *Acta Seismologica Sinica*, vol. 21, no. 3, pp. 258–266, 2008.

[13] W. F. Brace, A. S. Orange, and T. M. Madden, "The effect of pressure on the electrical resistivity of water—saturated crystalline rocks," *Journal of Geophysical Research*, vol. 70, no. 22, pp. 5669–5678, 1965.

[14] W. F. Brace and A. S. Orange, "Electrical resistivity changes in saturated rock under stress," *Science*, vol. 153, no. 3743, pp. 1525–1526, 1966.

[15] W. F. Brace and A. S. Orange, "Electrical resistivity changes in saturated rocks during fracture and frictional sliding," *Journal of Geophysical Research*, vol. 73, no. 4, pp. 1433–1444, 1968.

[16] W. F. Brace and A. S. Orange, "Further studies of the effects of pressure on electrical resistivity of rocks," *Journal of Geophysical Research*, vol. 73, no. 16, pp. 5407–5420, 1968.

[17] C. Y. Wang, R. E. Goodman, P. N. Sundaram, and H. F. Morrison, "Electrical resistivity of granite in frictional sliding: application to earthquake prediction," *Geophysical Research Letters*, vol. 2, no. 12, pp. 525–528, 1975.

[18] K. Kondo, T. Mashimo, and A. Sawaoka, "Electrical resistivity and phase transformation of hematite under shock compression," *Journal of Geophysical Research*, vol. 85, no. 2, pp. 977–982, 1980.

[19] C. Morrow and W. F. Brace, "Electrical resistivity changes in tuffs due to stress," *Journal of Geophysical Research*, vol. 86, no. 4, pp. 2929–2934, 1981.

[20] C. Pearson, J. Murphy, and R. Hermes, "Acoustic and resistivity measurements on rock samples containing tetrahydrofuran hydrates: laboratory analogues to natural gas hydrate deposits," *Journal of Geophysical Research*, vol. 91, no. 14, pp. 14132–14138, 1986.

[21] K. Skagius and I. Neretnieks, "Diffusivity measurements and electrical resistivity measurements in rock samples under mechanical stress," *Water Resources Research*, vol. 22, no. 4, pp. 570–580, 1986.

[22] F. Chen, J. Xiu, J. An, C. Liao, and D. Chen, "Detecting rupture precursors and determining the main fracture spread direction of rock with dynamic rock resistivity change anisotropy," *Acta Seismologica Sinica*, vol. 13, no. 2, pp. 234–237, 2000.

[23] F. Chen, J. Xiu, J. An, C. Liao, and D. Chen, "Research on dependence of resistivity changing anisotropy on microcracks extending in rock with experiment," *Acta Seismologica Sinica*, vol. 13, no. 3, pp. 331–341, 2000.

[24] F. Chen, C. Liao, and J. An, "Directional characteristic of resistivity changes in rock of original resistivity anisotropy," *Chinese Journal of Geophysics*, vol. 46, no. 2, pp. 381–395, 2003.

[25] F. Chen, C. Liao, and J. An, "Amplitude and anisotropy of apparent resistivity change in big models during shear and frictional slipping," *Chinese Journal of Geophysics*, vol. 46, no. 5, pp. 954–965, 2003.

[26] D. Fitterman, "Theoretical resistivity variations along stressed strike—slip faults," *Journal of Geophysical Research*, vol. 81, no. 26, pp. 4909–4915, 1976.

[27] D. L. Johnson and H. J. Manning, "Theory of pressure dependent resistivity in crystalline rocks," *Journal of Geophysical Research*, vol. 91, no. 11, pp. 11611–11617, 1986.

[28] K. Teisseyre, "Anisotropy of electric resistivity related to crack process before fracturing," *Acta Geophysica Polonica*, vol. 37, no. 2, pp. 185–192, 1989.

[29] K. Teisseyre, "Simulation of anisotropic changes of resistivity in primarily anisotropic rocks under load," *Acta Geophysica Polonica*, vol. 41, no. 3, pp. 316–333, 1991.

[30] K. Teisseyre, "Modeling of crack induced resistivity changes-applications to earthquake studies," *Publications Institute of Geophysics, Polish Academy of Sciences, Series A*, vol. 24, no. 274, pp. 1–63, 1997.

[31] Y. Yamazaki, "Electrical conductivity of strained rocks, 1, Laboratory experiments on sedimentary rocks," *Bulletin of the Earthquake Research Institute, University of Tokyo*, vol. 43, p. 783, 1965.

[32] Y. Yamazaki, "Electrical conductivity of strained rocks, 2, Further experiments on sedimentary rocks," *Bulletin of the Earthquake Research Institute, University of Tokyo*, vol. 44, p. 1553, 1966.

[33] D. Chen, F. Chen, and L. Wang, "Studies on resistivity of rock samples under uniaxial pressure-anisotropy of resistivity," *Acta Geophysica Sinica*, vol. 26, pp. 784–792, 1983 (Chinese).

[34] J. An, J. Xiu, F. Chen, and D. Chen, "Anisotropy studies of rock resistivity changes under uniaxial pressure and water replenishment," *Earthquake Research in China*, vol. 12, no. 3, pp. 300–306, 1996 (Chinese).

[35] I. Diaferia, M. Barchi, M. Loddo, D. Schiavone, and A. Siniscalchi, "Detailed imaging of tectonic structures by multiscale Earth resistivity tomographies: the Colfiorito normal faults (central Italy)," *Geophysical Research Letters*, vol. 33, no. 9, Article ID L09305, 2006.

[36] A. Roy and A. Apparao, "Depth of investigation in direct current methods," *Geophysics*, vol. 36, no. 5, pp. 943–957, 1971.

[37] L. S. Edwards, "A modified pseudosection for resistivity and IP," *Geophysics*, vol. 42, no. 5, pp. 1020–1036, 1977.

[38] F. Qian, Y. Zhao, and Y. Huang, "Calculation of the parameters of georesistivity anisotropy and case history of earthquake precursors," *Acta Seismologica Sinica*, vol. 9, no. 4, pp. 617–627, 1999.

8

Characterization of Fractured Reservoirs Using a Combination of Downhole Pressure and Self-Potential Transient Data

Yuji Nishi and Tsuneo Ishido

GSJ, National Institute of Advanced Industrial Science and Technology (AIST), Tsukuba Central 7, Tsukuba, Ibaraki 305-8567, Japan

Correspondence should be addressed to Tsuneo Ishido, ishido-t@aist.go.jp

Academic Editor: Laurence Jouniaux

In order to appraise the utility of self-potential (SP) measurements to characterize fractured reservoirs, we carried out continuous SP monitoring using multi Ag-AgCl electrodes installed within two open holes at the Kamaishi Mine, Japan. The observed ratio of SP change to pressure change associated with fluid flow showed different behaviors between intact host rock and fractured rock regions. Characteristic behavior peculiar to fractured reservoirs, which is predicted from numerical simulations of electrokinetic phenomena in MINC (multiple interacting continua) double-porosity media, was observed near the fractures. Semilog plots of the ratio of SP change to pressure change observed in one of the two wells show obvious transition from intermediate time increasing to late time stable trends, which indicate that the time required for pressure equilibration between the fracture and matrix regions is about 800 seconds. Fracture spacing was estimated to be a few meters assuming several micro-darcies (10^{-18} m^2) of the matrix region permeability, which is consistent with geological and hydrological observations.

1. Introduction

Geothermal reservoirs are frequently found in fractured rock formations which are otherwise nearly impermeable. The fractures serve as conduits for the geothermal fluids, and the relatively low-permeability country rock provides the reservoir storage capacity. The spacing between the discontinuities (faults/fractures/joints) is obviously an important parameter in any mathematical description of fluid flow through fissured rocks. Apart from major fault zones which is necessary to be modeled individually, the spacing (λ) of the discontinuities is usually small compared to the reservoir dimensions (L). Hence, the fissured rock mass can be treated as a continuum on an intermediate length scale (l), with $\lambda < l < L$ [1].

One continuum description of such a system is provided by the so-called "double-porosity model" (e.g., [2, 3]), in which the fractures and the porous matrix blocks are regarded as two separate but overlapping continua. But this approach is of limited utility in geothermal reservoir engineering mainly due to employment of analytical approximations for mass and energy exchange between the two

continua. Pruess and Narasimhan [4] developed the "MINC" (multiple interacting continua) model, which circumvents the difficulties associated with the estimation of mass and energy exchange between the fracture and matrix regions. The "MINC" method was successfully applied to explain, for example, "excess enthalpy" phenomena [5]; when a well is drilled into the fractured reservoir and fluid is withdrawn, the enthalpy of the stable fluid discharge is often anomalously high—sometimes the well discharges steam alone even though the reservoir is initially filled with almost liquid water. Pritchett and Garg [1] showed that two time constants are key parameters in characterizing two-phase flow in fractured reservoirs: the time required for pressure equilibration (τ_{pe}) and temperature equilibration (τ_{hc}) between the fracture and matrix regions, both of which are proportional to the square of fracture spacing (λ^2).

In the problem of cold water injection into a geothermal reservoir, the cold water advancement in a fractured reservoir that is represented as "MINC" double-porosity medium will be very different from that in a reservoir that can be represented by equivalent porous medium. In the fractured reservoir, the cold water will advance along the fracture zones, gradually extract heat from the adjacent rock matrix, and eventually arrive at the production wells. If the representative time scale t_R is shorter than τ_{hc}, the cold water has not fully heated up by then, which brings about undesired effects on heat recovery from decreasing fluid enthalpies. To achieve more complete heat recovery from the matrix rocks, we need to reduce the injection rate so as that the representative time scale is sufficiently longer than $\tau_{hc} \propto \lambda^2$ (e.g., [6, 7]).

It is highly desirable to know the time constants τ_{pe} and τ_{hc} in advance for prediction of two-phase flow behavior and/or cold water advancement in fractured geothermal reservoirs. Ishido and Pritchett [8] extended the so-called EKP-postprocessor [9] to apply it to fractured reservoirs represented by MINC media. They carried out pressure-transient simulations and calculated associated "self-potential transients" by using the extended EKP-postprocessor and showed that much more pronounced differences will be brought about in the self-potential transients between competing "fractured/MINC" and "porous-medium" descriptions of the same reservoir than is the case for pressure transients. They suggested that combining continuous pressure and SP measurements may therefore provide a means for better characterizing fractured geothermal reservoirs. This prediction motivated us to carry out flow tests at the Kamaishi Mine in Japan.

Self-potential (SP) observations associated with flow tests of boreholes have been conducted in various fields by now (e.g., [13–17]). Among them, SP observations designed for hydraulic characterization of groundwater aquifers were also conducted; the SP signals, which complement piezometric observations, were used to estimate the transmissivity of aquifers (e.g., [18–21]).

In the present study, we focus on the characterization of fractured reservoirs. First, SP transient signals expected for fractured reservoirs are explained based upon the results of numerical simulations [8, 10], and then SP data obtained from experiments at the Kamaishi Mine are described and interpretation of the data is discussed.

2. SP Transients in Fractured Reservoirs

2.1. Electrokinetic Coupling. The flow of a fluid through a porous medium will generate an electrical potential gradient (called the electrokinetic or streaming potential) along the flow path by the interaction of the moving pore fluid with the electrical double layer at the pore surface. This process is known as electrokinetic coupling. The general relations between the electric current density **I** and fluid volume flux **J** and the electric potential gradient $\nabla\phi$ and pore pressure gradient $(\nabla P - \rho\mathbf{g})$ forces are

$$\mathbf{I} = -L_{ee}\nabla\phi - L_{ev}(\nabla P - \rho\mathbf{g}), \tag{1}$$

$$\mathbf{J} = -L_{ve}\nabla\phi - L_{vv}(\nabla P - \rho\mathbf{g}), \tag{2}$$

where the L_{ij} are phenomenological coefficients (e.g., [22]). The first term on the right-hand side in (1) represents Ohm's law, and the second term in (2) represents Darcy's law.

The cross-coupling terms (with the L_{ev} and L_{ve} coefficients) represent the electrokinetic effect, $L_{ev} = L_{ve}$ according to Onsagar's reciprocal relations. See the tutorial of this special issue for more details (Jouniaux and Ishido, this issue).

Based upon a capillary model, the above coefficients may be written as follows (e.g., [20]):

$$L_{ev} = -\frac{\eta\varepsilon\zeta R_{ev}G}{\tau\mu}, \tag{3}$$

$$L_{ee} = \frac{\eta(\sigma + m^{-1}\Sigma_s)}{\tau}, \tag{4}$$

where η = porosity, ε = liquid-phase dielectric permittivity, ζ = zeta-potential, R_{ev} = "electrical relative permeability" for liquid/gas two-phase flow, G = correction factor which becomes less than unity only if the hydraulic radius is comparable to the thickness of the electrical double layer, τ = square of tortuosity ($\tau = t^2$), μ = liquid-phase viscosity, σ = electrical conductivity of pore fluid (two-phase mixture), m = hydraulic radius of pores and/or cracks, which equals half of radius and aperture for pores with circular and slit-like cross-sections, respectively, and Σ_s = surface conductance.

Equation (1) describes the total current density, composed of a drag (convection) current density $\mathbf{I}_{drag}$ caused by charges moved by fluid flow and a conduction current density $\mathbf{I}_{cond}$ caused by electric conduction; hence,

$$\mathbf{I} = \mathbf{I}_{cond} + \mathbf{I}_{drag}, \tag{5}$$

where

$$\begin{aligned}\mathbf{I}_{cond} &= -L_{ee}\nabla\phi, \\ \mathbf{I}_{drag} &= -L_{ev}(\nabla P - \rho\mathbf{g}).\end{aligned} \tag{6}$$

In the absence of external current sources, $\nabla \cdot \mathbf{I} = 0$, so from (5),

$$\nabla \cdot \mathbf{I}_{cond} = -\nabla \cdot \mathbf{I}_{drag}. \tag{7}$$

Equation (7) represents sources of conduction current that are required for the appearance of electrical potential.

In a homogeneous region with a homogeneous density (ρ), (7) can be written as

$$\nabla^2\phi = C\nabla^2(P - \rho gz), \tag{8}$$

where C is called the streaming potential coefficient, which is given as follows in case of liquid single-phase flow (and $G = 1$),

$$C = -\frac{L_{ev}}{L_{ee}} = \frac{\varepsilon\zeta}{(\sigma + m^{-1}\Sigma_s)\mu}. \tag{9}$$

If the pore pressure change occurs within a finite homogeneous volume, the following relation between changes in ϕ (streaming potential) and $(P - \rho gz)$ (pressure) prevails:

$$\Delta\phi = C\Delta(P - \rho gz). \tag{10}$$

2.2. EKP Postprocessor. The cross-coupling term in (2) may be safely neglected for typical geologic situations, and Darcy's law alone may be used to model the hydraulic problem; it is not necessary to solve (1) and (2) simultaneously. A "postprocessor" may then be used to calculate the drag current (I_{drag}) from the results of an unsteady thermohydraulic reservoir simulation.

The "EKP-postprocessor" [9, 23] simulates electric potentials caused by subsurface fluid flow by a two-step process. First, it calculates the distribution of L_{ev}, L_{ee}, and I_{drag} from the reservoir-simulation results using the same spatial grid used for the reservoir simulation calculation (called the RSV grid hereafter). Next, the postprocessor calculates the electric potential (ϕ) distribution by solving the above Poisson equation (7) within a finite-difference grid that is usually much greater in spatial extent than the RSV grid (hereafter called the SP grid).

Within that portion of the SP grid overlapped by the RSV grid, the distribution of electrical conductivity is obtained directly from RSV grid values. Elsewhere within the SP grid, the electrical conductivity distribution is user specified and time invariant. Ordinarily, boundary conditions on the potential are zero normal gradients (Neumann condition) on the ground surface (upper surface) and zero potential (Dirichlet condition) along the bottom and vertical sides of the SP grid. Equation (7) is solved numerically using a Gauss-Seidel iteration procedure incorporating intermittent automatic optimization of the overrelaxation factor.

2.3. Model for Drag Current in MINC Media. The model which Ishido and Pritchett [8] adopted to calculate the drag current density in "MINC" media [4] amounts to the following:

$$I_{drag} = I_f + I_m, \qquad (11)$$

with

$$
\begin{aligned}
I_f &= \left[\frac{\varepsilon \zeta \eta R_{ev} G(\nabla P - \rho \mathbf{g})}{\tau \mu} \right]_f \times \psi, \\
I_m &= \left[\frac{\varepsilon \zeta \eta R_{ev} G(\nabla P - \rho \mathbf{g})}{\tau \mu} \right]_m \times (1 - \psi),
\end{aligned}
\qquad (12)
$$

where I_{drag} = total drag current density vector, I_f = drag current density due to fracture zone effects, I_m = drag current density due to matrix region effects, ψ = fracture zone volume/total volume, and where subscript "f" denotes conditions in the fracture zone, and subscript "m" denotes "averaged" conditions in the matrix region (see [8] for further details). We detail now the SP transients related to a continuous injection, first for the MINC double-porosity media, and secondly for the reservoir model taking into account individual fractures and a borehole. We show that the characteristic behavior of SP predicted by the first model is present in the second model only if a skin zone with a lower streaming potential coefficient is assumed.

2.4. Pressure and SP Transients in MINC Media. Ishido and Pritchett [8] performed a pressure-transient simulation for a two-dimensional axisymmetric horizontal reservoir model. The formation is represented by an "MINC" double-porosity medium with the following properties: global permeability: $k = 10^{-14}$ m^2, fracture zone volume fraction: $\psi = 0.1$, fracture zone porosity: $\eta_f = 0.1$, matrix region porosity: $\eta_m = 0.1$, matrix region permeability: $k_m = 10^{-17}$ m^2, and fracture spacing: $\lambda = 10$ m. (In the present parameter setting, the fracture zone permeability is $k/\psi = 10^{-13}$ m^2, which is due to fractures occupying 10% volume of the fracture zone. The rest 90% volume is assumed to be of impermeable rock matrix.) The time required ($\tau_{pe} = \eta_m \mu C_t \lambda^2 / 10 k_m$) for pressure equilibration between the fracture and matrix regions is $\sim 10^4$ sec (here C_t: total system (fluid plus rock) compressibility). The initial thermodynamic state is uniform (temperature = 200°C and pressure = 10 MPa). For the corresponding SP calculations shown in Figure 1, the reservoir fluid's NaCl concentration is assumed to be 0.02 mol/L, and the formation conductivity L_{ee} is assumed to be 0.03 S/m (homogeneous).

Figure 1(a) shows semilog plots of changes in pressure and in SP due to continuous injection at 0.5 tons per hour per meter of reservoir thickness. The pressure transient at a point near the injection well shows behavior typical of a double-porosity medium; the late-time slope develops after the time required for pressure equilibrium within the matrix region τ_{pe} has elapsed.

The SP transient exhibits three segments. The drag current contribution through the matrix region is small at early times (up to ~ 0.01 day), so the slope is smaller than that at late times (after ~ 0.1 day), by the factor ψ (=0.1). At intermediate times, SP changes rapidly with increasing involvement of matrix region. The time τ_{pe} can be clearly identified at the intersection of the intermediate-time and late-time semilog straight lines (In Figure 1, the "observation" point is not located within the borehole, but ~ 5 meters away from the injection well. The reason for this is that in the case of "open hole," the SP change within the borehole does not show the typical behavior like that shown in Figure 1 since the pressure in the matrix region coincides with the borehole pressure even in early times as approaching the borehole. This topic will be explained in the next Section 2.5).

Figure 1(b) shows the ratio of SP changes to pressure changes for the results shown in Figure 1(a). In the case of the equivalent porous medium, relationship (10) is satisfied for the entire period, resulting in an almost constant ratio. In this plot, the difference between double-porosity and equivalent porous medium behavior is much more apparent and the time τ_{pe} is more evident than in a plot of SP change itself shown in Figure 1(a). The change-ratio plot has the additional advantage that, in real situations, pressure transient data suffer from fluctuations in the sandface flow-rate, so it is often difficult to discern the three segments such as those shown in Figure 1(a). By contrast, the ratio of SP change to pressure change is insensitive to flow-rate fluctuations, so a combination of pressure and SP measurements is expected to provide a more robust and reliable technique for fractured reservoir characterization.

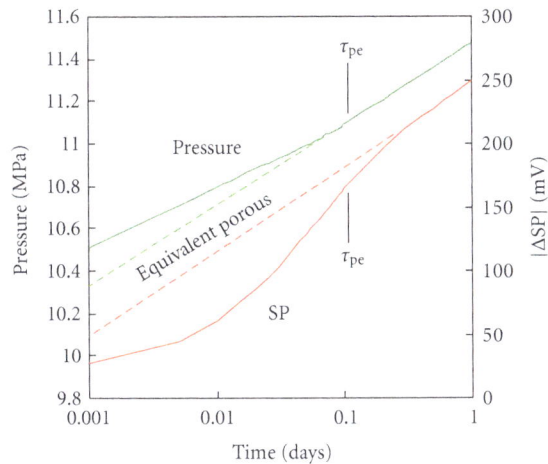

(a)

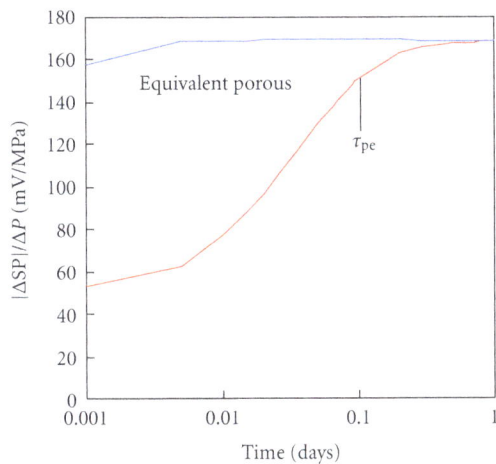

(b)

FIGURE 1: Results from the MINC double-porosity model. (a) Changes in pressure and SP at a point ~ 5 m away from the injection well (open hole) during injection test for fractured medium (after [8]). The pressure and SP changes for the equivalent porous medium are shown by broken curves. (b) Plot of the ratio of SP change to pressure change converted from the pressure and SP changes shown in (a).

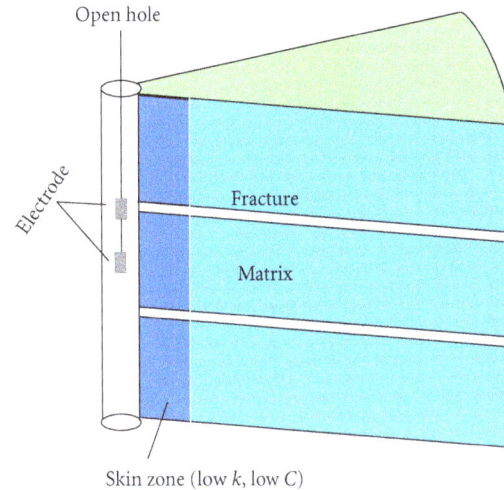

FIGURE 2: "Open hole" model used for numerical simulation of SP transient near a borehole (after [10]). In "open (skin)" case, a skin zone of low permeability and low magnitude of streaming potential coefficient is assumed for the matrix region.

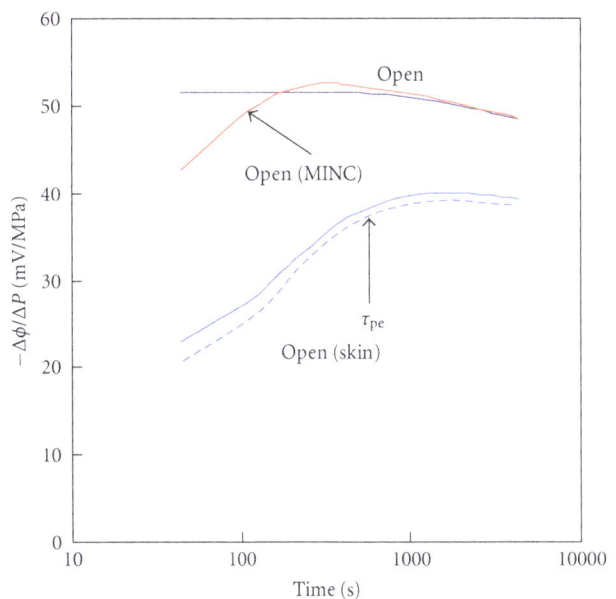

FIGURE 3: The ratio of SP change to pressure change as a function of time (after [10]). Results from the reservoir model, for (a) open hole homogeneous case "open" and (b) open hole with a skin-zone case "open (skin)"; (c) Result from the equivalent MINC double-porosity model "open (MINC)."

2.5. *Near-Field Effects.* In the calculations described in the previous section, the "near-field" effects around a borehole were not considered. Ishido et al. [10] constructed a reservoir model to treat a borehole and individual fractures explicitly instead of using the MINC double-porosity representation (Figure 2). The model is axisymmetric, eight meters thick, and of 1 km horizontal extent (radius). Five equally spaced horizontal fractures intersect the borehole located along the axis of symmetry (only two fractures are drawn schematically in Figure 2). Sufficiently fine block spacing was adopted near the borehole (radius of 0.075 meter) to represent the well casing. Fine block spacing was also used for the host rock (matrix) region close to the fracture zone so as to resolve the high electrical potential gradients there. Here, only the

results for open hole cases are explained. See [10] for the results of cased wells.

The formation properties are: fracture zone permeability: $k_f = 10^{-12}$ m^2, fracture zone thickness = 0.01 m, fracture zone porosity: $\eta_f = 0.5$, host rock porosity: $\eta_m = 0.01$, host rock permeability: $k_m = 10^{-18}$ m^2, and fracture spacing: $\lambda = 1$ m. The time required (τ_{pe}) for pressure equilibration between the fracture and host rock regions is ~ 500 sec. The initial thermodynamic state is uniform (temperature = 45°C

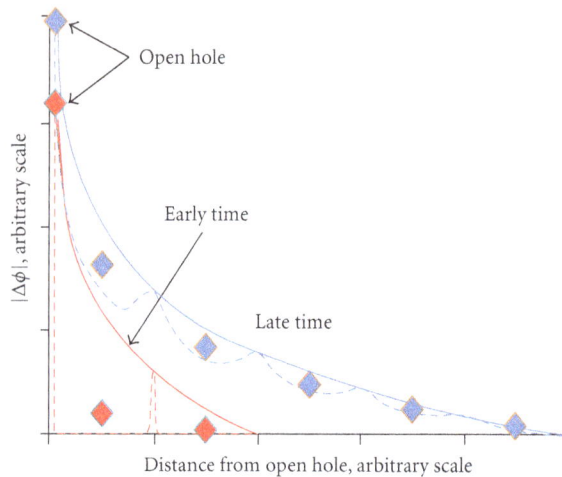

FIGURE 4: Results from the reservoir model taking into account individual fractures and a borehole. Schematic diagram showing the distribution of "microscopic" potential in the fracture (solid lines) and matrix (broken lines) regions in the reservoir, which is equal to change in local pressure multiplied by the streaming potential coefficient C (here homogeneous C is assumed for simplicity). "Macroscopic" potential (♦) calculated for MINC-medium blocks corresponds to an averaged potential over the fracture and matrix regions (after [10]).

and pressure = 10 MPa). Fluid production takes place from the borehole with a constant pressure drawdown of 1 MPa. For the corresponding SP calculations, the reservoir fluid's NaCl concentration is assumed to be 0.005 mol/L, and the streaming potential coefficient is uniform throughout the fracture and host rock regions.

In the "open hole" cases, the SP change that would be measured by electrode(s) installed within the borehole (Figure 2) is calculated, and then the ratio of SP change to pressure change is plotted as a function of time (Figure 3). As Figure 3 shows, the "open" ratio is almost constant with time, which is similar to the "equivalent porous medium" behavior shown in Figure 1(b) and does not exhibit any characteristic fractured reservoir behavior.

But if a skin zone in which the streaming potential coefficient is much smaller than that of the outer matrix region is present, the typical double-porosity behavior appears in the plot ("open (skin)" curves in Figure 3). Although the SP change magnitude is independent of the location of the electrode for the case without a skin zone, the SP change is slightly smaller at the electrode located at the matrix region for the case with skin zone as shown by the broken line in Figure 3.

In Figure 3, also shown (in red) is the result for a case in which the reservoir is represented by "equivalent" MINC double-porosity medium. In this "open (MINC)" case, the SP/pressure change ratio exhibits similar behavior as that of "open" case. This is because the present scheme used by the EKP postprocessor to calculate SP in an MINC medium gives an "averaged" potential over the fracture and matrix regions.

Figure 4 illustrates the spatial distribution of "microscopic" potential, which is the streaming potential coefficient

multiplied by local pressure change, in the fracture and matrix regions around an open hole. Along the wall of the open hole, the local pressures in both the fracture and matrix regions are equal to the borehole pressure, so the microscopic potentials in both the regions converge as the borehole is approached. This is the reason why the SP change is in proportion to the pressure change within the borehole in the "open" case. However, in the "open (skin)" case, substantial drag current is not induced in the matrix region, while a large pressure gradient remains within the skin zone. So, SP change within the borehole is brought about solely by the drag current induced in the fracture zone, resulting in smaller SP change magnitude at early times.

3. Field Experiments at the Kamaishi Mine

The Kamaishi Mine had been one of the largest mines in Japan, which produced copper and iron ore over 130 years since 1857 (see the inset of Figure 5 for its location). The total amount of ore mined during this period is about 70 million tons. The galleries of ~140 km in total length were made mainly in hard rocks composed of Kitakami Paleozoic granitic rocks, Mesozoic granitic intrusions, and skarn ore deposits formed at the contacts of the intrusions. During the last more than two decades, the mine has been diversifying from mining into underground research sites for developing new methods of geological and geophysical studies and for rock mechanics, hydrology, and so forth (e.g., [24]). In this section, we describe the experiments performed by inducing a fluid flow from horizontal wells, which causes pressure disturbance along the entire borehole both in the fracture and host rock matrix regions.

3.1. Outline of Measurements. We carried out flow tests and pressure and SP measurements in two open holes (KF-1 and KF-3) which were drilled nearly horizontally from the wall of one of the levels into the surrounding granodiorite body (both wells were drilled in the direction ~15° from the direction of the tunnel on the horizontal plane as shown in Figure 5). Both wells maintain stable pressures of about five bars under shut-in conditions, so that flow tests may be carried out by simply opening and closing the wellhead valves. After preliminary experiments in 2005 and 2006 [25], we installed twelve custom-made silver-silver chloride electrodes in each of the two wells in 2007 [26].

To reduce flow effects on measuring electrodes, each silver-silver chloride electrode was installed in a container made of hard plastic tube, the upstream and downstream ends of which were closed and open (via sponge), respectively (Figure 6). The voltages between each electrode and the reference (the location of which is shown in Figure 5) and the pressure and flow rate of the two wells were recorded with two data loggers Campbell Scientific's CR5000. The valve operation of wells KF-1 and KF-3 was automatically controlled, and all measuring equipments were powersupplied by a few car batteries, which were fully charged at maintenance time. One unit of the experiment was carried out in two days; KF-1 was opened for one hour on the first

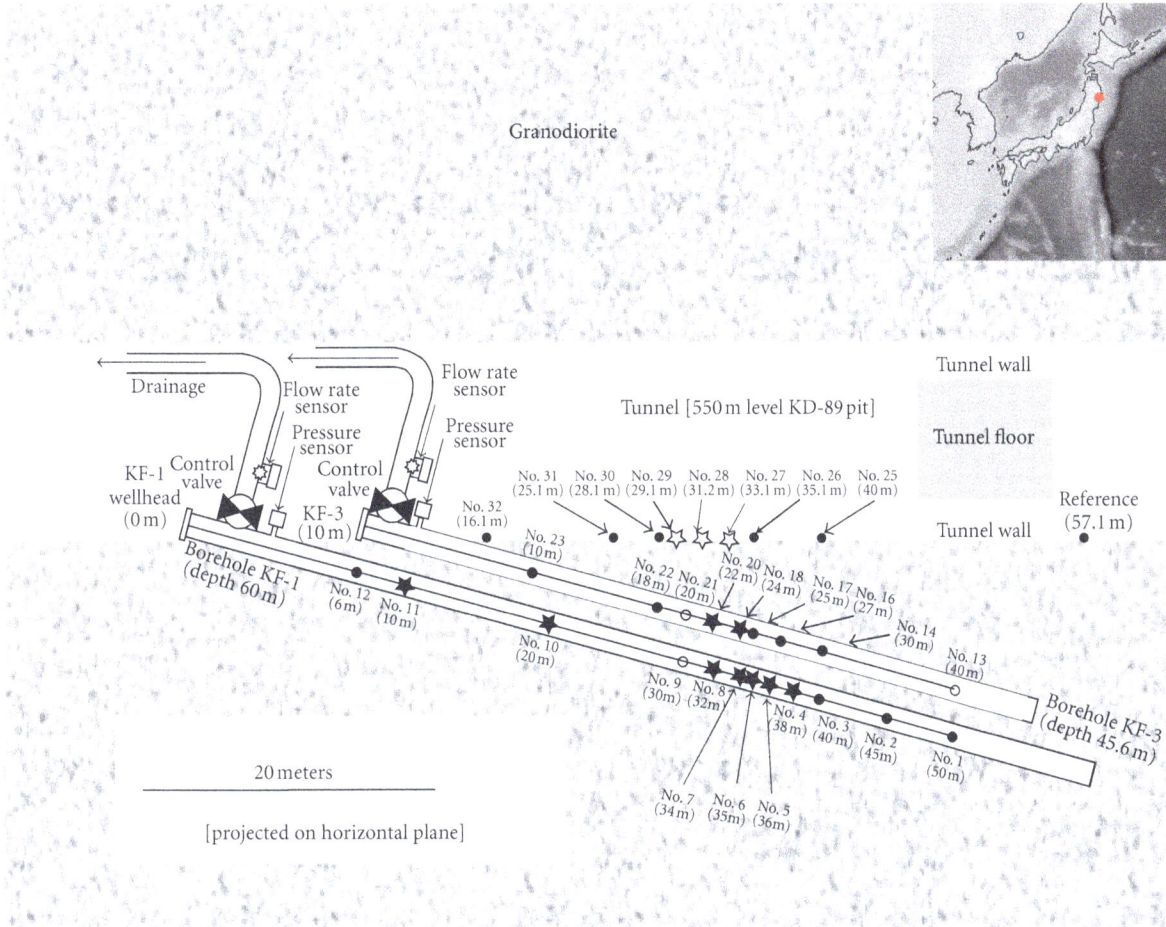

FIGURE 5: Plan view of experimental setup of pressure and SP measurements at the Kamaishi Mine in April–June 2007. Ag-AgCl electrodes No. 1 to 12 and No. 13 to 24 were installed in wells KF-1 and KF-3, respectively. Ag-AgCl electrodes were also installed on the tunnel floor at eight points No. 25 to 32 and at the reference point. Locations of the borehole electrodes near permeable fractures which showed distinguishable behaviors from those of other electrodes are shown by solid asterisks. Locations of tunnel floor electrodes which showed SP changes corresponding to the flow from KF-1 or KF-3 are shown by open asterisks. Locations of other electrodes are shown by solid or open circles (open circles show the locations of electrodes which were sometimes unstable during the period from April to June 2007).

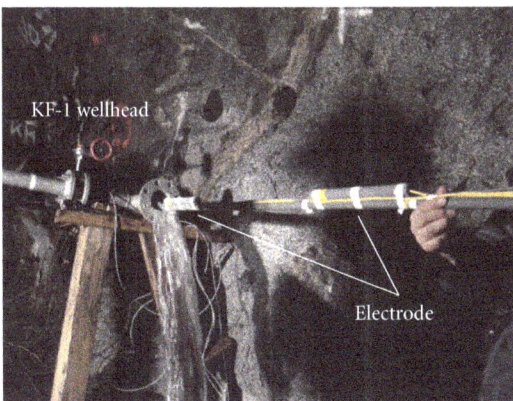

FIGURE 6: Photo of installation of custom-made Ag-AgCl electrode which is contained in a plastic tube into well KF-1.

day, and then KF-3 was opened for one hour on the second day. We repeated this procedure as many as possible with various valve openings in April through June 2007.

Examples of the test results are shown in Figure 7. In Figure 7(a), the results are shown for an experiment when well KF-1 was flowing on May 13. The two wells are connected to each other through permeable fractures, so KF-3 pressure also substantially decreased. Corresponding to the pressure decreases, SP in both wells KF-1 and KF-3 increased several millivolts. Concerning the SP changes on the tunnel (level) floor, their appearance is restricted to an interval of ~5 meters near the fracture zones; only three electrodes nos. 27, 28, and 29 (see Figure 5 for the locations) showed a few millivolts decrease and increase corresponding to the start and stop of the flow, respectively (in Figure 7, the data of nos. 27 and 29 showing substantial changes and the data of no. 25

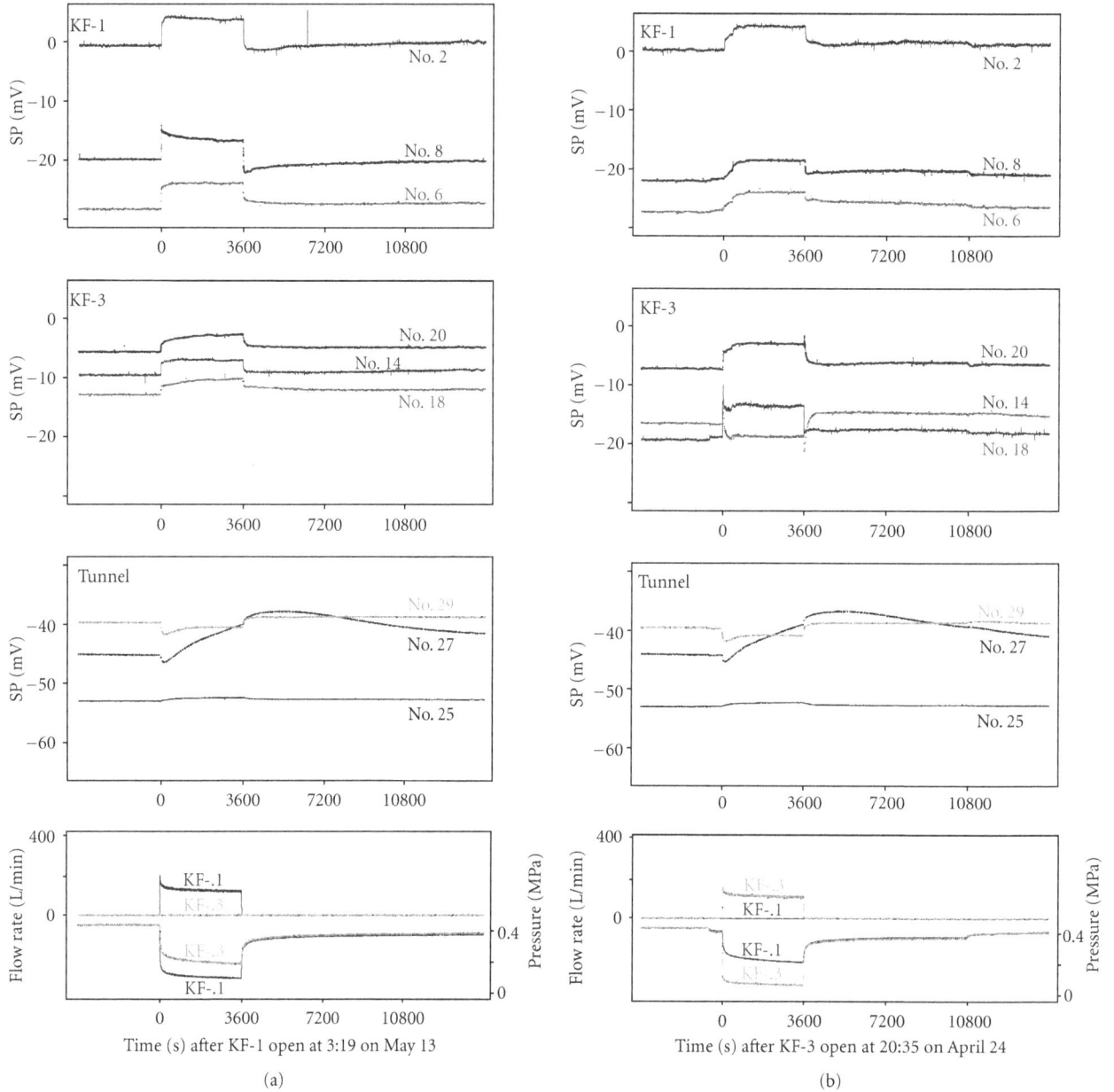

FIGURE 7: Recorded data during one cycle of valve operation. Changes in SP in wells KF-1 and KF-3 and on the tunnel floor are shown for the valve opening of KF-1 (a) and KF-3 (b). Also shown are the recorded pressure and flow rate of wells KF-1 and KF-3 (bottom).

showing only a tiny fluctuation are shown). In Figure 7(b), the results are shown for an experiment when well KF-3 was flowing on April 24. Similar SP changes to those when KF-1 was flowing were observed.

As seen in Figure 7, some of the electrodes installed in the flowing well showed spike-like changes at the start and stop of the flow, which seem to correlate with sudden flow rate changes and influence the successive SP changes. So hereafter, we focus on the SP data measured in "observation" wells, that is, KF-3 and KF-1 corresponding to the flow of KF-1 and KF-3 respectively, which are thought to be less influenced by the flow effects on the electrodes. In the next section, three KF-3

observations associated with KF-1 valve opening (70%) for one hour on April 23, May 3, and May 13 and three KF-1 observations associated with KF-3 valve opening (70%) for one hour on April 24, May 4, and May 14 will be shown. Among these, the data from experiments on April 24 and May 13 are the same as those shown in Figure 7.

3.2. Results. Figure 8 shows three records (with one-second sampling interval) from each of four electrodes in KF-3 (observation well) associated with KF-1 valve opening on April 23, May 3, and May 13. As seen in the figure, the

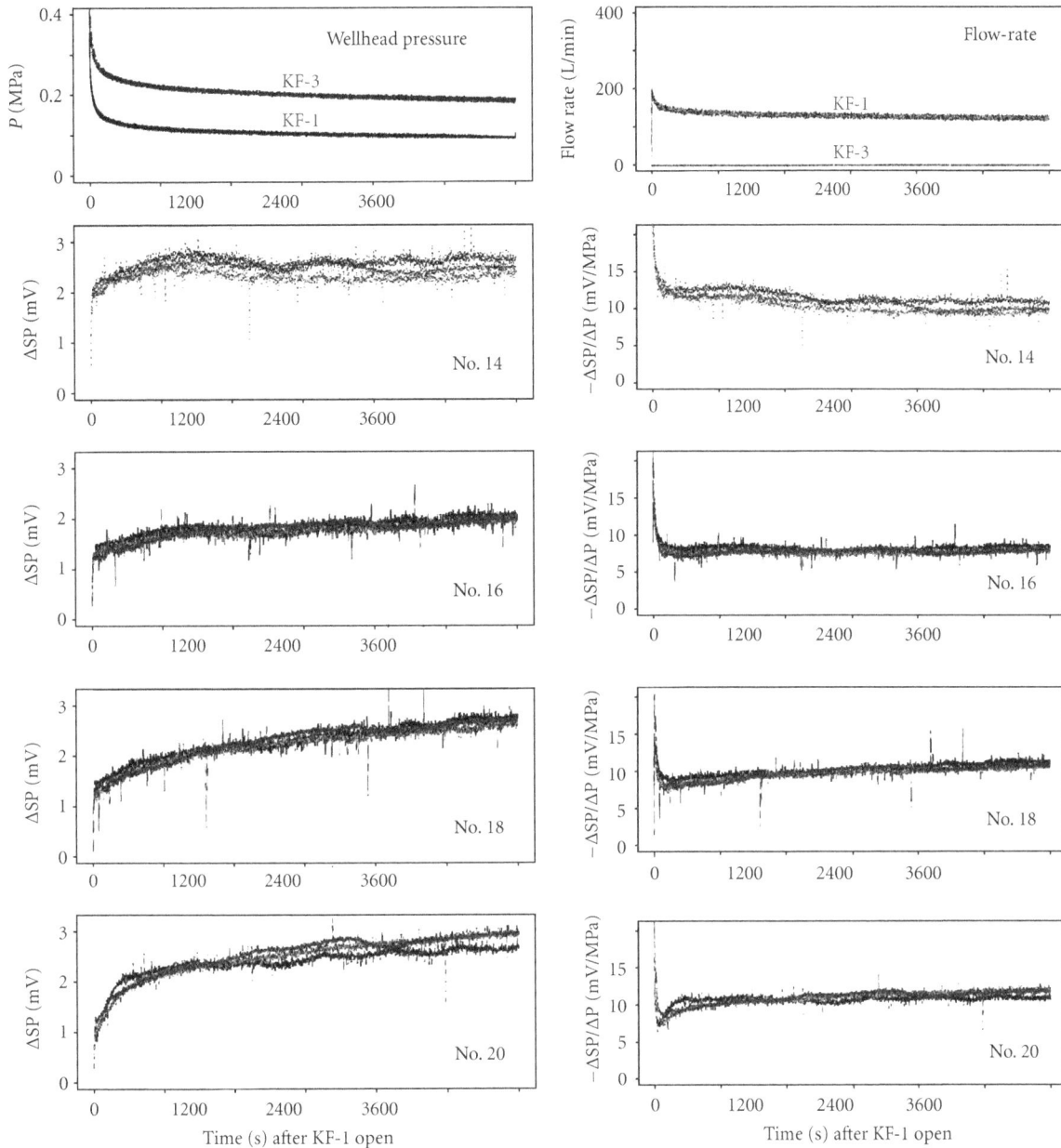

FIGURE 8: Changes in SP in KF-3 (observation well) for three repeated KF-1 valve operations. Ratio of SP change to pressure change is also shown on the right. Both of ΔSP and ΔSP/ΔP are fairly reproducible.

differences among the three records are not so small compared to those of pressure and flow rate histories, but overall trends are fairly reproducible both for SP change (ΔSP) and the ratio of SP change to pressure change (ΔSP/ΔP). The ΔSP/ΔP corresponds to the streaming potential coefficient, which is estimated to be around −10 mV/MPa from the records at later times. During the first ~100 seconds, the magnitude of ΔSP/ΔP rapidly decreased from the initial values of around 20 mV/MPa to ~10 mV/MPa. This rapid change is thought to be caused by (1) electrode drifts induced by the flow start, which were small in the observation well but not negligible compared to the small signals at early times and (2) uncertainty involved in the evaluation of SP

values just before the valve opening. We usually determined "static" SP values from the average values over intervals between 25 and 15 minutes before the valve opening. In Figure 9, comparison between the results using this average "static" value and an instantaneous "static" value just at the valve opening is shown. The differences between the two curves are relatively minor even for earlier times than 100 seconds. However, in cases such as the experiment on April 24 shown in Figure 7, a small pressure down occurred before 10 minutes before the flow start, which was caused by a faint unexpected opening of the automatically controlled wellhead valve when power was supplied for operational preparation. In such cases, SP was disturbed from 10 minutes before the

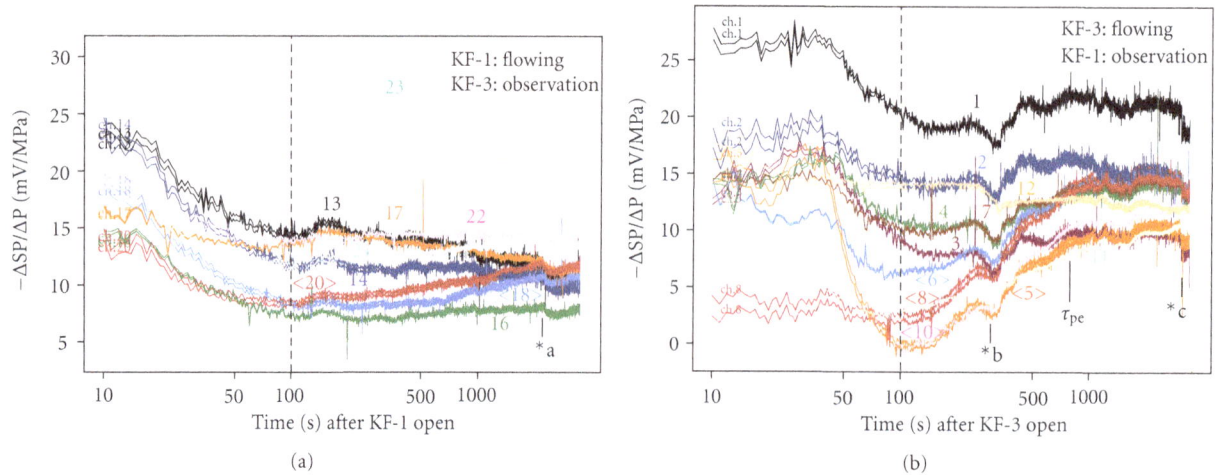

FIGURE 9: Ratio of SP change to pressure change as a function of time for KF-3 observation on May 13, 2007 (a) and KF-1 observation on May 14, 2007 (b). Two records for each electrode are drawn with different "static" SP values: one is with an average value over interval between 25 and 15 minutes before the valve opening and another is with an instantaneous value just at the valve opening at time zero. The differences between the two curves are relatively small even for earlier times than 100 seconds in the results shown here. At times pointed by $*a$, $*b$, and $*c$, all electrodes in wells KF-1 and KF-3 were suffered from a shift or a short-term disturbance. If the "double-porosity" model is applied to the KF-1 observation, the time required for pressure equilibration between the fracture and matrix regions is estimated as $\tau_{pe} \sim 800$ seconds.

flow start at $t = 0$, and the difference between the two results with "static" values determined from the instantaneous (at $t = 0$) and average (between $t = -25$ and -15 minutes) values became significant for early times. While the SP changes remained below ~1 mV, these uncertainties would easily bring about factor 2 to 3 uncertainty in $\Delta SP/\Delta P$ values. So we will disregard the first 100 seconds of $-\Delta SP/\Delta P$ plots in the following discussion.

In Figures 9 and 10, $\Delta SP/\Delta P$ is plotted as a function of logarithm of time. Figure 10 shows the records of selected electrodes for three KF-3 observations on the left and three KF-1 observations on the right. As seen in the figure, the results for KF-3 observations are fairly reproducible. Three different behaviors are recognized: the first one shows a stable or small decreasing trend observed at electrodes located at host rock zone without permeable fractures (no. 14 and no. 16), the second one shows an increasing trend observed at electrodes located near permeable fractures (no. 18 and no. 20), and the third one shows a high magnitude at early times and decreasing trend after ~400 seconds observed at the shallowest electrode no. 23.

In contrast to the KF-3 observations, the reproducibility is not good for the KF-1 observations. This is mainly due to a shift (equivalent to ~1 mV positive SP change) appearing around 300–400 seconds for all electrodes except the shallowest electrode no. 12 in the first and second observations. A "V"-shaped change appeared also around 300–400 seconds for all electrodes in the third observation. Such shift or rapid change did not appear for the electrodes installed on the tunnel floor (eight electrodes nos. 25–32), and the occurrence time was not correlated with time of a day (the first, second, and third ones were at 20:40 on April 24, at 12:08 on May 4, and at 3:35 on May 14, resp.), and a possibility of external noise source due to working

activity of the mine is low. In addition, no anomalous change is present in pressure and flow rate histories in the three observations (see Figure 7 for April 24 data). We have not understood the cause of these anomalous changes yet. However, comparing the earlier records than 300–400 seconds of the three observations, the changing patterns are quite similar to each other. This is also true for the later records than 300–400 seconds.

Records of all electrodes for the third KF-1 observation are shown in Figure 9. The behaviors can be divided into three groups; the difference between the earlier and later portions than 300–400 seconds is relatively small for the first group to which electrodes located at host rock zone without permeable fractures belong (nos. 1, 2, 3, and 12). The magnitude is small at early times, increasing up to ~800 seconds, and stable afterwards for the second group to which electrodes located near permeable fractures belong (nos. 5, 6, 8, and 10). The third group (electrodes nos. 4, 7, and 11) shows an intermediate behavior between those of the first and second groups. The observed $\Delta SP/\Delta P$ is almost constant with time in the host rock zone but exhibits temporal variation near the fractures, which looks like the "double-porosity behavior" predicted by Ishido and Pritchett [8]. If their prediction is applied, the time required for pressure equilibration (τ_{pe}) between the fracture and matrix regions will be identified at the intersection of the intermediate time increasing and late-time stable trends seen in the change-ratio plots; τ_{pe} is estimated to be about 800 seconds for KF-1 observations.

3.3. Interpretation. Figure 11 shows the resistivity distribution around wells KF-1 and KF-3 [11]. The main permeable fracture zone located around 30 meters depth of KF-1

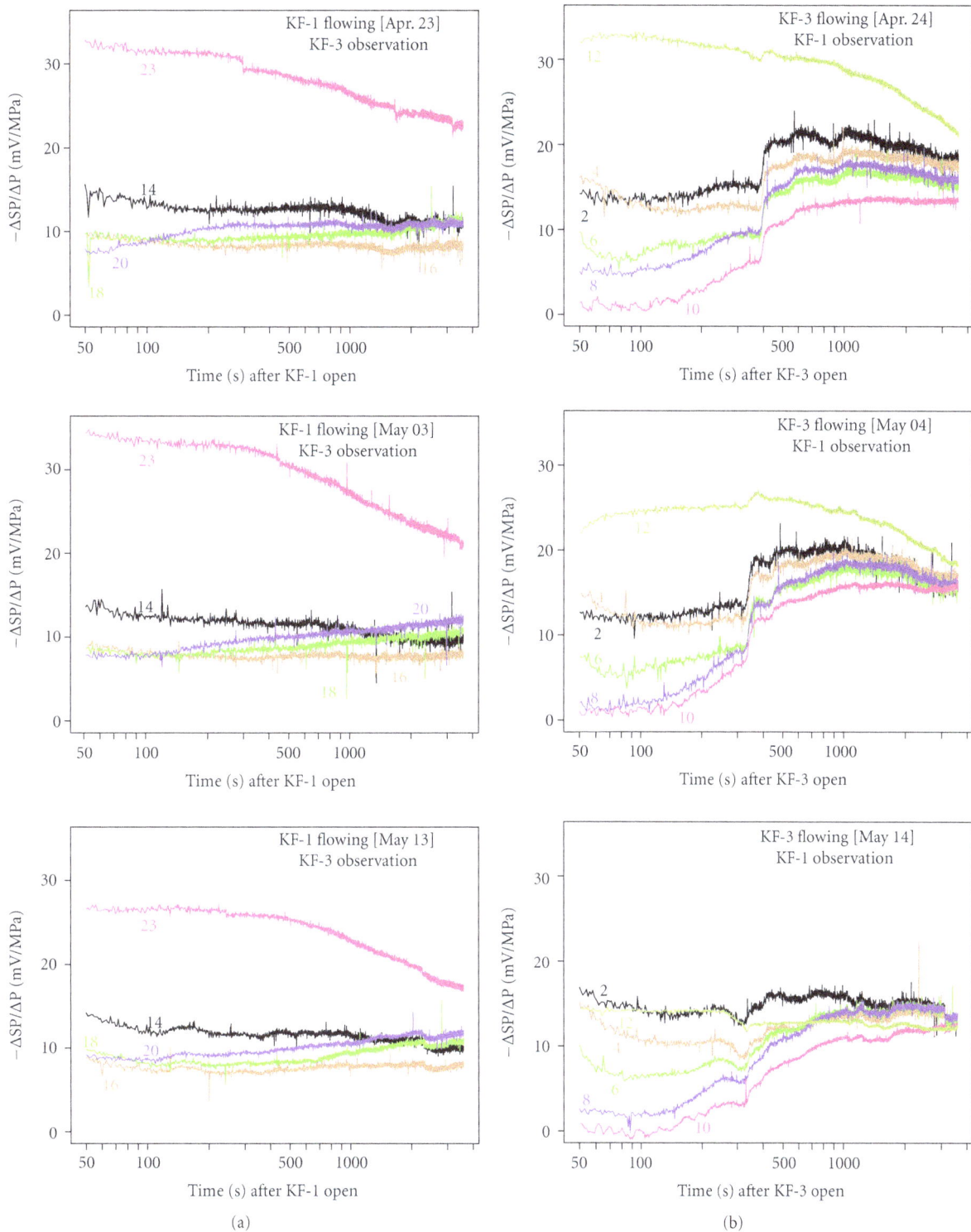

FIGURE 10: Ratio of SP change to pressure change as a function of time for KF-3 observations associated with three KF-1 valve opening (a) and for KF-1 observations associated with three KF-3 valve opening (b). The start of pressure drop is at time zero.

[12] is clearly delineated as a relatively low-resistivity zone (~800 Ωm) within high-resistive (~4000 Ωm) rocks. The extent of this low-resistive zone is well correlated with low-velocity zone revealed by seismic tomography measurements [27]. The locations of KF-1 electrodes of the first, second, and third groups mentioned above are indicated by circle, yellow-filled asterisk, and white-filled asterisk, respectively, in Figure 11. The locations of KF-3 electrodes showing increasing trend and other KF-3 electrodes are indicated by white-filled asterisk and circle, respectively, in Figure 11.

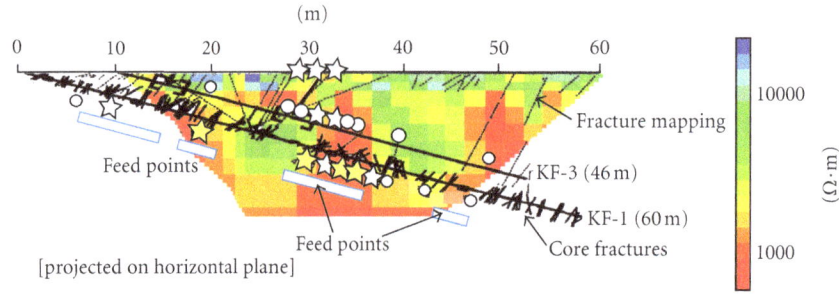

FIGURE 11: Electrical resistivity distribution around wells KF-1 and KF-3 (after [11]). Also shown are traces of major fractures, directions of core fractures, and locations of fluid feed points for KF-1 (e.g., [12]). Locations of KF-1 electrodes (nos. 5, 6, 8, and 10) showing typical fractured rock behavior in Figures 9 and 10 are denoted by yellow-filled asterisks, and those of KF-3 electrodes (nos. 18, 20) showing increasing trend are shown by white-filled asterisks. Locations of electrodes showing almost constant ΔSP/ΔP with time are shown by circles.

Different ΔSP/ΔP behaviors observed at low-permeability host rock and permeable fractured rock regions can be explained by a conceptual model shown in Figure 12. After valve opening of KF-1 (KF-3), pressure drawdown immediately propagates to nearby observation well KF-3 (KF-1) through permeable fractures. Due to substantial pressure drawdown, radial fluid flow toward the wellbore also takes place in host rocks surrounding the observation well (point "A" in Figure 12), which makes ΔSP/ΔP equal to the streaming potential coefficient (C) from early times. In fractured rocks, pressure within the matrix region remains near the initial pressure at early times. So large pressure gradient is created in the matrix region near the fractures and/or the wellbore (point B in Figure 12). As shown in Section 2.5, the magnitude of ΔSP/ΔP is modeled to be quite large from early times and constant with time thereafter if the matrix region's C is assumed to be homogeneous. But if C in the near surface skin zone is assumed significantly smaller in magnitude than that in the interior of the matrix region, $|\Delta$SP/ΔP$|$ is modeled to remain at quite small value at early times.

As shown in Figure 11, quite large contrast of resistivity is present between the main fracture and the surrounding host rock regions. This might bring about difference in the streaming potential coefficient between the fracture and host rock regions. We need to take into account the presence of such heterogeneity in addition to that from EDZ (excavation-damaged zone), which is thought to have ~1 m thickness around the tunnel wall (e.g., [24]), in future studies.

If we apply the "double-porosity" model mentioned in Section 2 to the KF-1 observations shown in Figure 9, the fracture spacing is estimated to be a few meters based on the τ_{pe} value of ~800 seconds and the matrix rock permeability of several microdarcies (10^{-18} m^2) deduced by testing core samples. This is in the range of spacing of permeable fractures estimated for the fracture zone of KF-1 from detailed geological and hydrological observations (e.g., [12, 28]). The ratio of SP change to pressure change corresponds to the streaming potential coefficient C and is $-7 \sim -20$ mV/MPa as seen in Figure 9 for the nonfractured host rock region. This value is in the range of C measured for an intact granite sample in dilute solutions [29]. As for

the fracture zone, the final asymptotic value is also about -10 mV/MPa. This suggests that the contribution of drag current through the matrix region dominates under steady-state conditions.

We measured the zeta potential (-20 mV) for a crushed sample of granodiorite rock cored near the wellheads of KF-1 and KF-3 under room temperatures using the water discharged from the wells ($T \sim 11$°C, pH~9 and electrical conductivity ~0.85×10^{-4} S/m at sampling site; major chemical components are Na$^+ \sim 6$, Ca$^{++} \sim 9$, Cl$^- \sim 2.5$, HCO$_3^- \sim 33$, and SO$_4 \sim 7$ mg/L). If we assume the zeta potential of in situ intact host rock is also around -20 mV, we can estimate the in situ streaming potential coefficient by using (9). Unfortunately, we do not know the surface conductivity, but its contribution must be significant since the in situ bulk conductivity ($1/800 \sim 1/4000$ S/m) is even higher than the pore fluid conductivity itself. If we use the observed bulk conductivity in (9), the streaming potential coefficient is given as $C = (\eta/\tau)(\varepsilon\zeta/\mu)/L_{ee}$. By substituting $\zeta = -0.02$ V, $\varepsilon = 83.6 \times 8.85 \times 10^{-12}$ F/m, $\mu = 0.00127$ Pa-s, and $L_{ee} = 2.5 \times 10^{-4}$ S/m into this equation, $C = -(\eta/\tau) \times 4.7 \times 10^4$ mV/MPa. If we assume term η/τ ($=F^{-1}$, here F is the electrical formation factor) to be 0.0002, which is thought to be in the range of F^{-1} for intact crystalline rocks (e.g., [29]), C becomes about -10 mV/MPa, which is comparable to the observed "in situ" values: $-7 \sim -20$ mV/MPa. Using this value for F^{-1}, we can also estimate the surface conductivity ($m^{-1}\Sigma_s$) in (4) or (9) to be around 1.25 S/m.

4. Concluding Remarks

We have carried out continuous pressure and SP monitoring in KF-1 and KF-3 wells at the Kamaishi Mine by inducing a fluid flow from these wells, which caused pressure disturbance in the surrounding formation along the entire borehole. The observed ratio of SP change to pressure change associated with the fluid flow showed different behaviors between intact host rock and fractured rock regions. In open holes, the appearance of double-porosity behavior is modeled to depend on whether or not a skin zone or some heterogeneity in the streaming potential coefficient is present. It may be that skin zones will be found in most open

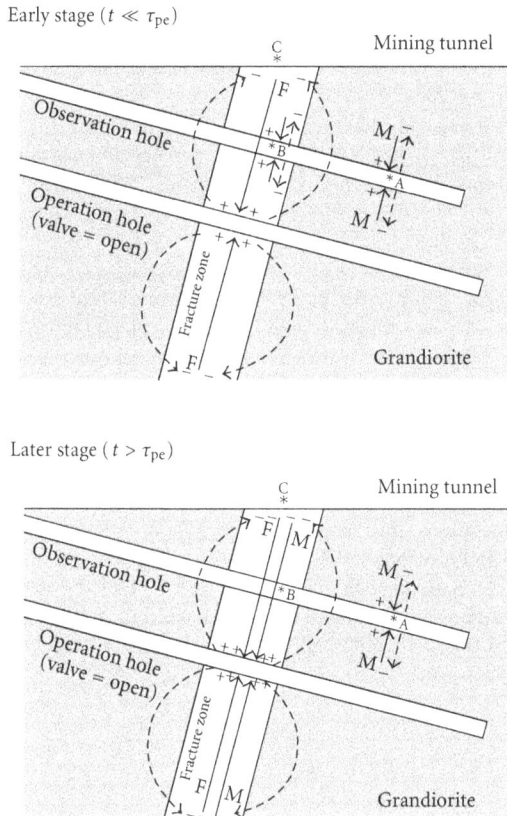

FIGURE 12: A conceptual model for different behaviors of SP changes at the country rock (A) and fractured rock (B) regions, and on the tunnel wall/floor (C). The drag current (solid line) flows only through the fracture (F) region at early times and through both of the fracture (F) and matrix (M) regions at later times than the time required for pressure equilibrium between the fracture and matrix regions. The conduction current (broken line) flows to compensate the resulting charge separation. The model is drawn for a case that KF-3 and KF-1 are observation and flowing wells, respectively; but the model is also applicable to a case that KF-1 and KF-3 are observation and flowing wells, respectively.

hole completions, so measurements of this type in open holes are likely to be useful to deduce the time required for pressure equilibrium between the fractures and the rock matrix.

When the observation is carried out using cased wells, the hydraulic communication between the borehole and the formation should be restricted to the fracture region. Direct hydraulic contact between the borehole and matrix region is prevented by solid casing and cementing [10]. To detect microscopic $\Delta SP/\Delta P$, an electrode array installed outside the insulated casing is desirable. Such observations were reported from an oil field [17]. Downhole SP measurement is thought to be a promising monitoring technique in various applications (e.g., [30, 31]). How to equip observation and/or production/injection wells with appropriate electrodes is very important for these applications. To detect macroscopic $\Delta SP/\Delta P$ such as predicted by Ishido and Pritchett [8], the conductive casing itself (with electrical

continuity extending over a distance longer than the typical fracture spacing) can be used as an electrode. Surface SP measurements (e.g., [14, 21]) detected SP changes which were generated at the reservoir depths and transferred to the earth's surface through the conductive well casing. Thus surface SP measurement around a wellhead is also promising to detect macroscopic $\Delta SP/\Delta P$ if the hydraulic communication between the borehole and the formation is restricted to the fracture region.

Acknowledgments

The present study was supported by basic research fund from The Institute for Geo-Resources and Environment, AIST. The authors greatly acknowledge the generous courtesy received from Kamaishi Mine Co., Ltd. and Nittetsu Mining Consultants Co., Ltd. during the field experiments. They express special thanks to the Kamaishi city where they stayed during the field surveys. They also wish the city recovers as soon as possible from the tsunami damages caused by the 2011 off the Pacific Coast of Tohoku Earthquake. The associate editor L. Jouniaux and two anonymous reviewers are acknowledged for helpful comments to improve the paper.

References

[1] J. W. Pritchett and S. K. Garg, "On similitude, heat conduction, and two-phase flow in fractured porous media," in *Proceedings of the 15th Stanford Workshop on Geothermal Reservoir Engineering*, 1990.

[2] J. E. Warren and P. J. Root, "The behavior of naturally fractured reservoirs," *Society of Petroleum Engineers Journal*, vol. 3, pp. 245–255, 1963.

[3] H. Kazemi, "Pressure transient analysis of naturally fractured reservoirs with uniform fracture distribution," *Society of Petroleum Engineers Journal*, vol. 9, pp. 451–462, 1969.

[4] K. Pruess and T. N. Narasimhan, "A practical method for modeling fluid and heat flow in fractured porous media," in *Proceedings of the the Reservoir Simulation Symposium of the Society of Petroleum Engineers*, New Orleans, La, USA, 1982, Also see SPE Journal, pp.14–26, 1985.

[5] K. Pruess and T. N. Narasimhan, "On fluid reserves and the production of superheated steam from fractured, vapor-dominated geothermal reservoirs," *Journal of Geophysical Research*, vol. 87, no. 11, pp. 9329–9339, 1982.

[6] J. W. Pritchett, "Efficient numerical simulation of nonequilibrium mass and heat transfer in fractured geothermal reservoirs," in *Proceedings of the 22nd Stanford Workshop on Geothermal Reservoir Engineering*, pp. 287–293, 1997.

[7] T. Ishido, *Geothermal Reservoir Engineering*, Japan Geothermal Energy Association, 2002.

[8] T. Ishido and J. W. Pritchett, "Characterization of fractured reservoirs using continuous self-potential measurements," in *Proceedings of the 28th Stanford Workshop on Geothermal Reservoir Engineering*, pp. 158–165, 2003.

[9] T. Ishido and J. W. Pritchett, "Numerical simulation of electrokinetic potentials associated with subsurface fluid flow," *Journal of Geophysical Research B*, vol. 104, no. 7, pp. 15247–15259, 1999.

[10] T. Ishido, Y. Nishi, and J. W. Pritchett, "Application of self-potential measurements to geothermal reservoir engineering: characterization of fractured reservoirs," in *Proceedings of the 35th Stanford Workshop on Geothermal Reservoir Engineering*, 2010.

[11] S. Takakura, Y. Nishi, and T. Ishido, "Resistivity monitoring during an air injection experiment to fractured rocks," in *Proceedings of the 123rd Meeting of SEGJ Extended Abstract*, pp. 225–228, 2010.

[12] T. Negi, Y. Yoneda, and T. Senba, "Application of streaming potential method for detection of fractures in granite rock (part II)," in *Proceedings of the 97th Meeting of SEGJ Extended Abstract*, pp. 274–278, 1997.

[13] V. V. Bogoslovsky and A. A. Ogilvy, "Deformation of natural electric fields near drainage structures," *Geophysical Prospecting*, vol. 21, no. 4, pp. 716–723, 1973.

[14] T. Ishido, H. Mizutani, and K. Baba, "Streaming potential observations, using geothermal wells and in situ electrokinetic coupling coefficients under high temperature," *Tectonophysics*, vol. 91, no. 1-2, pp. 89–104, 1983.

[15] M. Darnet, A. Maineult, and G. Marquis, "On the origins of self-potential (SP) anomalies induced by water injections into geothermal reservoirs," *Geophysical Research Letters*, vol. 31, no. 19, article L19609, 2004.

[16] M. Darnet, A. Maineult, and G. Marquis, "On the origins of self-potential (SP) anomalies induced by water injections into geothermal reservoirs," *Geophysical Research Letters*, vol. 31, no. 19, article L19609, 2004.

[17] M.-Y. Chen, B. Raghuraman, I. Bryant, and M. Supp, "Streaming potential applications in oil fields," in *Proceedings of SPE Annual Technical Conference and Exhibition*, vol. 2, pp. 24–27, 2006, Paper SPE 102106.

[18] E. Rizzo, B. Suski, A. Revil, S. Straface, and S. Troisi, "Self-potential signals associated with pumping tests experiments," *Journal of Geophysical Research B*, vol. 109, no. 10, article B10203, 2004.

[19] S. Straface, C. Fallico, S. Troisi, E. Rizzo, and A. Revil, "An inverse procedure to estimate transmissivity from heads and SP signals," *Ground Water*, vol. 45, no. 4, pp. 420–428, 2007.

[20] A. Maineult, E. Strobach, and J. Renner, "Self-potential signals induced by periodic pumping tests," *Journal of Geophysical Research B*, vol. 113, no. 1, Article ID B01203, 2008.

[21] M. Darnet, G. Marquis, and P. Sailhac, "Hydraulic stimulation of geothermal reservoirs: fluid flow, electric potential and microseismicity relationships," *Geophysical Journal International*, vol. 166, no. 1, pp. 438–444, 2006.

[22] T. Ishido and J. Muzutani, "Experimental and theoretical basis of electrokinetic phenomena in rock-water systems and its applications to geophysics," *Journal of Geophysical Research*, vol. 86, no. 3, pp. 1763–1775, 1981.

[23] T. Ishido and J. W. Pritchett, "Numerical simulation of electrokinetic potentials associated with subsurface fluid flow," in *Proceedings of the 21st Stanford Workshop on Geothermal Reservoir Engineering*, pp. 143–149, 1996.

[24] K. Yokoi, Y. Noguchi, H. Choh, and S. Hamabe, "Geophysical methods applied in underground investigation programs at the Kamaishi Mine," *Busuri-Tansa*, vol. 44, no. 6, pp. 350–361, 1991.

[25] Y. Nishi, T. Ishido, and T. Negi, "Field experiments to characterize hydrological properties of fractured rock using continuous borehole self-potential measurement," *Transactions of Geothermal Resources Council*, vol. 30, pp. 835–838, 2006.

[26] Y. Nishi, T. Ishido, and T. Negi, "Continuous borehole self-potential measurement—a new approach to characterize hydrological properties of fractured rock," *Butsuri-Tansa*, vol. 61, pp. 285–299, 2008.

[27] Y. Yoneda, H. Takahara, N. Nakamura, S. Akiyama, T. Moriya, and T. Negi, "Investigation of geological environment of deeper granitic rock at the Kamaishi Mine (H9)," Report PNC TJ 1380 98-002, Power Reactor and Nuclear Fuel Development, 1998.

[28] Y. Yoneda and T. Negi, "Geological surveys of fracture zone of deeper granitic rock at the Kamaishi Mine (H7)," PNC TJ 1380 96-001, Power Reactor and Nuclear Fuel Development, 1996.

[29] T. Ishido and N. Matsushima, "Streaming potential measured for an intact granite sample at temperatures to 200°C," in *Proceedings of the 32nd Stanford Workshop on Geothermal Reservoir Engineering*, 2007.

[30] J. W. Pritchett and T. Ishido, "Numerical simulation of underground electrical signals caused by hydrofracturing operations," in *Proceedings of the World Geothermal Congress*, Bali, Indonesia, 2010.

[31] J. H. Saunders, M. D. Jackson, and C. C. Pain, "Fluid flow monitoring in oilfields using downhole measurements of electrokinetic potential," *Geophysics*, vol. 73, no. 5, pp. E165–E180, 2008.

9

Observations of an 11 September Sahelian Squall Line and Saharan Air Layer Outbreak during NAMMA-06

J. W. Smith,[1,2] A. E. Reynolds,[2,3] A. S. Pratt,[4] S. Salack,[5] B. Klotz,[6] T. L. Battle,[1] D. Grant,[7] A. Diop,[5] T. Fall,[5] A. Gaye,[5] D. Robertson,[8] M. S. DeLonge,[9] and S. Chan[10]

[1] *Program in Atmospheric Sciences, Howard University, Washington, DC 20059, USA*
[2] *NASA Goddard Space Flight Center, Greenbelt, MD 20771, USA*
[3] *Atmospheric Science Group, Department of Geosciences, Texas Tech University, Lubbuck, TX 79409, USA*
[4] *Earth Resources Technology, Inc., Laurel, MD 20707, USA*
[5] *Laboratory of Atmospheric Physics—Simeon Fongong, Cheikh Anta Diop University, Dakar, Senegal*
[6] *Rosenstiel School of Marine and Atmospheric Science, Cooperative Institute for Marine and Atmospheric Science Studies, University of Miami, Miami, FL 33149, USA*
[7] *Environmental Protection Agency, Region 4, Atlanta, GA 30303, USA*
[8] *Department of Mechanical Engineering, Howard University, Washington, DC 20059, USA*
[9] *Department of Environmental Science, Policy, and Management, University of California Berkeley, Berkeley, CA 94720, USA*
[10] *Department of Environmental Sciences, University of Virginia, Charlottesville, VA 22903, USA*

Correspondence should be addressed to J. W. Smith, jonathan.smith@nasa.gov

Academic Editor: Alessandra Giannini

The 2006 NASA-African Monsoon Multidisciplinary Analyses (NAMMA-06) field campaign examined a compact, low-level vortex embedded in the trough of an AEW between 9–12 September. The vortex triggered a squall line (SL) in southeastern Senegal in the early morning of 11 September and became Tropical Depression 8 on 12 September. During this period, there was a Saharan Air Layer (SAL) outbreak in northwestern Senegal and adjacent Atlantic Ocean waters in the proximity of the SL. Increases in aerosol optical thicknesses in Mbour, Senegal, high dewpoint depressions observed in the Kawsara and Dakar rawinsondes, and model back-trajectories suggest the SAL exists. The close proximity of this and SL suggests interaction through dust entrainment and precipitation invigoration.

1. Introduction

Westerly propagating squall lines (SLs) are frequent phenomena in the West African Sahel during the Northern Hemisphere summer [1, 2]. This SL formed and propagated ahead of the trough region of an African Easterly Wave (AEWs) [3, 4]. The development of SLs involves a combination of atmospheric conditions including conditional instability, low-level convergence, and vertical wind shear between relatively cool, moist, monsoon lower tropospheric air and warmer, drier Harmattan winds driven by the 650 hPa AEJ [3, 5].

The structure of SLs in West Africa was observed and analyzed during the Global Atmospheric Research Program's (GARP's) Atlantic Tropical Experiment (GATE) and Convection Profonde Tropicale (COPT 81) field campaigns [3, 5–8]. Squall lines typically have a convective leading edge that produces a short period of intense rainfall. Observations show that intense precipitation extended 5 to 6 km above the surface (see reflectivity values of ≥ 50 dBZ in [7], Figure 8, [8], Figure 7). Observations and modeling studies suggest that SLs in the Sahel were embedded in the troughs of AEWs [3, 9, 10]. The axis of the AEW trough often contains two distinct mesoscale features: a low-level Saharan heat low to the north of the AEJ and a circulation at 650 to 700 hPa around 10°N [11–14]. The trough of the AEW highlighted as "Wave 7" of the 2006 NASA-African Monsoon Multidisciplinary Analyses (NAMMA-06) [15] took on these coupled dynamical features.

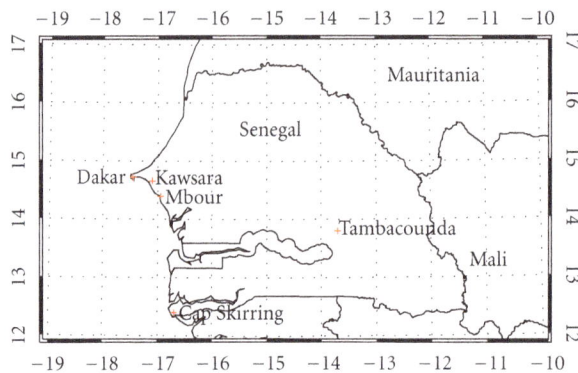

FIGURE 1: Map of cities and villages in Senegal where meteorological data relevant to this study was obtained.

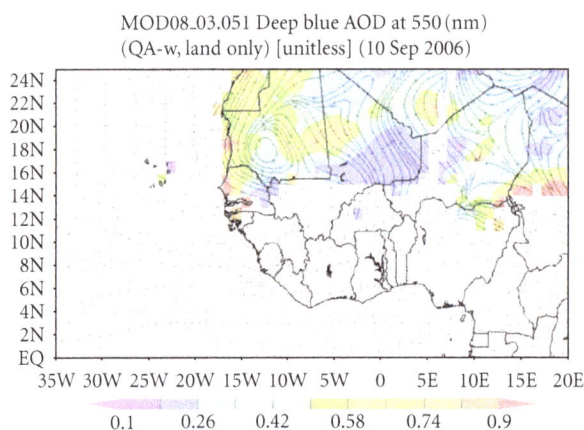

FIGURE 2: ECMWF Reanalyses streamlines and MODIS-Terra 550 nm AOT at 1200 UTC 10 September 2006.

The NAMMA-06 campaign extended from 15 August through 30 September 2006. The NAMMA-06 Campaign's goals were to find differences between AEWs that produce hurricanes and those that do not, the specific role of the Saharan Air Layer (SAL) has on AEWs and tropical cyclones, and to examine the role of cloud microphysics on precipitation amounts and contribution. The NAMMA-06 study accomplished the goals by observing seven AEWs and SAL outbreaks [15]. The SL in this study was one of two that developed with an AEW propagating westward out of Mali during the 9–11 September 2006 period and was a precursor to Hurricane Helene. Moreover, NAMMA-06 included an investigation of five SAL outbreaks and associated cloud microphysical properties [16].

The primary goals of this paper are to use data from the NAMMA-06 period to examine observations of an 11 September SL, SAL outbreak, and its interaction with the SL. Data locations and sources are discussed in Section 2 and the synoptic analysis and observations in Section 3. Observations of the 11 September SAL outbreak are presented in Section 4. A brief synthesis of the SL and SAL interaction is given in Section 5. Section 6 gives a summary of the major results and ideas for future work.

2. Data Locations and Sources

The locations of sites within Senegal where meteorological data were collected near the SL and SAL outbreak are presented in Figure 1. The European Center for Medium Range Weather Forecasting (ECMWF) 40-year (ERA-40) Reanalysis data are used to examine the synoptic-scale and mesoscale conditions in the vicinity of the SL. The ERA-40 Reanalysis covers the years of 1957–2010 and has a spatial and temporal resolution of approximately $1° \times 1°$ and six hours, respectively [17]. The data and locations include flux tower measurements in Kawsara (14.7°N, 17.1°W) (Figure 1), surface observations at the Cap Skirring Airport (GOGS) (12.4°N, 16.7°W) (Figure 1), rawindsonde data from Kawsara, Leopold Sedar Senghor International Airport in Dakar (GOOY) (14.7°N, 17.5°W), and Tambacounda Airport (GOTT) (13.8°N, 13.7°W) (Figure 1). The National Centers for Environmental Prediction/Climate Prediction Center (NCEP/CPC) 4 km globally merged Infrared cloud top brightness temperature dataset [18] are used to evaluate the location and intensity of the SL convection. The merged IR product is also a Tropical Rainfall Measuring Mission (TRMM) ancillary dataset. Additional storm-scale data are

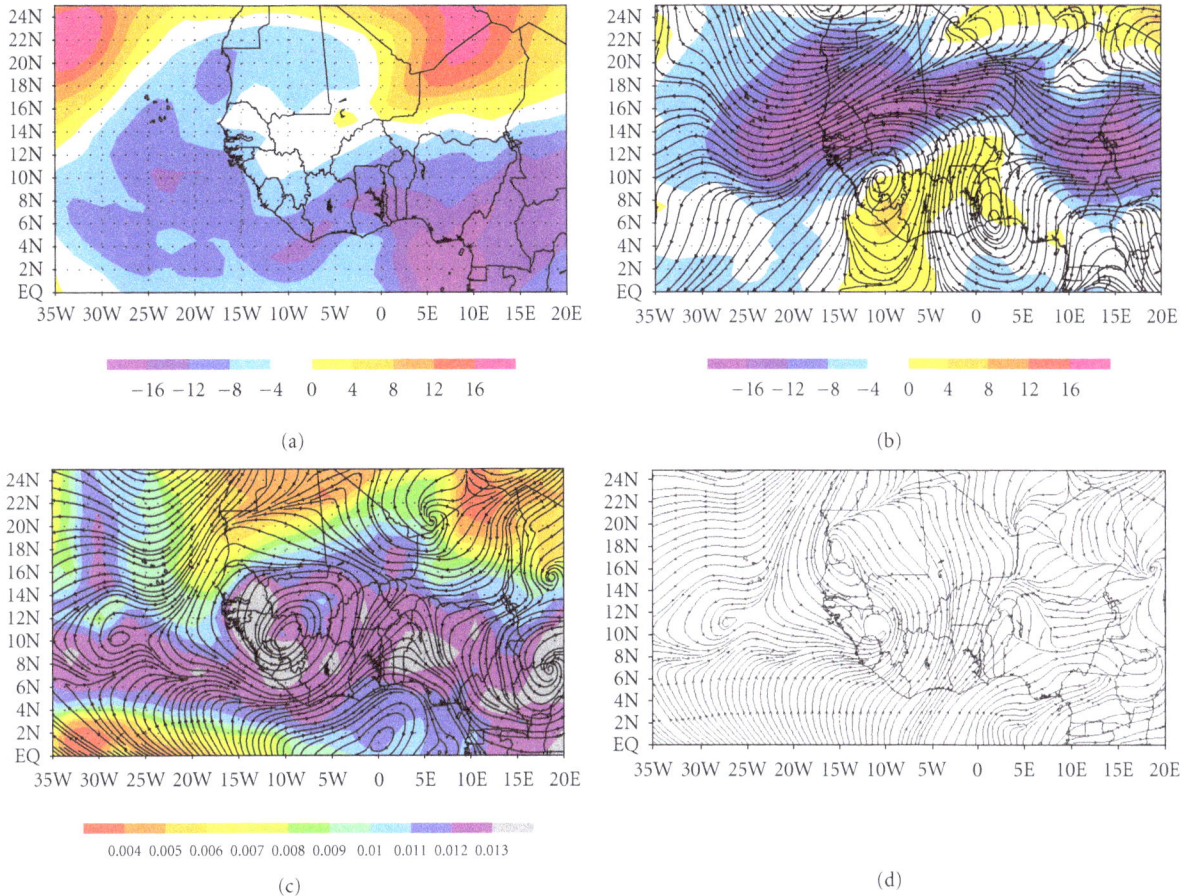

FIGURE 3: (a) ECMWF Reanalysis zonal winds $(m\,s^{-1})$ at 0000 UTC 11 September for 200 hPa. (b) Composite ECMWF Reanalysis zonal winds $(m\,s^{-1})$ and streamlines at 0000 UTC 11 September for 650 hPa. (c) Composite ECMWF Reanalysis specific humidity $(kg\,kg^{-1})$ and streamlines at 0000 UTC 11 September for 850 hPa. (d) ECMWF Reanalysis streamlines at 0000 UTC 11 September for 925 hPa.

provided by the constant altitude plan position indicator (CAPPI) and radar height indicator (RHI) cross-sections from the NPOL radar (located in Kawsara), as well as the United Kingdom Meteorological Service Arrival Time Difference (UKMET ATD) cloud-to-ground lightning data set. The Aerosol Robotic Network (AERONET) Aerosol Optical Thickness (AOT) data from Mbour, Senegal (14.39°N, 16.95°W), are used to indicate the presence of Saharan dust aerosols in the storm environment. Back-trajectories from the Hybrid Single Particle Lagrangian Integration Trajectory (HYSPLIT) Model and MODIS Terra satellite images over West Africa and the adjacent eastern Atlantic Ocean are examined to highlight the interaction between the SL and the SAL. Taken together, these data provide a robust and detailed picture of the environmental conditions for this case.

3. Squall Line Environmental Characteristics

3.1. Synoptic View. The 10 September 2006 MODIS-Terra 550 nm AOT (Figure 2) displays AOT values that exceed 0.9 in western Senegal. These values stretch northward into Mauritania. The 925 hPa flow from around a heat low in Southern Mauritania at 1200 UTC 10 September suggests

that air from dust aerosols in the Sahara are advecting into Senegal as early as 18 hours before the SL arrives.

The 0000 UTC analysis on 11 September 2006 depicts several large-scale dynamical features over West Africa. The Tropical Easterly Jet (TEJ) can be seen at 200 hPa extending from central Africa westward along the Gulf of Guinea coast, and then northwestward just offshore of Guinea-Bissau and Senegal (Figure 3(a)). This jet helped support upper-level divergence, which aided in repeated squall line development with this AEW.

A closed circulation exists within the AEW trough axis at 650 hPa, with the core of the AEJ just to its north (Figure 3(b)). This closed circulation is also noted at 850 hPa (Figure 3(c)) and 925 hPa (Figure 3(d)). A sharp gradient in specific humidity can be seen over Senegal at 850 hPa, separating the dry SAL regime from the much moister (specific humidity > 0.012 kg kg^{-1}) regime near the vicinity of the incipient squall line. A Saharan heat low is also present at 925 hPa (Figure 3(d)) over southern Mauritania; the flow around the heat low advects dust aerosols south and east toward central and eastern Senegal. This general circulation pattern is characteristic of some AEWs [11–14]. The same general circulation features were observed for this AEW with the Global Data Assimilation System analyses [15].

(a)

(b)

(c)

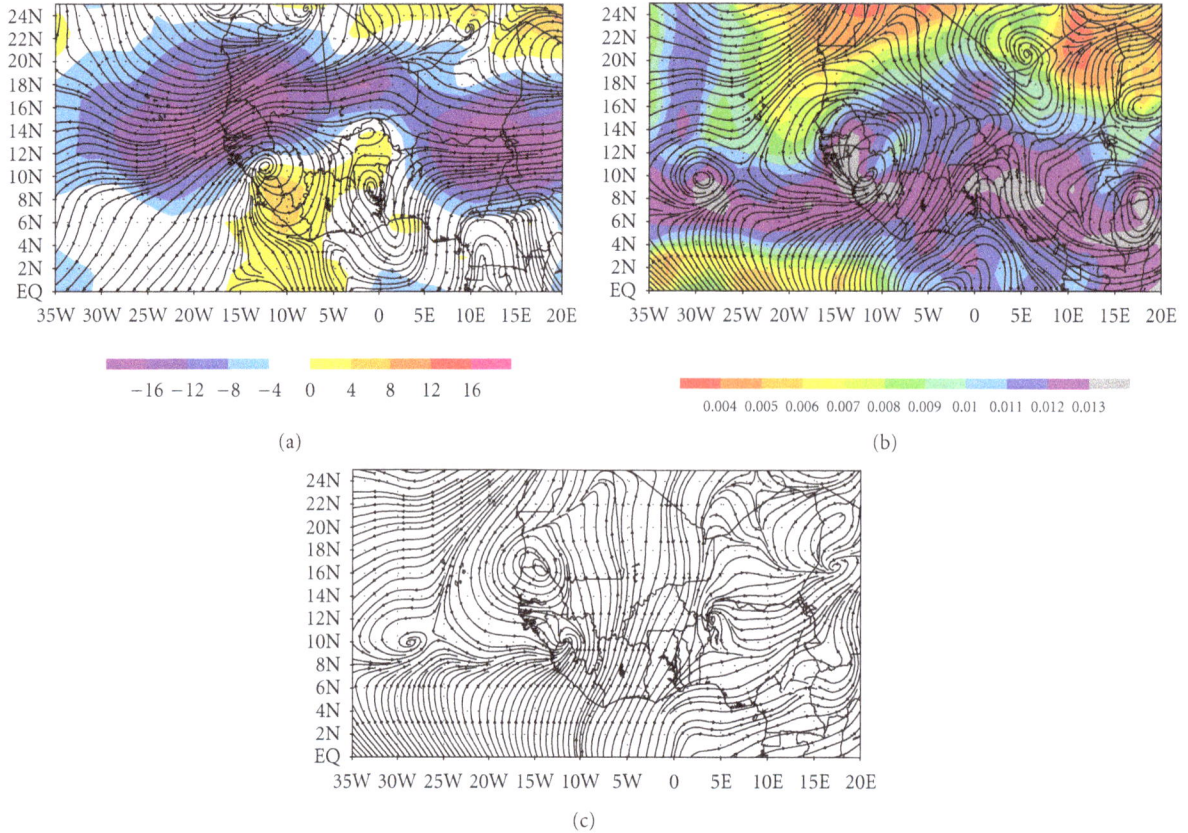

FIGURE 4: (a) Composite ECMWF Reanalysis zonal winds $(m\,s^{-1})$ and streamlines at 0600 UTC 11 September for 650 hPa. (b) Composite ECMWF Reanalysis at 0600 UTC 11 September for 850 hPa, specific humidity $(kg\,kg^{-1})$ and streamlines. (c) Composite ECMWF Reanalysis streamlines at 0600 UTC 11 September for 925 hPa.

At 0600 UTC (Figures 4(a)–4(c)), the low-level vortex associated with the AEW at 850 hPa and 925 hPa is located over southern Guinea (12°W, 10°N). These features, along with the collocated 650 hPa AEW vortex and 925 hPa heat low over southern Mauritania, are propagating westward. The drier SAL regime has pushed further to the south (Figure 5(b)), while dust continues to be advected into the storm environment by the 925 hPa circulation. At 1200 UTC (Figure 5(a)–5(d)), the vortex at 850 hPa has moved further west with wind speeds increasing as suggested by the streamlines. The Saharan heat low (Figure 5(d)) moves southwestward to the northern coast of Senegal and the compact streamlines at 650 hPa suggest that the circulation is also strengthening. This circulation is centered 75–100 km east of the coast of Guinea and continues moving westward (Figure 5(b)). The position of these coupled features shows that the northern end of the AEW trough axis is moving faster than its southern end. Moreover, the 650 hPa circulation is located northwest of the 850 hPa circulation, indicating that the AEW circulation is tilted from southeast to northwest with height (Figure 5(c)).

3.2. Satellite and Surface Observations. The merged IR satellite observations (Figure 6) show the development and evolution of the convection associated with the AEW. This

particular AEW had a history of producing numerous squall lines; one such squall line can be seen moving offshore at 0000 UTC 11 September [15]. A second SL (the purpose of this study) is initiated in southeastern Senegal at 0200 UTC (Figure 6(b)). As this SL developed IR, its cloud-top temperatures cool to 200–220 K south of the 14°N latitude line (Figures 6(c)–6(f)); there is also an expansion of the cloud canopy associated with this system. The cooling cloud top temperatures are indicative of the SL intensifying; the cloud tops are coolest at 0600 UTC (Figure 6(d)). This system moved offshore between 0900–1000 UTC.

Pregust front/SL conditions in Kawsara and Cap Skirring (Figure 7(a)) show a steady west (270°) to west-northwest (300°) wind at both locations throughout the morning, providing moist low-level onshore flow into the region. The wind speeds range from 3.0 to 8.0 m s^{-1}. Meanwhile, the SL continued to propagate westward into a thermodynamic regime conducive for the maintenance of strong convection. Convective Available Potential Energy (CAPE) value of 3237 J kg^{-1} at 0000 UTC is observed at Dakar and 4443 J kg^{-1} at 0549 UTC at Kawsara (Table 1). At both locations, the CAPE exceeds one of the highest COPT 81 values of 2810 J kg^{-1} [9].

There was a marked temperature drop from 28.0°C at 0900 to 22.0°C at 1100 UTC at Cap Skirring (Figure 7(a))

(a)

(b)

(c)

(d)

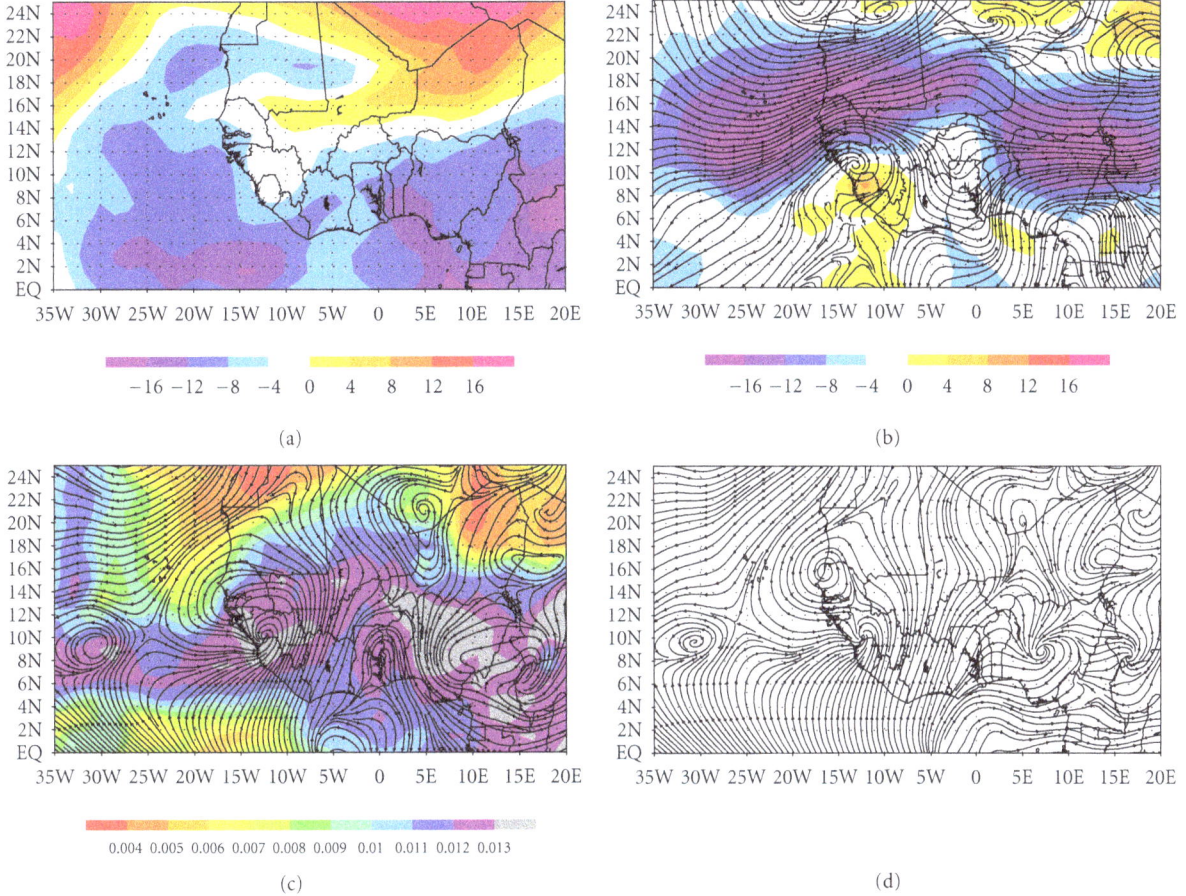

FIGURE 5: (a) ECMWF Reanalysis zonal winds (m s^{-1}) at 1200 UTC 11 September for 200 hPa. (b) Composite ECMWF Reanalysis zonal winds (m s^{-1}) and streamlines at 1200 UTC 11 September for 650 hPa. (c) Composite ECMWF Reanalysis specific humidity (kg kg^{-1}) and streamlines at 1200 UTC 11 September for 850 hPa. (d) Composite ECMWF Reanalysis streamlines at 1200 UTC 11 September for 925 hPa.

TABLE 1: Depiction of the CAPE and CIN for Dakar, Kawsara, and Tambacounda, Senegal, prior to the SL propagating into the vicinity of these cities.

Locations	Time UTC	CAPE J kg^{-1}	CIN J kg^{-1}
Dakar (14.7°N, 17.5°W)	0000	3237	−269
Kawsara (14.7°N, 17.1°W)	0549	4433	−128
Tambacounda (13.8°N, 13.7°W)	0000	1112	−190

associated with the squall line passage. In Kawsara, the temperature dropped from 30.2°C at 1130 UTC to 24.8°C by 1330 UTC (Figure 7(b)). Between the hours of 0900 and 1100 UTC, the pressure rises from 1011 to 1014 hPa in Cap Skirring after the SL passage and from 1002 to 1006 hPa in Kawsara after the gust front passed. These observations are typical of squall lines and suggest a mesohigh followed the SL and gust front [7, 8]. In both locations, the pressure dropped quickly as the SL approaches and then rose afterward due to the post-SL mesohigh. Moreover, at both Kawsara and Cap Skirring, the wind shifts to an easterly direction and the speed decreases to 1.0–4.0 m s^{-1} after SL passage.

3.3. Radar and Lightning Observations. The northern end of the heavy convection associated with the SL passes through Cap Skirring and to the south of Kawsara between 0900–1200 UTC. Surveillance reflectivity scans (Figures 8(a)–8(d)) depict heavy precipitation (reflectivity greater than 45 dBZ) propagating over Cap Skirring. By contrast, no precipitation falls over Kawsara from the squall line; nevertheless, a narrow band of relatively weak reflectivity values depict the gust front moving through this region as well. While the squall struggled to build northward, the gust front pushed as far as 50–60 km to the north and northeast of the heavier convection.

The CAPPI images in the top panels of Figures 9(a)–9(c) depict the westward progression of the SL off the coast of Senegal. The radar scans focus on the northern portion of the SL in western Senegal. These observations coupled with the surface data in Figure 7 and the surveillance scans in Figure 8 provide additional evidence that a dry gust front passes through Kawsara during the late morning of 11 September 2006.

The dashed lines on the CAPPI images are the horizontal locations of RHI cross-section plots in the bottom panels of Figures 8(a)–8(c). A reflectivity contour of 6 dBZ reaches

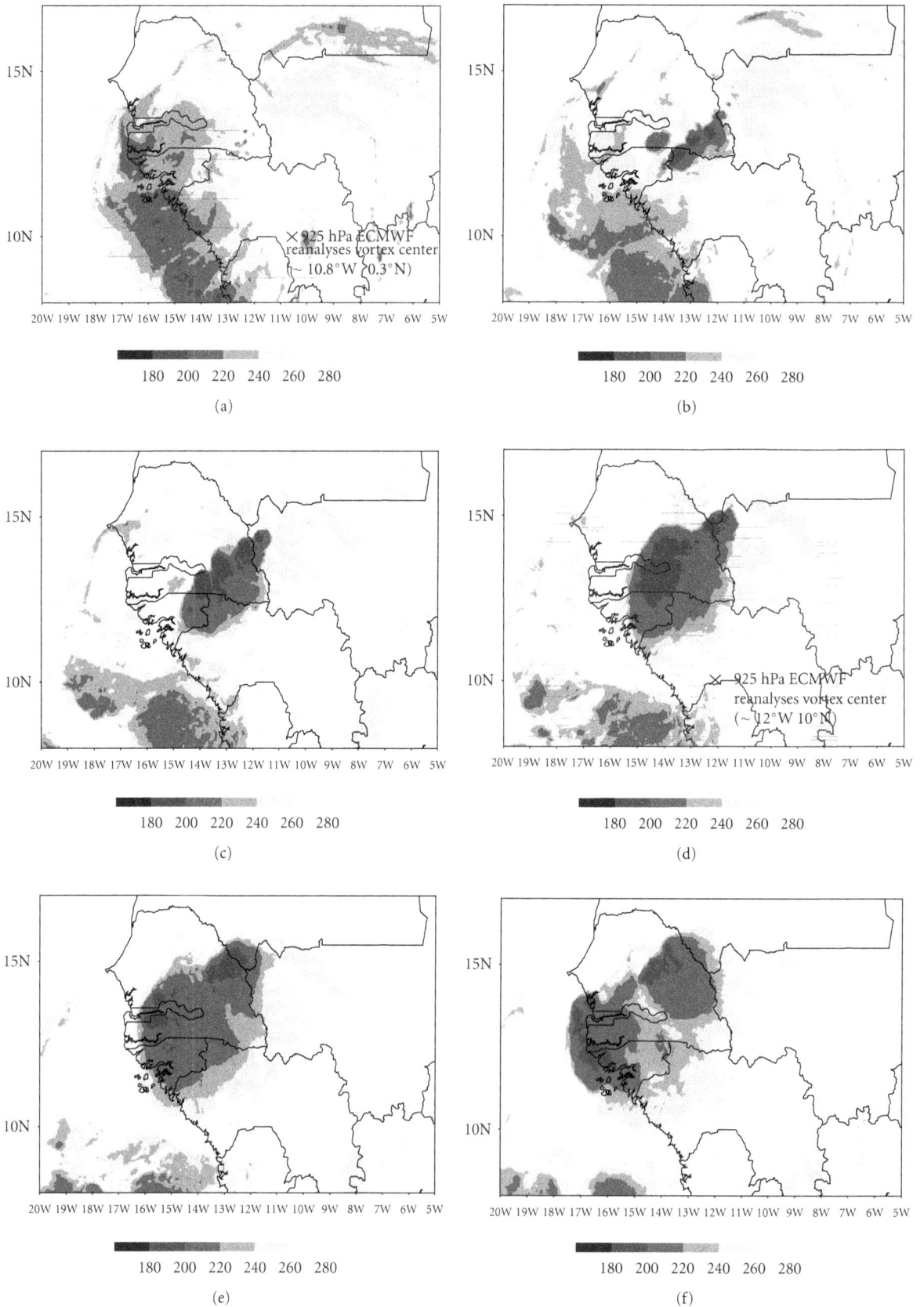

FIGURE 6: Continued.

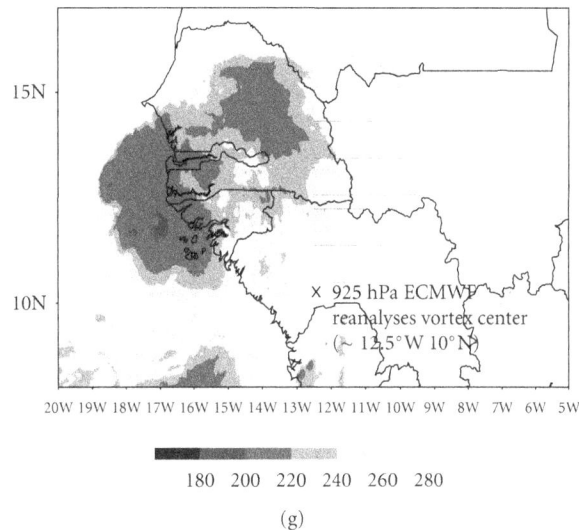

(g)

FIGURE 6: (a) Depiction of TRMM globally merged IR cloud top temperatures (K) on 11 September 2006 for 8°N to 17°N and 20°W to 5°W for 0000 UTC. The X denotes the 0000 UTC 925 hPa ECMWF Reanalyses Vortex Center. (b) Depiction of TRMM MERG IR cloud top temperatures (K) on 11 September 2006 for 8°N to 17°N and 20°W to 5°W for 0200 UTC. (c) Depiction of TRMM MERG IR cloud top temperatures (K) on 11 September 2006 for 8°N to 17°N and 20°W to 5°W for 0400 UTC. (d) Depiction of TRMM MERG IR cloud top temperatures (K) on 11 September 2006 for 8°N to 17°N and 20°W to 5°W for 0600 UTC. The X denotes the 0600 UTC 925 hPa ECMWF Reanalyses Vortex Center. (e) Depiction of TRMM globally merged IR cloud top temperatures (K) on 11 September 2006 for 8°N to 17°N and 20°W to 5°W for 0800 UTC. (f) Depiction of TRMM MERG IR cloud top temperatures (K) on 11 September 2006 for 8°N to 17°N and 20°W to 5°W for 1000 UTC. (g) Depiction of TRMM MERG IR cloud top temperatures (K) on 11 September 2006 for 8°N to 17°N and 20°W to 5°W for 1200 UTC. The X denotes the 1200 UTC 925 hPa ECMWF Reanalyses Vortex Center.

up to nearly 15 km AGL, which is consistent with higher cloud top heights at 0900 UTC (see Figure 6). However, while the 48 dBZ contour extends horizontally for 10 km, it only reached an altitude of 3 km AGL. This horizontally extensive but shallow convection is consistent with the general lack of convective intensity seen in this SL at 0900 UTC. At this time, there is no trailing stratiform rain region associated with this SL. Surveillance scans to the south (see Figure 8) further reveal this underdeveloped stratiform region.

As the squall line transitions from a continental to maritime regime, the convection associated with it became shallower and less intense (Figure 9(b)). The reduced amount of CAPE offshore relative to land most likely contributed to the decrease in convective intensity; dropsondes released offshore on 12 September indicated maximum CAPE values of approximately 2500 J kg^{-1} (not shown), considerably less than the values observed over land the previous day. A study investigating an mesoscale convective system moving over West Africa on 31 August 2006 also noted less favorable thermodynamic and dynamic conditions offshore that contributed to convective weakening [19]. Although there were a few small cells with 54 dBZ contours at 4 km AGL observed in the horizontal cross section, the vertical cross section taken through the leading edge shows the 42 dBZ contour reaching only slightly above 4 km AGL. A secondary reflectivity maximum at approximately 47 to 50 km distance along the dashed line cross section is also observed. This suggests that decaying cells from the convective leading edge move rearward once they have reached a mature stage [20]. However, at 1000 UTC (Figure 9), the stratiform rain region

is even less cohesive with the leading edge to the west than it was at 0900 UTC. This lack of cohesion of the stratiform region is uncharacteristic of SLs observed in GATE and COPT 81 [3, 5–8]. Figure 9(c) shows that the northern end of the SL is approximately 50 km southwest of the NPOL radar location at 1100 UTC.

Figure 10 depicts the UK-ATD lightning data [21, 22] between 0400–1200 UTC on 11 September 2006. Lightning flashes show that the SL extended southward from approximately 14.5°N to 12.5°N from 0400 to 0600 UTC as shown by purple crosses. Lightning associated with the SL moved westward from 0600 to 0800 UTC as shown by blue circles. Between 0800–1000 UTC, the lightning flashes associated with the SL as shown by red squares approached the West African coastline. The lightning flashes associated with the SL moved westward into the Atlantic between 1000–1200 UTC shown by the yellow triangles. The region of lightning flashes to the northeast of the SL persisted throughout this time period.

4. Observations of Saharan Air Layer Outbreak

A significant SAL outbreak, one of five observed between 15 August and 15 September 2006 [23], preceded the SL on 11 September 2006. SAL outbreaks are defined as the onset of a well-mixed, warm, dry, and dusty air mass 2.0 to 6.0 km deep that originates from Saharan Africa and propagates westward toward the Eastern Atlantic Ocean [24]. This air mass (bounded by temperature inversions at its base and top)

(a)

(b)

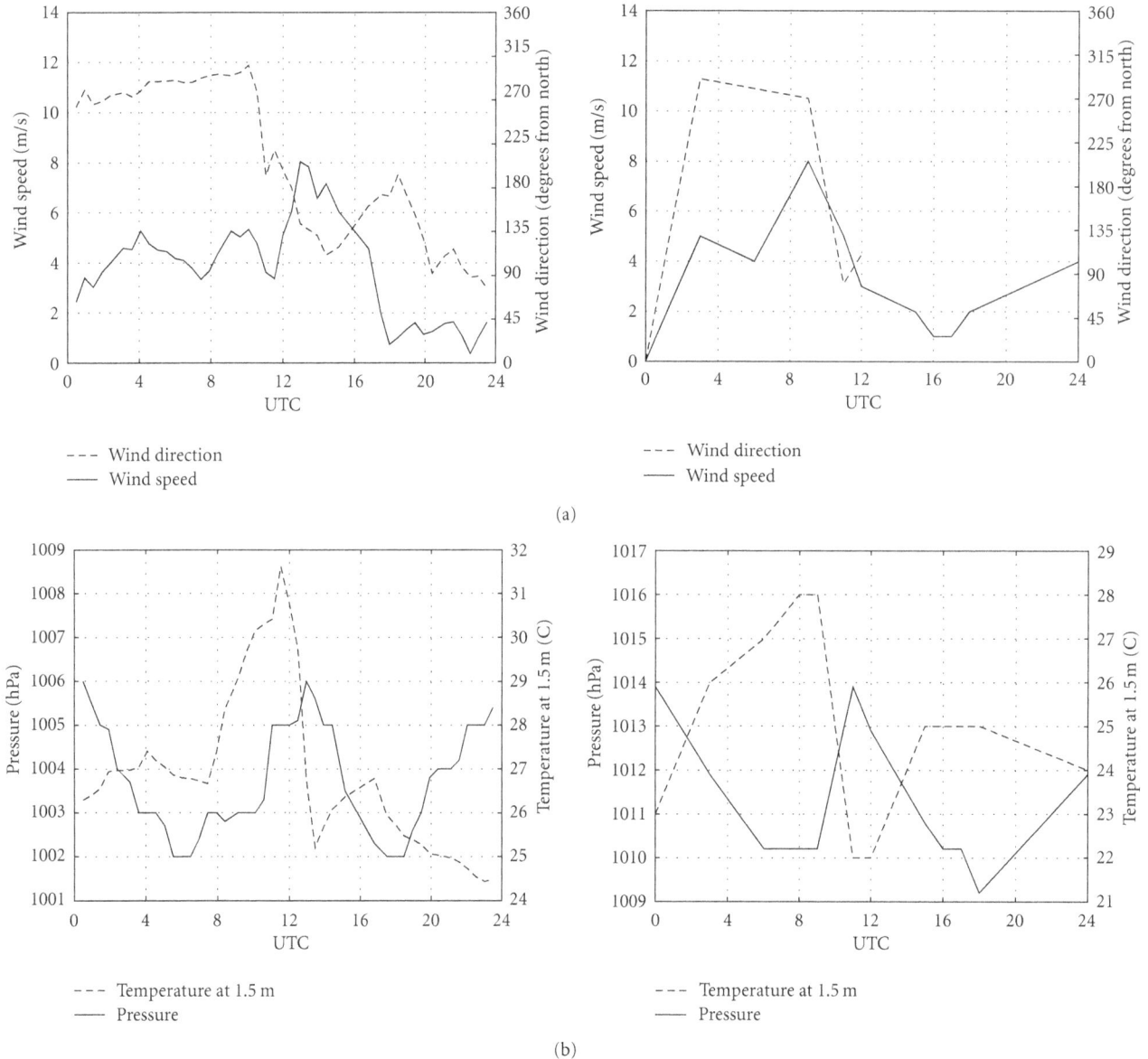

FIGURE 7: (a) Double y-axis plots containing an 11 September 2006 time series of wind speed in m s^{-1} and wind direction in degrees from north for Kawsara and Cap Skirring, Senegal. (b) Double y-axis plots containing an 11 September 2006 time series of pressure in hPa denoted by the solid line and 1.5 m temperature in °C denoted by the dashed line.

is usually 5 to 6°C warmer than the more humid and shallow (e.g., 1.2 to 1.8 km) marine boundary layer air below [25–27]. Typically the AEJ forms around 650 hPa on the southern edge of the SAL primarily due to the temperature contrast between the warm SAL to the north and the relatively cooler marine air to its south [24, 28, 29].

The ECMWF Reanalyses of 850 hPa specific humidity on 11 September (see Figures 3(c), 4(b), and 5(c)) indicate a tongue of low specific humidity (values less than 0.009 kg kg^{-1}) extending south into coastal and northern Senegal. The low moisture measurements are indicative of drier air and dust intruding from the Saharan Desert into northern Senegal. The moisture steadily increases to the

south and east over southeast Senegal, Mali, and Guinea-Bissau.

Rawindsonde data also show the presence of the SAL in the vicinity of the SL on 11 September 2006. At 0000 UTC, the SAL (though weak and shallow) is present at Tambacounda, Senegal (Figure 11(a)) and (more prominently) in Dakar (Figure 11(b)) and at 0549 UTC in Kawsara (Figure 11(c)). The depth of the SAL is denoted by the vertical extent of the SAL between its lower and upper temperature inversions. At Dakar (Figure 11(b)), the SAL extends from 935 hPa to 580 hPa, while at Kawsara (Figure 9(c)) it starts at 900 hPa and is capped at 600 hPa. At both locations (Figures 11(b) and 11(c)), there is a robust dewpoint depression on

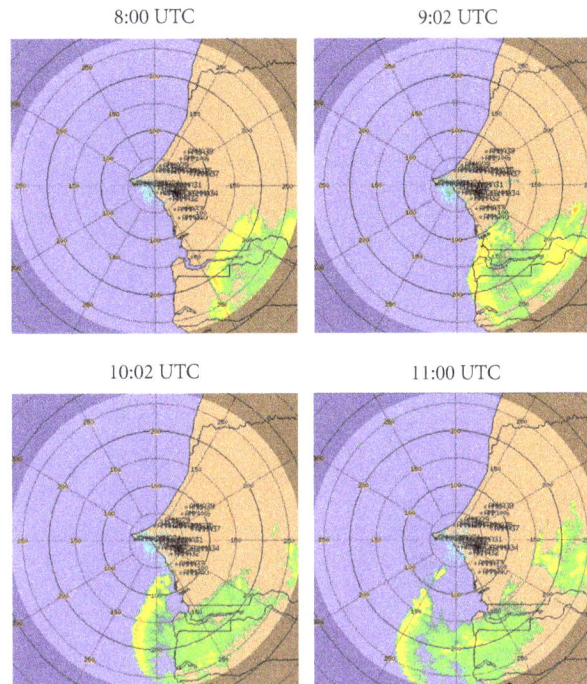

FIGURE 8: Surveillance scans of radar reflectivity (dBZ) time series for 0800 UTC, 0902 UTC, 1002 UTC, and1100 UTC on 11 September 2006. The black circles designate the 100, 150, 200, and 250 km radial distances from the NPOL radar.

the order of 15 to 20°C in the layer of 850 to 800 hPa. The depth and atmospheric stability of the SAL in Dakar and Kawsara prevented the squall line from building this far north. At the same time, the rather weak and shallow SAL at Tambacounda did not impede the squall line here.

Post-SL rawindsonde (Figure 12(a)) data indicate a moist profile (modified by the squall line passage just before 0800 UTC) at 1200 UTC for Tambacounda. At both Dakar (1200 UTC, Figure 12(b)) and Kawsara (1402 UTC, Figure 12(c)), a deep SAL layer remains after the squall has moved to the west of this region. The SAL exists between 920 hPa and 610 hPa at Dakar. The highest dewpoint depression at Dakar and Kawsara (Figure 12(c)) is around 875 hPa where the dewpoint was 7°C while the temperature is 26°C.

At Mbour (see location in Figure 1), Figures 13(a)–13(d) depict aerosol optical thickness (AOT) measurements from the Aerosol Robotic Network (AERONET) Dakar station. Low AOT values below 0.30 are found on 08 September (Figure 13(a)), and a steady increase is observed over subsequent days (Figures 13(b)–13(d)). By 10 September (Figure 13(c)), AOT values are approximately 0.65 and increase to 0.71 during the early morning of 11 September. The high AOT values on 10-11 September are indicative of the dustiness associated with the SAL. The AOT values, in concert with the wind flow at 925 hPa and 850 hPa, suggest that the dry, dusty SAL is being entrained into the squall line inflow.

Two-day back-trajectories from the HYSPLIT Model at Dakar, Kawsara, and Tambacounda (Figure 14) illustrate the origins of the air mass at various altitudes over western Senegal on 11 September. The air mass at 3.0 km in altitude in Tambacounda at 1200 UTC 10 September originated out of the Sahara; by contrast, the origin of the air mass at 0.5 km and 1.5 km originates over the Eastern Atlantic basin and northern Senegal, respectively (Figure 14(a)). Dakar and Kawsara exhibit airflow from the Sahara at both 1.5 km and 3.0 km at 0000 UTC 11 September (Figure 14). This deeper airflow out of the Sahara explains the greater vertical extent of the SAL and stronger temperature inversion at its base over Dakar and Kawsara as compared to Tambacounda. These conditions also contributed to enhanced atmospheric stability, which lead to weakening of the northern portion of the squall line.

5. Discussion

The SL was generated by an AEW in southeastern Senegal at 0000 UTC on 11 September 2006 (Figure 6). A SAL outbreak was building into the western Senegal and the adjacent Atlantic Ocean waters. The SL approached the coast and was several hundred kilometers ahead of the AEW vortex by 0900 UTC. The southern two-thirds of the SL maintained its strength. This is noted by its NPOL reflectivity and the robust lightning flashes observed by WWLLN south of 14°N. The northern third weakened in the presence of abundant dust noted by the lack of lightning north of 14°N (Figure 10), the Kawsara and Dakar radiosondes (Figures 11(a) and 11(b), 12(a) and 12(b)), increasingly high AOT at Mbour (Figure 13), no recorded rainfall by the AMMA22 and AMMA23 rain gauges (not shown), and precipitation-free boundary in Kawsara during the 0900 to 1200 UTC hour. Moreover, HYSPLIT modeling (Figure 14) provides

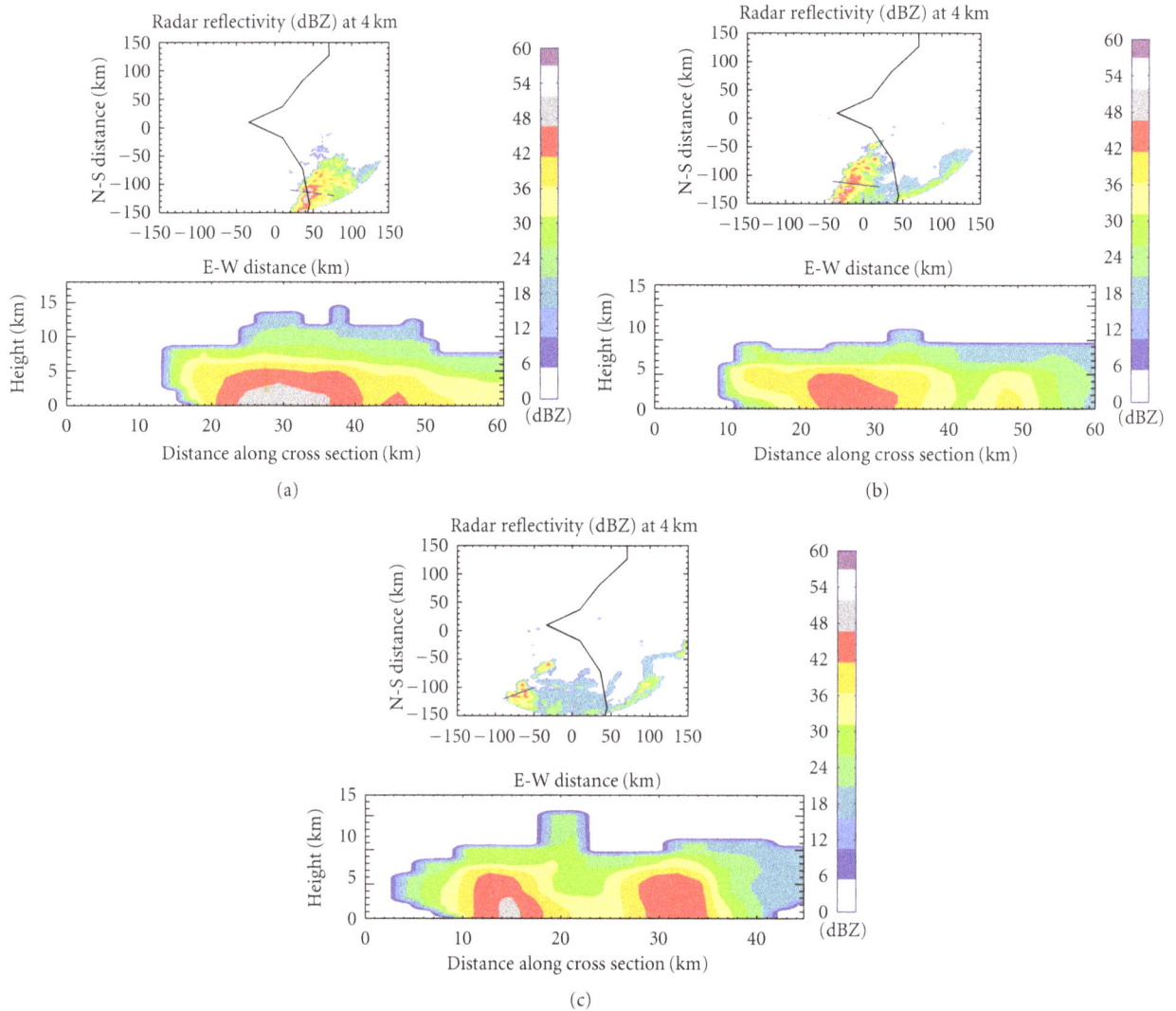

FIGURE 9: (a) A horizontal constant altitude plan position indicator (CAPPI) image at 4.0 km altitude and radar height indicator RHI plots of the NPOL radar reflectivity (dBZ) along the leading convective line on 11 September 2006 at 0900 UTC. The negative sign in the CAPPI images indicates distance in the westward direction. The dashed line in the CAPPI image represents the location of the cross section. In the RHI images, ordinate axis is height z in km and the azimuthal is the distance in km along the cross-section. (b) A horizontal constant altitude plan position indicator (CAPPI) image at 4.0 km altitude and radar height indicator RHI plots of the NPOL radar reflectivity (dBZ) along the leading convective line on 11 September 2006 at 1000 UTC. The negative sign in the CAPPI images indicates distance in the westward direction. The dashed line in the CAPPI image represents the location of the cross section. In the RHI images, ordinate axis is height z in km and the azimuthal is the distance in km along the cross-section. (c) A horizontal constant altitude plan position indicator (CAPPI) image at 4.0 km altitude and radar height indicator RHI plots of the NPOL radar reflectivity (dBZ) along the leading convective line on 11 September 2006 at 1100 UTC. The negative sign in the CAPPI images indicates distance in the westward direction. The dashed line in the CAPPI image represents the location of the cross section. In the RHI images, ordinate axis is height z in km and the azimuthal is the distance in km along the cross-section.

evidence that particles in coastal Senegal at 0800 to 1000 UTC originated from the Sahara implying that Saharan dust was present in this region.

The reflectivity values in Figure 8 indicate that the southern part of the SL weakened as it propagated off the coast, particularly by 1100 UTC; however, the reflectivity is the only observation that reveals the southern portion of the SL weakens. This is similar to the transition the 31 August SL in DeLonge et al. [19]. No solid conclusion can be made on the SL and AEW transition to Hurricane Helene. There are not aircraft observations or TRMM overpasses over this SL on 11 September to make the connection. More aircraft- and surface-based measurements are needed to tie this AEW and SL to other coastally transitioning storms. The next set of aircraft observations are not until 12 September 2006. At this time, the AEW develops into Hurricane Helene by then (Jenkins et al. [23] and Zawislak and Zipser [15]).

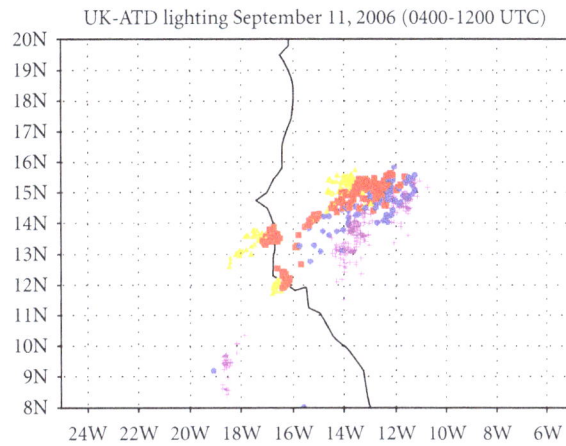

FIGURE 10: An image of cloud-to-ground lightning flashes on 11 September 2006 between 0400 and 1200 UTC. Purple crosses are for CG flashes from 0400 to 0600 UTC; blue circles are for flashes from 0600 to 0800 UTC; red squares are for flashes from 0800 to 1000 UTC; yellow triangles are for flashes from 1000 to 1200 UTC.

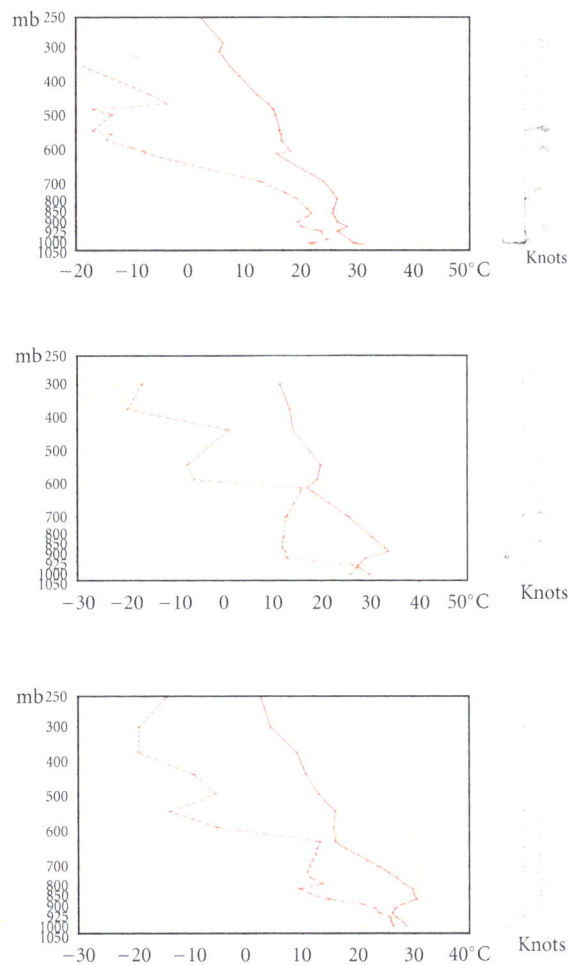

FIGURE 11: Rawindsonde of temperature and dewpoint and wind profiles (kts) for Tambacounda at 1200 UTC on 10 September, Dakar at 0000 UTC on 11 September, and Kawsara at 0530 UTC on 11 September 2006. The solid line is temperature, the dashed line is dewpoint, and both are given in degrees C. The soundings are plotted every 50 hPa.

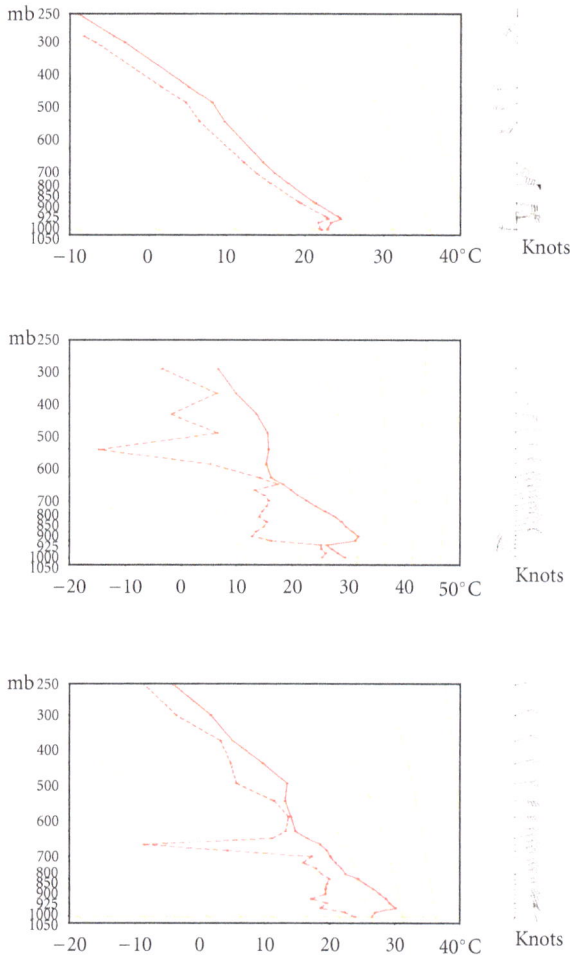

FIGURE 12: Same as Figure 9 but at Tambacounda at 1200 UTC, Dakar at 1200 UTC, and Kawsara at 1402 UTC on 11 September 2006.

6. Conclusion

The NAMMA-06 campaign provided measurements for coastal Senegal. These observations provided insight into the dynamics and thermodynamics of the SL and SAL. The ECMWF Reanalyses show that the AEW and SL were in a favorable environment for continued strengthening and maturing. These observations included ground-based measurements such as flux tower measurements, radiosonde launches, and AERONET. There was remote sensing data from the NPOL radar, TRMM MERG, and MODIS satellite products. These observations revealed that the westerly propagating SL and the SAL in the lower troposphere interacted. The interaction was evident through the weakening of the northern portion of the SL. The HYSPLIT back-trajectory model gives indication of impending dust. Even though CAPE was high and an SL-related boundary passed through Kawsara, there was not precipitation on its northern end.

There are avenues for additional work on this paper. Additional plausible ideas for future research include aircraft observations that might identify the role SAL aerosols have

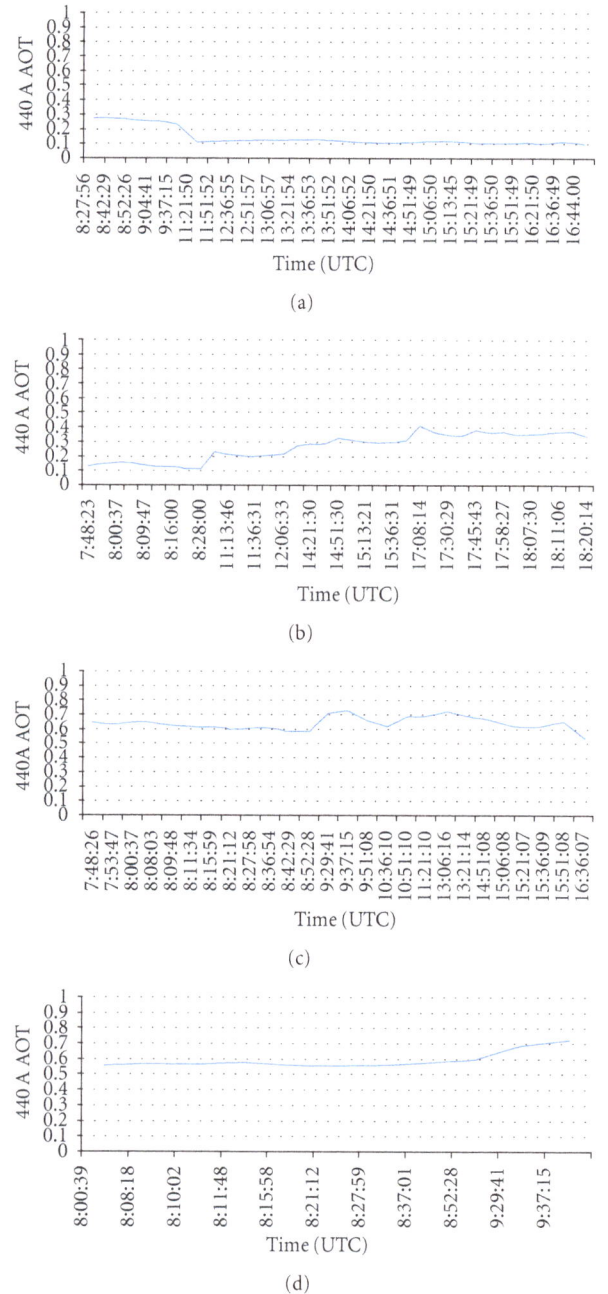

FIGURE 13: (a) A time series plot of 440 Angstrom AOT for 8 September 2006 at Mbour, Senegal. (b) A time series plot of 440 Angstrom AOT for 9 September 2006 at Mbour, Senegal. (c) A time series plot of 440 Angstrom AOT for 10 September 2006 at Mbour, Senegal. (d) A time series plot of 440 Angstrom AOT for 11 September 2006 at Mbour, Senegal.

on clouds, precipitation, and developing tropical systems. Despite there being observations and arguments to support the negative interaction of the two (e.g., Wong and Dessler [30]), there is evidence that dust and aerosol entrainment into convective systems acts as cloud condensation nuclei subsequently forming precipitation and banding. Wong and Dessler [31] concluded that Saharan dust increased the number of cumulus cloud particles in the North Atlantic

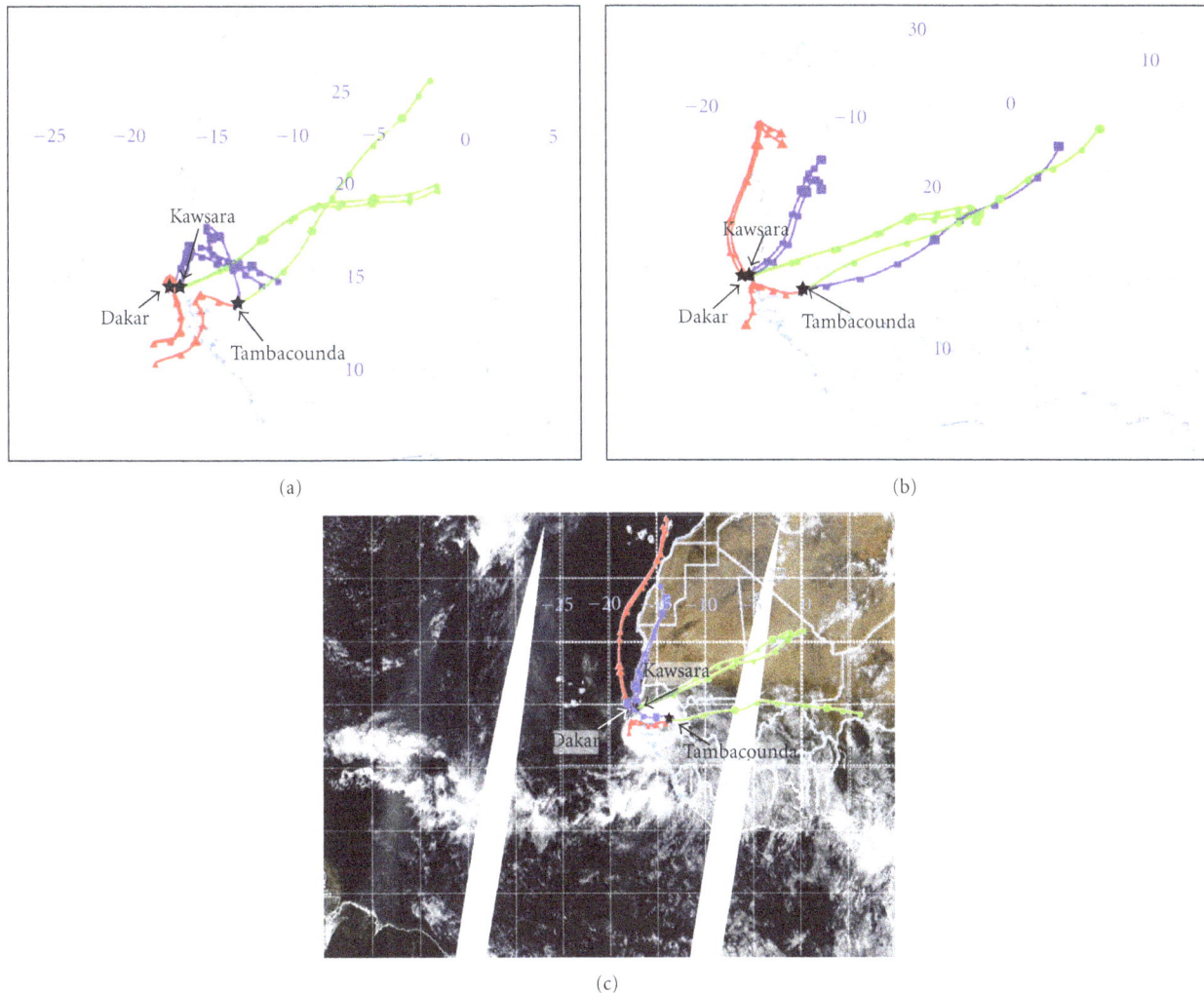

(a)

(b)

(c)

FIGURE 14: (a) The 48-hour HYSPLIT back-trajectory valid for 1200 UTC 10 September 2006. (b) The 48-hour HYSPLIT back-trajectory valid for 0000 UTC 11 September 2006. (c) The 1150 UTC Terra MODIS Visible Composite Satellite and 48-hour HYSPLIT back-trajectories valid 1200 UTC, 11 September 2006. Back-trajectories are for air parcels at 0.5 km AGL (red), 1.5 km AGL (blue), and 3.0 km AGL (green).

in the summer of 2006. Rainbands of 2006 Tropical Storm Debby were invigorated by dust (Jenkins et al. [23]). The precipitation-free gust front that passed through Kawsara is an interesting phenomena to explore.

Exploring these topics include high-resolution mesoscale model simulations of AEWs, SLs, and SAL outbreaks, which combine cloud microphysics, aerosols, and precipitation. Sensitivity tests of microphysical, convective, planetary boundary layer, and aerosol schemes could be done for the Western and Central African continent and the Eastern Atlantic Ocean. Answers to these questions would provide rich insight into SL, AEW, and SAL formation, characteristics, and interactions and to how crucial they are to forecasting in the Eastern Tropical Atlantic Ocean.

Acknowledgments

The authors thank G. S. Jenkins, Ph. D., Professor in the Howard University Department of Physics and Astronomy and the Program in Atmospheric Sciences for constructive dialogue and suggestions on this paper. They thank B. Emory of Goddard Space Flight Center for his critique of the text. The Dakar AERONET data was provided by D. Tanre. This study was funded by National Science Foundation (NSF) Grant OISE-0553959 and NASA Grant no. NNX06AC78G.

References

[1] S. W. Payne and M. M. McGarry, "The relationship of satellite inferred convective activity to easterly waves over West Africa and the adjacent ocean during phase III of GATE," *Monthly Weather Review*, vol. 105, pp. 413–420, 1977.

[2] A. Gaye, A. Viltard, and P. de Félice, "Squall lines and rainfall over Western Africa during summer 1986 and 87," *Meteorology and Atmospheric Physics*, vol. 90, no. 3-4, pp. 215–224, 2005.

[3] R. A. Houze, "Structure and dynamics of a tropical squall line system," *Structure and Dynamics of a Tropical Squall Line System*, vol. 105, pp. 1540–1567, 1977.

[4] C. Thorncroft and K. Hodges, "African easterly wave variability and its relationship to Atlantic tropical cyclone activity," *Journal of Climate*, vol. 14, no. 6, pp. 1166–1179, 2001.

[5] D. P. Rowell and J. R. Milford, "On the generation of African squall lines," *Journal of Climate*, vol. 6, no. 6, pp. 1181–1193, 1993.

[6] E. J. Zipser, "Mesoscale and convective-scale downdrafts as distinct components of SL circulation," *Monthly Weather Review*, vol. 105, pp. 1568–1589, 1977.

[7] M. Chong, P. Amayenc, G. Scialom, and J. Testud, "A tropical squall line observed during the COPT 81 experiment in West Africa: part 1: kinematic structure inferred from dual-Doppler radar data," *Monthly Weather Review*, vol. 115, no. 3, pp. 670–694, 1987.

[8] J. P. Chalon, G. Jaubert, F. Roux, and J. P. Lafore, "The West African squall line observed on 23 June 1981 during COPT 81: mesoscale structure and transports," *Journal of the Atmospheric Sciences*, vol. 45, no. 19, pp. 2744–2763, 1988.

[9] F. Roux, "The West African squall line observed of 23 June 1981 during COPT 81: kinematics and thermodynamics of the convective region," *Journal of the Atmospheric Sciences*, vol. 45, no. 3, pp. 406–426, 1988.

[10] J. Schwendike and S. C. Jones, "Convection in an African easterly wave over West Africa and the eastern Atlantic: a model case study of Helene (2006)," *Quarterly Journal of the Royal Meteorological Society*, vol. 136, no. 1, pp. 364–396, 2010.

[11] T. N. Carlson, "Synoptic histories of three African disturbances the developed into Atlantic hurricanes," *Monthly Weather Review*, vol. 97, pp. 256–276, 1969.

[12] T. N. Carlson, "Some remarks on African disturbances and their progress over the tropical Atlantic," *Monthly Weather Review*, vol. 97, pp. 716–726, 1969.

[13] R. W. Burpee, "The origin and structure of easterly waves in the lower troposphere of North Africa.," *Journal of the Atmospheric Sciences*, vol. 29, 1972.

[14] R. J. Reed, D. C. Norquist, and E. E. Recker, "The structure and properties of African wave disturbances as observed during Phase III of GATE," *Monthly Weather Review*, vol. 105, pp. 317–333, 1977.

[15] J. Zawislak and E. J. Zipser, "Observations of seven African easterly waves in the east Atlantic during 2006," *Journal of the Atmospheric Sciences*, vol. 67, no. 1, pp. 26–43, 2010.

[16] E. J. Zipser, C. H. Twohy, S.-C. Tsay et al., "The Saharan air layer and the fate of African Easterly Waves: NASA's AMMA 2006 field program to study tropical cyclogenesis: NAMMA," *Bulletin of the American Meteorological Society*, vol. 8, pp. 1137–1156, 2009.

[17] P. Kållberg, A. Simmons, S. Uppala, and M. Fuentes, "The ERA-40 archive," Tech. Rep., ECMWF, 2007, http://www.ecmwf.int/publications/library/ecpublications/_pdf/era/era40/ERA40_PRS17_rev1.pdf.

[18] J. E. Janowiak, R. J. Joyce, and Y. Yarosh, "A real-time global half-hourly pixel-resolution infrared dataset and its applications," *Bulletin of the American Meteorological Society*, vol. 82, no. 2, pp. 205–217, 2001.

[19] M. S. DeLonge, J. D. Fuentes, S. Chan et al., "Attributes of mesoscale convective systems at the land-ocean transition in Senegal during NASA African Monsoon Multidisciplinary Analyses 2006," *Journal of Geophysical Research D*, vol. 115, no. 10, Article ID D10213, 2010.

[20] R. A. Houze, S. A. Rutledge, M. I. Biggerstaff, and B. F. Smull, "Interpretation of Doppler weather radar displays of mid-latitude mesoscale convective systems," *Bulletin—American Meteorological Society*, vol. 70, no. 6, pp. 608–619, 1989.

[21] J. Nash, "Progress in introducing new technology sites for the Met Office long range lightning detection system," Tech. Rep. 2.9, Paper 2.9 WMO, WMO/TD-No.1265.

[22] S. Keogh, E. Hibbett, J. Nash, and J. Eyre, "The Met Office Arrival Time Difference (ATD) system for thunderstorm detection and lightning location," Tech. Rep. 488, Met Office, Numerical Weather Prediction: Forecasting Research, 2006.

[23] G. Jenkins, P. Kucera, E. Joseph et al., "Coastal observations of weather features in Senegal during the African monsoon multidisciplinary analysis special observing period 3," *Journal of Geophysical Research D*, vol. 115, no. 18, Article ID D18108, 2010.

[24] V. M. Karyampudi and T. N. Carlson, "Analysis and numerical simulations of the Saharan air layer and its effect on easterly wave disturbances," *Journal of the Atmospheric Sciences*, vol. 45, no. 21, pp. 3102–3136, 1988.

[25] B. Marticorena and G. Bergametti, "Two-year simulations of seasonal and interannual changes of the Saharan dust emissions," *Geophysical Research Letters*, vol. 23, no. 15, pp. 1921–1924, 1996.

[26] I. Koren and Y. J. Kaufman, "Direct wind measurements of Saharan dust events from Terra and Aqua satellites," *Geophysical Research Letters*, vol. 31, no. 6, pp. L06122–4, 2004.

[27] C. H. Twohy, S. M. Kreidenweis, T. Eidhammer et al., "Saharan dust particles nucleate droplets in eastern Atlantic clouds," *Geophysical Research Letters*, vol. 36, no. 1, Article ID L01807, 2009.

[28] T. N. Carlson and J. M. Prospero, "The large-scale movement of Saharan air outbreaks over Northern Equatorial Atlantic," *Journal of Applied Meteorology*, vol. 11, pp. 283–297, 1972.

[29] J. P. Dunion and C. S. Velden, "The impact of the Saharan Air Layer on Atlantic tropical cyclone activity," *Bulletin of the American Meteorological Society*, vol. 85, no. 3, Article ID L12804, pp. 353–365, 2004.

[30] G. S. Jenkins and A. Pratt, "Saharan dust, lightning and tropical cyclones in the eastern tropical Atlantic during NAMMA-06," *Geophysical Research Letters*, vol. 35, no. 12, Article ID L12804, 2008.

[31] S. Wong and A. E. Dessler, "Suppression of deep convection over the tropical North Atlantic by the Saharan Air Layer," *Geophysical Research Letters*, vol. 32, no. 9, Article ID L09808, 4 pages, 2005.

10

Preliminary Results of Marine Electromagnetic Sounding with a Powerful, Remote Source in Kola Bay off the Barents Sea

Valery Grigoryev,[1] **Sergey Korotaev,**[2,3] **Mikhail Kruglyakov,**[3,4] **Darya Orekhova,**[3] **Yury Scshors,**[3] **Evgeniy Tereshchenko,**[1] **Pavel Tereshchenko,**[5] **and Igor Trofimov**[2]

[1] *Kola Science Centre, Polar Geophysical Institute, Russian Academy of Science, Murmansk,*
15 Khalturina Street, Murmansk 183010, Russia
[2] *Geoelectromagnetic Research Centre of Schmidt Institute of Physics of the Earth, Russian Academy of Sciences,*
P.O. Box 30, Troitsk, Moscow Region 142190, Russia
[3] *Kurchatov Institute, Moscow, 1 Akademika Kurchatova Square, Moscow 123182, Russia*
[4] *Moscow State University, Moscow, GSP-1 Leninskie Gory, Moscow 119991, Russia*
[5] *Institute of Terrestrial Magnetism, Ionosphere, and Radiowave Propagation, Russian Academy of Sciences,*
St. Petersburg Branch, 1 Mendeleevskaya Linia, St. Petersburg 199034, Russia

Correspondence should be addressed to Darya Orekhova; ordaal@gmail.com

Academic Editor: Michael S. Zhdanov

We present an experiment conducted in Kola Bay off the Barents Sea in which new, six-component electromagnetic seafloor receivers were tested. Signals from a powerful, remote super-long wave (SLW) transmitter at several frequencies on the order of tens Hz were recorded at the six sites along a profile across Kola Bay. In spite of the fact that, for technical reasons, not all the components were successfully recorded at every site, the quality of the experimental data was quite satisfactory. The experiment resulted in the successful simulation of an electromagnetic field by the integral equation method. An initial geoelectric model reflecting the main features of the regional geology produced field values that differed greatly from the experimental ones. However, step-by-step modification of the original model considerably improved the fit of the fields. Thereby specific features of the regional geology, in particular the fault tectonics, were able to be corrected. These preliminary results open the possibility of inverse problem solving with more reliable geological conclusions.

1. Introduction

Recently marine electromagnetic methods have become a valuable tool for seafloor mapping (e.g., [1, 2]). Among them, methods based on the natural (magnetotelluric) field offer advantages over those based on controlled source electromagnetic methods (CSEM) for lower-crust and upper-mantle mapping, while the latter offer advantages for upper-crust mapping. Although in general terms the audiomagnetotelluric method has been successfully applied to upper-crust survey too, such is not the case for the auroral zone, where the necessary plane-wave condition is roughly violated. A number of types of seafloor electromagnetic receivers measuring four (horizontal electric and magnetic) or five

(the same plus vertical magnetic) components of the field, with various types of control sources dropped into the sea have been implemented ([3–8]). However, it is known that measurement of the sixth component (vertical electric) can be of much interest, because it is particularly sensitive to the insulating crust structures ([9]). On the other hand, it can be advantageous to use the powerful stationary control source for land mapping (e.g., [10]) and it is hoped that employment of such a powerful land-based source will also be beneficial in shelf mapping.

In this paper, we present the experiment that made this possibility a reality. The first six-component electromagnetic seafloor receivers have been tested in Kola Bay off the Barents Sea. Signals from a distant, powerful super-long wave (SLW) transmitter were recorded at six sites along a profile across the bay. The results have been preliminarily interpreted by simulation of the electromagnetic field by the integral equation method.

2. Experiment

This pioneer experiment on seafloor measurements of the electromagnetic field emitted by the powerful land-based SLW transmitter was performed in September 2011. The source represented a grounded horizontal line current (electric bipole) approximately 60 km long located on the Kola Peninsula and oriented along the latitude parallel. The field was emitted at frequencies of 41 Hz, 62 Hz, 82 Hz, and 144 Hz, by 200 A sinusoidal current every fifteen minutes and was measured by the six-component receivers (Figure 1) sited on the floor of the bay. The amplitude of the antenna current was recorded with reference to GPS time.

The receivers were equipped with three orthogonal induction magnetic field sensors with low-noise amplifiers and with three orthogonal electric field sensors. The measured analog signals were received by the six-channel, 16 bit ADC, where they were converted to digital signals and saved for further analysis on a FLASH memory. Accuracy of the conversion was provided by pass-through calibration of the measurement channels using specialized metrology equipment. The receiver technical characteristics are follows.

Frequency range: 0.01 Hz–200 Hz.

Dynamic range of the measured signals: 72 dB.

The magnetic channels sensitivity in frequency band 0.01 Hz–200 Hz: above 0.5 pT at the signal-noise ratio 3/1.

The intrinsic noise of magnetic channels is

1000 fT at $f = 0.1$ Hz,
100 fT at $f = 10$ Hz,
100 fT at $f = 100$ Hz.

The electric channels sensitivity in frequency band 0.01 Hz–200 Hz: above 10 nV/m at the signal-noise ratio 3/1.

Operating depth-up to 500 m.

Before starting and stopping the recording, the verification of high-precision temperature-stabilized receiver clock with GPS time was produced. This ensured data synchronization with the correct time. During each measurement session values of receiver azimuth, roll, and pitch were fixed on a FLASH memory every minute by the receiver orientation unit. The coordinates of the station dive point were fixed by the navigation system of the boat. After measurements the station lifting was performed by the command from the ship through an acoustic channel. The reading of data is carried out after lifting station on the boat. So the antenna current data and the values of the field on the seabed were synchronous.

FIGURE 1: Six-component seafloor receiver before deployment.

The signal was separated from natural and man-made noise using the spectral analysis of measurement results from Welch's method [11]. As a result, the amplitudes of the six components of the field as well as the phase differences between them were gleaned. The signal-noise ratio for all the components exceeded 30 dB.

As the receiver orientation was random after descent on the floor, the gleaned data needed to be converted into a single-axis coordinate system, taking into account the values of pitch, roll, and yaw of the receiver units.

The new amplitudes of the orthogonal components from the measured ones, A_i and A_k, with phase difference ϕ after turning in one plane with angle ψ, are as follows:

$$\widetilde{A}_i = \sqrt{A_i^2\cos^2\psi - 2A_iA_k \sin\psi \cos\psi \cos\varphi + A_k^2\sin^2\psi},$$
$$\widetilde{A}_k = \sqrt{A_k^2\cos^2\psi + 2A_iA_k \sin\psi \cos\psi \cos\varphi + A_i^2\sin^2\psi}. \quad (1)$$

The new phase difference between them is

$$\widetilde{\varphi} = \varphi + \arcsin\left(\frac{A_k \sin\psi \sin\varphi}{\widetilde{A}_i}\right) + \arcsin\left(\frac{A_i \sin\psi \sin\varphi}{\widetilde{A}_k}\right). \quad (2)$$

To present the measurement results, the Cartesian axis system was used with the x-axis oriented along the geographical meridian facing North and the y-axis oriented along the latitude facing East. The z-axis was oriented vertically, facing up. Calculations (1)–(2) were consequently applied in three planes, taking into account the magnetic declination $14° 56'$ E in the area of observations when turning in the XY plane. As a result, we assembled the amplitude values of the field components and the phase differences in a single geographic axis system.

The observation sites were located along two sides of the waterway at a depth of 36 to 85 meters—four on the southern side of the waterway and two on the northern side

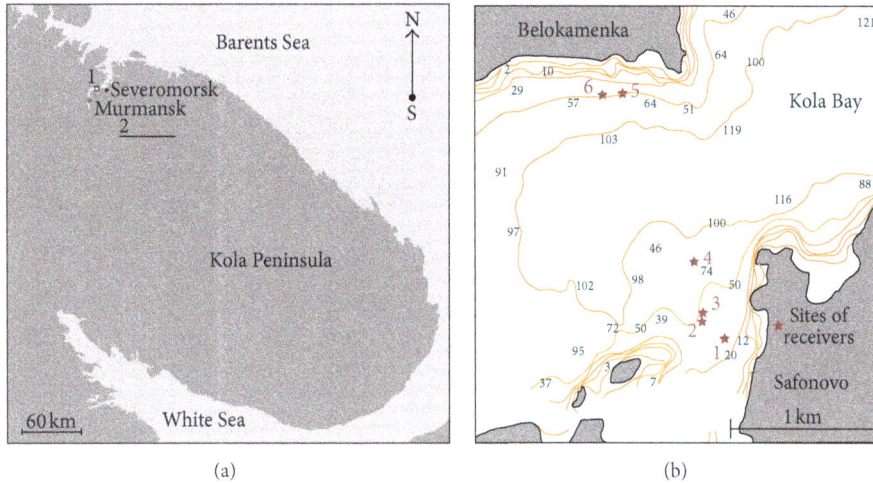

FIGURE 2: Region of electromagnetic sounding. (a) Area of observations, 1, and the source bipole, 2. (b) Observation sites.

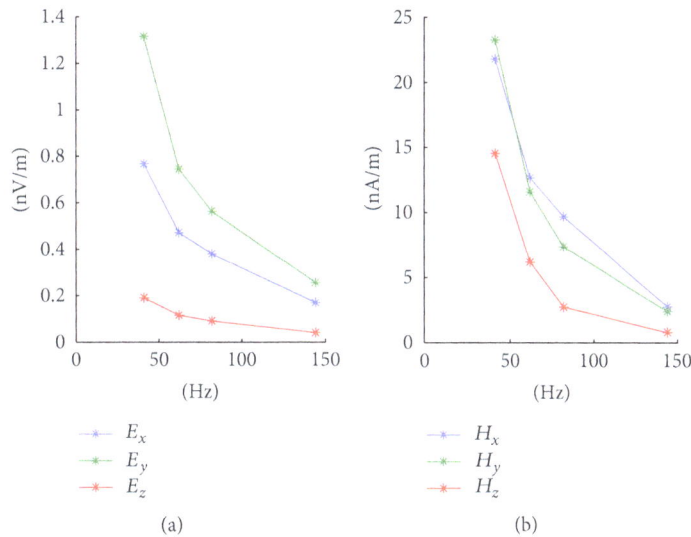

FIGURE 3: Amplitudes for current reduced to 1 A in the source: (a) electric, V/m and (b) magnetic, A/m components, plotted against frequency, Hz.

(Figure 2) because of the navigation peculiarities in Kola Bay. The receivers were deployed on the seafloor at least thirty minutes before transmission started. The receivers buoyed to the sea surface thirty minutes after transmission stopped.

All of the horizontal and vertical components of the electromagnetic field at all the frequencies were detected at all the observation sites. As a result, data for the magnetic components were obtained at every site except site 3, where the receiver unit had not been fixed on the bay slope and changed its orientation during the session, resulting in noise from the induction sensors. In view of the unreliability of some electric component measurements at sites 2, 5, and 6, electric field data for these sites could not be transformed into a geographical axis system. However, data for single components of the field can be used for interpretation.

Figure 3 presents an example of the plot of the reduced electric (a) and magnetic (b) fields at different frequencies at site 1. The frequency dependence is quite natural.

To take into account the possible space variability of seawater conductivity, we also carried out its vertical and horizontal profiling with an oceanic probe. The conductivity proved to be the same at all the sites of measurement and independent of the depth except for a thin surface layer. It was equal to 3.5 S/m.

3. The Main Features of Regional Geology and the Initial Geoelectric Model

Kola Bay is a submeridional fjord situated on the coast of the Barents Sea in the northwestern part of the Kola

FIGURE 4: A fragment of the geological map of the Kola Peninsula by Mitrofanov [18]. 92-leucogranites, granodiorites; 102-metavolcanics; 104-granodiorites, tonalites, plagiogranites; 106-micaceous, garnet-mica gneisses. Red line: the observation profile.

Peninsula. Geological and geoelectric description of the area and the fjord itself is given in many works ([12–17]). It passes through two regions: Murmansk Craton and Kola-Norwegian Province, in area of distribution of resistive Archean granitoids and gneisses (a fragment of the geological map by Mitrofanov [18] is given in Figure 4). The average thickness of quaternary sediments (generally glacial and marine) is 50–100 m and reaches 200 m at the entrance of the fjord. In accordance with the cranked curves of the fjord, it is divided into three parts: northern, middle, and southern bends. The northern part is the deepest with depths up to 300 m; in the middle part, where the observation profile is, the depths vary from 35 to 130 m. Kola Bay is also complicated by underwater rapids and branches in the form of bays. The formation of such a complex structure is determined by a system of northwest- and northeast-trending faults, glacial exaration, and irregular postglacial uplift.

On the basis of the publications cited above, the initial geoelectric model was constructed (Figure 5). The background of the model is a two-layered resistive medium: the conductivity of the upper layer is 10^{-4} S/m with a thickness of 2 km; the conductivity of the lower half-infinite layer is 10^{-5} S/m. The conductivity of all the faults is the same, 1 S/m (red in Figure 5); the water conductivity is 3.2 S/m (brown in Figure 5); the sediment conductivity is 1 S/m. The widths of the faults change from 2 to 4 km. The sediments follow the curve of the fjord banks, and their thickness changes from 50 to 200 m at the entrance of the fjord.

4. Method of Simulation

The simulation was based on the 3D integral equation method. The use of this method in electromagnetic geophysics problems is founded on the following model of the complex conductivity distribution. The whole space is

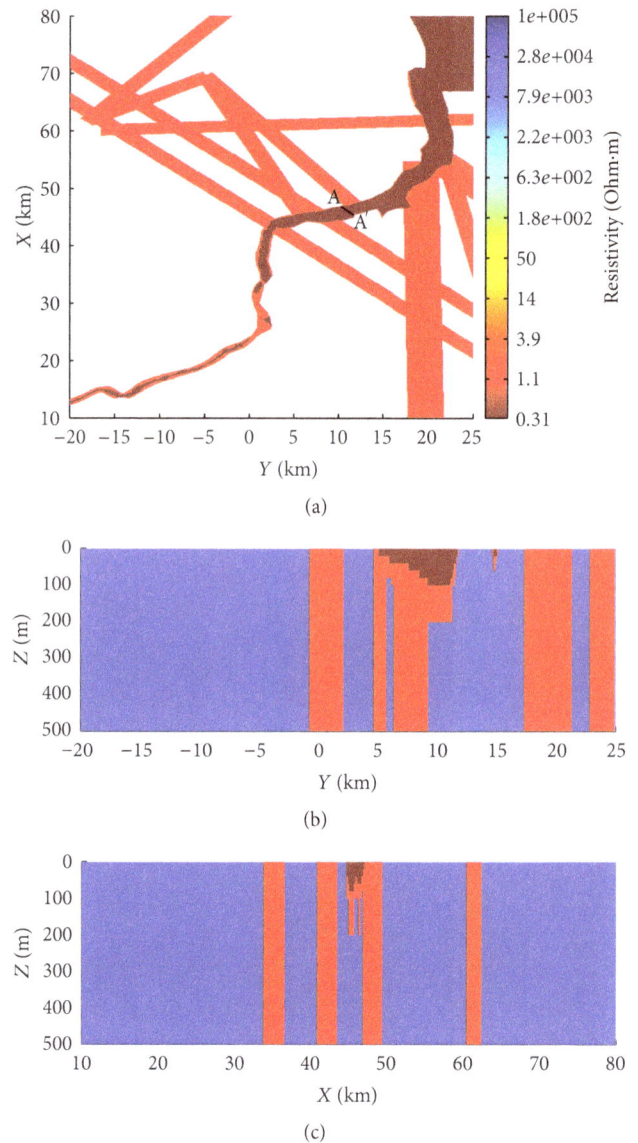

FIGURE 5: The initial geoelectric model, (a) XY section at $z = 0$ m, (b) YZ section at $x = 45.36$ km, and (c) XZ section at $y = 11.19$ km. The line AA' represents the profile of observations corresponding to the sites from 5 to 1 in Figure 2(b).

divided into two parts: the normal section and the anomaly. The normal section (background) means the set of unlimited horizontal homogenous layers and two homogeneous half-spaces, where the complex conductivity is constant in each layer and half-space. The conductivity of the upper half-space, which simulates the air, is zero. The anomaly is the 3D volume, where the conductivity is different from that in the normal section. Notice that if we use the integral equation method, we have to specify the conductivity in the whole space. Magnetic permeability is constant μ_0 in the whole space.

If conductivity σ_n of the normal section is defined, the electric and magnetic Green's tensors G_E and G_H can be

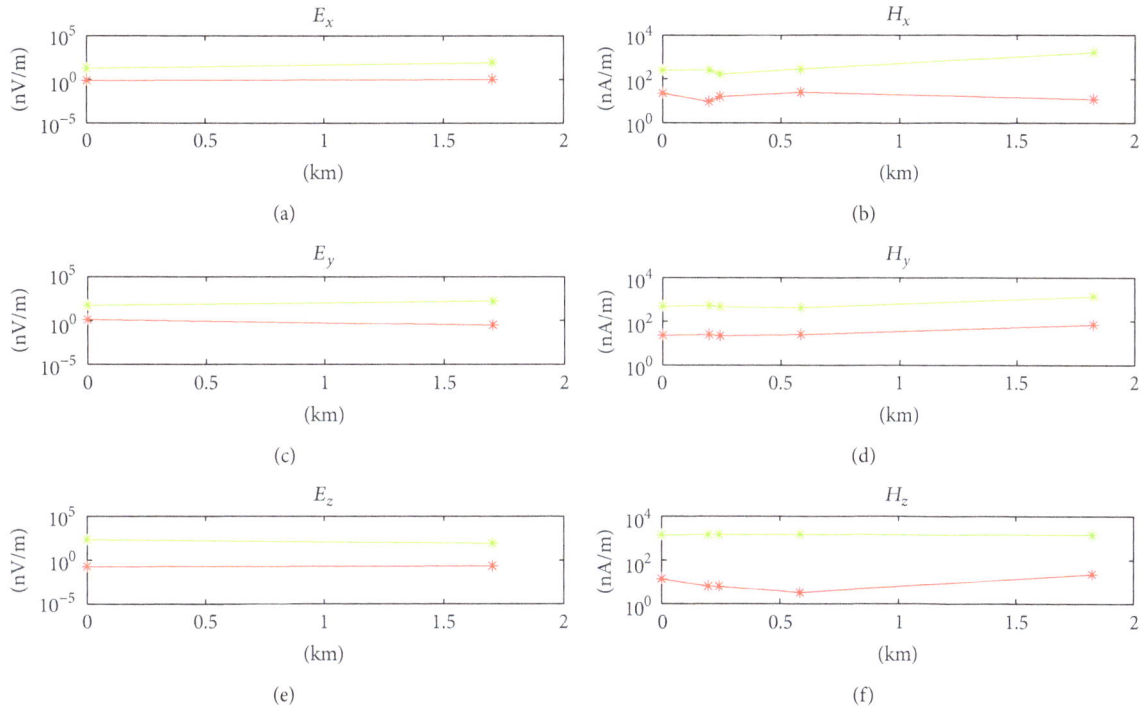

FIGURE 6: Amplitudes of the field components (reduced to 1 A in the source) for the initial geoelectric model at 41 Hz (green lines) as compared with the observed amplitudes (red lines).

computed. This way the following expressions for electrical E and magnetic H fields may be written ([19]):

$$
\begin{aligned}
E(M) &= i\omega\mu_0 \iiint_V \left(\sigma_a(M_0) - \sigma_n(M_0)\right) \\
&\quad \times G_E(M, M_0) E(M_0) dV_{M_0} + E_0(M), \\
H(M) &= i\omega\mu_0 \iiint_V \left(\sigma_a(M_0) - \sigma_n(M_0)\right) \\
&\quad \times G_H(M, M_0) E(M_0) dV_{M_0} + H_0(M),
\end{aligned}
\tag{3}
$$

where ω is the circular frequency, V is the anomalous volume, $\sigma_a(M_0)$ is the anomalous conductivity at the point M_0, and E_0 and H_0 are the primary electrical and magnetic fields. The primary field means the field, which is induced at this point in the normal section by the same source.

If M in the first expression in (3) is inside V, then this expression becomes a 3D singular integral equation of the second kind:

$$
\begin{aligned}
E(M) - i\omega\mu_0 \iiint_V \left(\sigma_a(M_0) - \sigma_n(M_0)\right) G_E(M, M_0) \\
\times E(M_0) dV_{M_0} = E_0(M).
\end{aligned}
\tag{4}
$$

If (4) is solved, its solution E is placed under the integral in (3). Then it becomes possible to compute the electrical and magnetic fields at any point in the space.

The integral equation (4) is solved by the collocation method. The anomalous volume V is divided into the set of

rectangular cells V_k, $k = 1, \ldots, N$. The electrical field E is approximated by the constant E^k in V_k, which is the field in the center of V_k. This way we get the following system of linear equations:

$$
\begin{aligned}
E^n - i\omega\omega_0 \sum_{k-1}^{N} E^k \iiint_{V_k} \left(\sigma_a(M_k) - \sigma_n(M_k)\right) G_E(M, M_0) dV_{M_0} \\
= E_0^n,
\end{aligned}
\tag{5}
$$

where M_k is the center of the V_k, $E_0^k = E_0(M_k)$. The system (5) is solved by the generalized minimal residual method.

We start with the very large anomaly extending more than 100 km horizontally in both directions and 2 km in the vertical direction, like the one which was modeled in our previous study ([10]). The computational experiments show that it is not necessary to use that large anomaly, because all the effects in the resulting field are induced by the local objects. These experiments also show that the effects of the geometry of the Kola Bay bottom are very strong, so it is necessary to use the right model of the bathyorography. This way we have to use cells with a very small size: 20 m in the horizontal direction and 5 m in the vertical one. The resulting model consists of 11 million cells. The computations have been conducted on the high-performance computer Skif-Tchebyshev, located at Moscow State University, using fully parallelized 3D forward modeling software PIE3D from the CEMI consortium of the University of Utah.

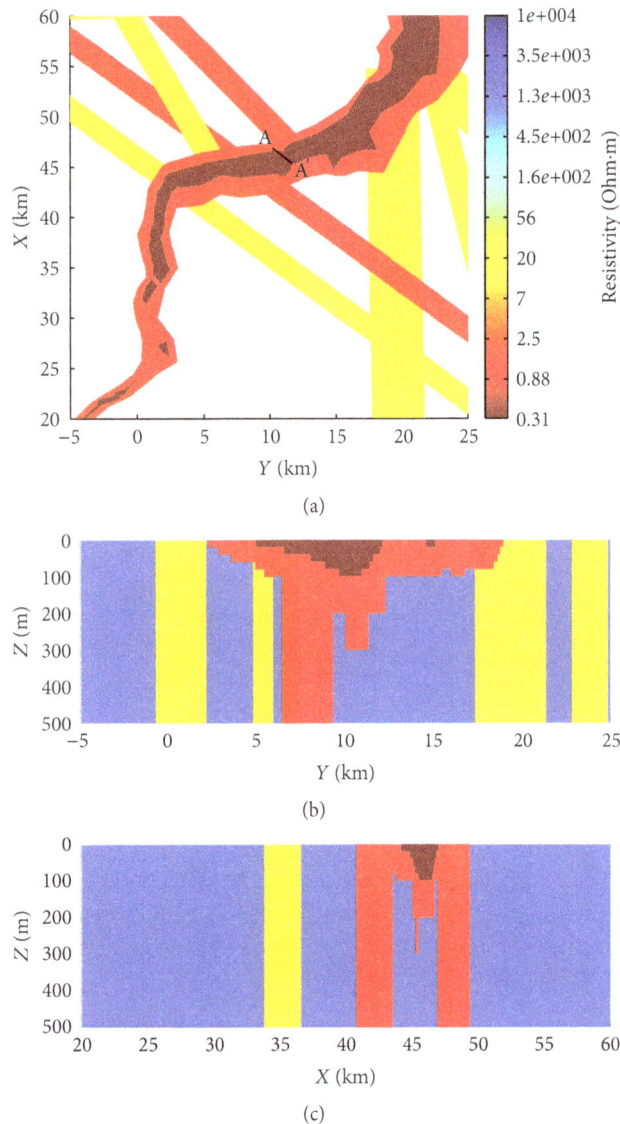

TABLE 1

| | Average relative misfit at 41 Hz | | | | | |
	E_x	E_y	E_z	H_x	H_y	H_z
Initial model	58.8146	156.3671	692.1128	28.6447	24.3338	214.7193
Final model	−0.0266	1.6131	2.8070	0.6004	0.4969	−0.3303

FIGURE 7: Geoelectric model resulting from trial and error, (a) XY section at $z = 0$ m, (b) YZ section at $x = 45.36$ km, and (c) XZ section at $y = 11.19$ km. The line AA′ represents the profile of observations corresponding to the sites from 5 to 1 in Figure 2(b).

5. Preliminary Interpretation

All six components of the field have been computed for the initial geoelectric model (Figure 5). In spite of the fact that there were amplitudes and phases at four frequencies, at the primary stage of investigation we limited ourselves by comparison of the experimental and modeled fields only by the amplitudes and mainly at the lowest frequency. The thing is, phase is too sensitive parameter, which is important at the stage of solving the inverse problem, but which is difficult to employ at the stage of preliminary rough model matching. On the other hand, due to the exposure of an insulating basement around Kola Bay, very thin conductors like bogs or small rivers (which are practically impossible to take into account) strongly influence the field at high frequencies. Therefore at

the preliminary rough model matching stage, we pay the most attention to modeling at the most reliable, lowest frequency (41 Hz). Such an approach follows from the experience of land electromagnetic sounding with the same source ([10]).

When the observed field amplitudes were compared with the computed ones for the initial geoelectric model (Figure 6), it was apparent that they differed widely both by their shape and level (up to three orders of value). Therefore the initial model had to be considerably corrected.

We undertook step-by-step modification of the original model, varying the normal cross section, positions, and conductivity of the faults and geometry of the Kola Bay alluvial belt. Thus, as our original model was rather large and correspondingly every step took considerable time, we also tested the possibility of decreasing the horizontal sizes of the model so that amplitude differences between "large" and "small" models at observation would not exceed about 1%.

We tested a total of about 50 models. One of the best variants is shown in Figures 7 and 8. Of course, the fit of the field components is rather imperfect, but obviously far better than in Figure 6. The average relative misfits for the EM-field amplitudes at 41 Hz are presented in Table 1. At frequencies of 62 and 82 Hz, the fit is only slightly worse than that presented in Figure 8 at 41 Hz and evidently departs from fit only at the highest frequency, 144 Hz.

Thus the model presented in Figure 7 can be a good starting point for strict solving of the inverse problem. Although this model is not final, it allows the inference of some qualitative geophysical and geological conclusions.

  (i) The normal cross section attained an order of magnitude more conductive than it was believed before.

 (ii) Kola Bay is surrounded by a sedimentary alluvial belt. Evidently it is the result of regional postglacial uplift. The sedimentary belt proved to be asymmetric in accordance with normal right displacement of the ancient river.

(iii) Conductivity of the faults proved essentially different. Probably the larger conductivity of a couple of faults seen in Figure 7 is related to contemporary tectonic activity.

6. Conclusions

This pioneer experiment with a new six-component electromagnetic seafloor receiver and with a distant, powerful SLW transmitter was carried out in Kola Bay off the Barents Sea. The receivers were deployed at six sites along a profile across the bay. Not all six components were successfully

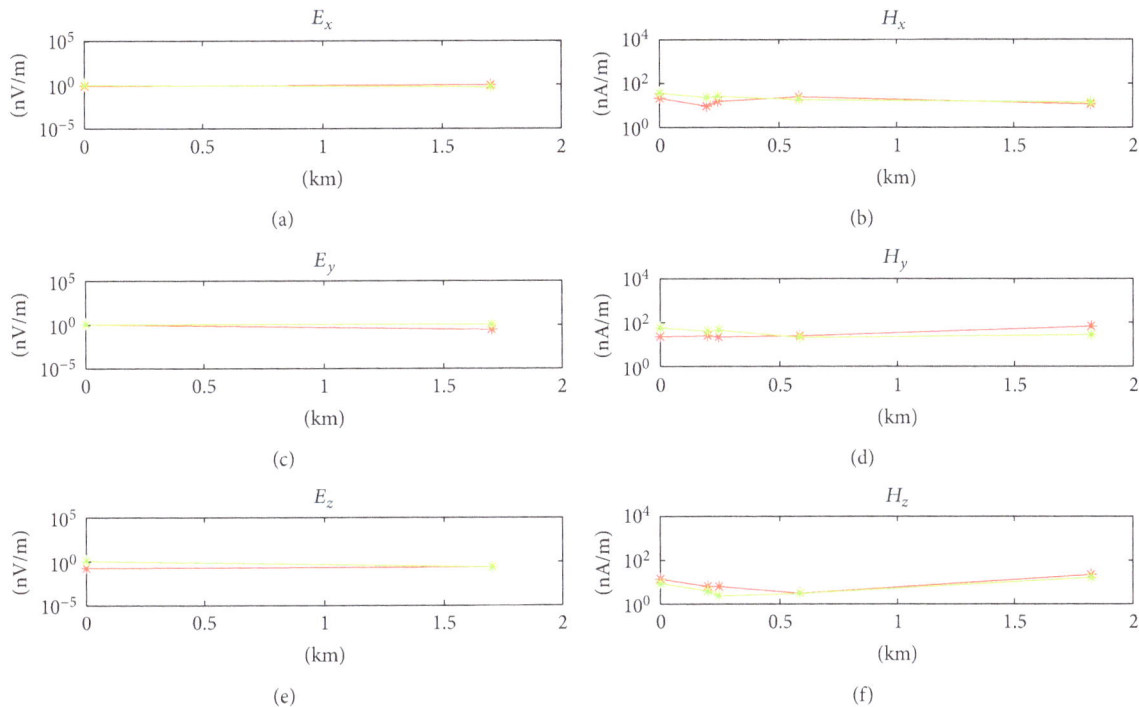

FIGURE 8: Amplitudes of the field components (reduced to 1 A in the source) for geoelectric model resulting from trial and error at 41 Hz (green lines) as compared with the observed amplitudes (red lines).

measured at all the sites in particular, due to the mechanical unreliability of electric antennas. Nevertheless, the quality of the experimental data turned out to be quite satisfactory.

The data have been preliminarily interpreted by a trial-and-error method with simulation of the electromagnetic field by an integral equation method. The resulting geoelectric model differs from the initial one, and this difference reveals new features of regional geology, in particular of fault tectonics. The resulting model is also the appropriate starting point for strict solving of the inverse problem with subsequently more comprehensive geological interpretation.

Acknowledgments

The work was supported by RFBR (Grant 11-05-12015). The authors acknowledge the University of Utah's Consortium for Electromagnetic Modeling and Inversion (CEMI) for providing 3D forward modeling software PIE3D.

References

[1] K. Key, "Marine electromagnetic studies of seafloor resources and tectonics," *Surveys in Geophysics*, vol. 33, no. 1, pp. 135–167, 2011.

[2] K. A. Weitemeyer, S. Constable, and A. M. Tréhu, "A marine electromagnetic survey to detect gas hydrate at Hydrate Ridge, Oregon," *Geophysical Journal International*, vol. 187, pp. 45–62, 2011.

[3] R. L. Evans, M. C. Sinha, S. C. Constable, and M. J. Unsworth, "On the electrical nature of the axial melt zone at 13∘N on the East Pacific Rise," *Journal of Geophysical Research B*, vol. 99, no. 1, pp. 577–588, 1994.

[4] S. Constable and C. S. Cox, "Marine controlled-source electromagnetic sounding 2. The PEGASUS experiment," *Journal of Geophysical Research B*, vol. 101, no. 3, pp. 5519–5530, 1996.

[5] L. MacGregor and M. Sinha, "Use of marine controlled-source electromagnetic sounding for sub-basalt exploration," *Geophysical Prospecting*, vol. 48, no. 6, pp. 1091–1106, 2000.

[6] L. MacGregor, M. Sinha, and S. Constable, "Electrical resistivity structure of the Valu Fa Ridge, Lau Basin, from marine Controlled-Source electromagnetic sounding," *Geophysical Journal International*, vol. 146, no. 1, pp. 217–236, 2001.

[7] S. Ellingsrud, T. Eidesmo, S. Johansen, M. C. Sinha, L. M. MacGregor, and S. Constable, "Remote sensing of hydrocarbon layers by seabed logging (SBL): results from a cruise offshore Angola," *Leading Edge*, vol. 21, no. 10, pp. 972–982, 2002.

[8] M. C. Sinha, P. D. Patel, M. J. Unsworth, T. R. E. Owen, and M. R. G. Maccormack, "An active source electromagnetic sounding system for marine use," *Marine Geophysical Researches*, vol. 12, no. 1-2, pp. 59–68, 1990.

[9] M. N. Berdichevsky, O. N. Zhdanova, and M. S. Zhdanov, *Marine Deep Geoelectrics*, Nauka, Moscow, Russia, 1989.

[10] E. P. Velikhov, V. F. Grigor'v, M. S. Zhdanov et al., "Electromagnetic sounding of the Kola Peninsula with a powerful extremely low frequency source," *Doklady Earth Sciences*, vol. 438, no. 1, pp. 711–716, 2011.

[11] F. J. Harris, "On the use of windows for harmonic analysis with the discrete Fourier transform," *Proceedings of the IEEE*, vol. 66, no. 1, pp. 51–83, 1978.

[12] E. P. Velikhov, Ed., *Geoelectrical Studies with a Powerful Current Source on the Baltic Shield*, Nauka, Moscow, Russia, 1989.

[13] A. A. Kovtun, S. A. Vagin, I. L. Vardanjants, E. L. Kokvina, and N. I. Uspenskiy, "Magnetotelluric study of the structure of the crust and mantle of the eastern part of the Baltic Shield," *Izvestiya—Physics of the Solid Earth*, no. 3, pp. 32–36, 1994.

[14] A. A. Zhamaletdinov, "Graphite in the Earth's crust and electrical conductivity anomalies," *Izvestiya—Physics of the Solid Earth*, vol. 32, no. 4, pp. 272–288, 1996.

[15] L. L. Vanjan and N. I. Pavlenkova, "Layer of low velocity and high electrical conductivity at the base of the upper crust of the Baltic Shield," *Izvestiya—Physics of the Solid Earth*, no. 1, pp. 37–45, 2002.

[16] E. V. Spiridonov, *Paleoseismodislocations on the coast of the Barents Sea [Ph.D. thesis]*, MSU, Moscow, Russia, 2007.

[17] E. A. Kovalchuk and E. V. Shipilov, "The first data about the structure and lithology of the section of the Kola Fjord sediments," in *Proceedings of the International Scientific Conference on the 100th Anniversary of D. G. Panov*, pp. 157–160, SSC Academy of Sciences, Rostov-on-Don, Russia, 2009.

[18] F. P. Mitrofanov, Ed., *Geological Map of the Kola Region*, Scale 1:500000. 2001. Apatity.

[19] M. S. Zhdanov, *Geophysical Inverse Theory and Regularization Problems*, Elsevier, Amsterdam, The Netherlands, 2002.

11

A Prototype System for Time-Lapse Electrical Resistivity Tomographies

Raffaele Luongo,[1,2] **Angela Perrone,**[2] **Sabatino Piscitelli,**[2] **and Vincenzo Lapenna**[2]

[1] *DIFA, Università della Basilicata, Contrado Macchia Romana, 85100 Potenza, Italy*
[2] *IMAA-CNR, Contrada Santa Loja Zona Industriale, Tito Scalo, 85055 Potenza, Italy*

Correspondence should be addressed to Raffaele Luongo, raffaele.luongo@imaa.cnr.it

Academic Editor: Pantelis Soupios

A prototype system for time-lapse acquisition of 2D electrical resistivity tomography (ERT) and time domain reflectometry (TDR) measurements was installed in a test site affected by a landslide in Basilicata region (southern Italy). The aim of the system is to monitor in real-time the rainwater infiltration into the soil and obtain information about the variation of the water content in the first layers of the subsoil and the possible influence of this variation on landslide activity. A rain gauge placed in the test site gives information on the rainfall intensity and frequency and suggests the acquisition time interval. The installed system and the preliminary results are presented in this paper.

1. Introduction

Landslides are complex geological phenomena depending on many factors. In order to study these factors, to understand the triggering mechanisms of the movement, and to monitor its dynamic evolution it is necessary to apply a multidisciplinary approach. The rainwater infiltration into the soil and the increase of pore water pressure in the vadose zone can be considered one of the main causes of shallow landslides triggering. Usually, the standard techniques used to measure the water content of the soil and the water table levels in areas of potential instability are the TDR method and the piezometric measurements, respectively. These techniques, while allowing to obtain direct information of the considered parameter, provide only 1D information. Considering that landslides are volumetric phenomena it is a clear need to experiment new investigation techniques which can provide at least 2D hydrological information. It would be better if this information could be continuous in time.

Recently, the literature reports many examples of application of indirect (geophysical) methods for the study and the estimate of water content in the first layers of the subsoil. Among these, the electrical resistivity tomography (ERT), usually applied to obtain information about the geometrical features of the landslides and estimate the thickness of the slide material [1–8], has been tested to obtain information on the temporal and spatial patterns of water infiltration processes [9–15].

The aim of this work is to present a prototype system planned to obtain time-lapse 2D ERT and TDR measurements in a landslide area located in Basilicata region (southern Italy). The system was planned with the aim to estimate the variation of electrical resistivity and soil moisture values in a long period and to obtain information about the influence of precipitations and seasonal changes on them. Very preliminary results allowed us to verify the functioning and confidence of the system, to decide the acquisition time interval, and to obtain information on the yearly variation of resistivity values.

The system was developed in the frame of MORFEO (Monitoraggio e Rischio da Frana mediante dati EO) project funded by the Italian Agency Space (ASI) and finalized to the activities of the Italian Civil Protection Department (DPC) in the landslide risk management.

2. Test Site Description

The planned monitoring system was installed in a landslide area located in Picerno territory at the western side of

FIGURE 1: Study area location (from Google Earth, modified).

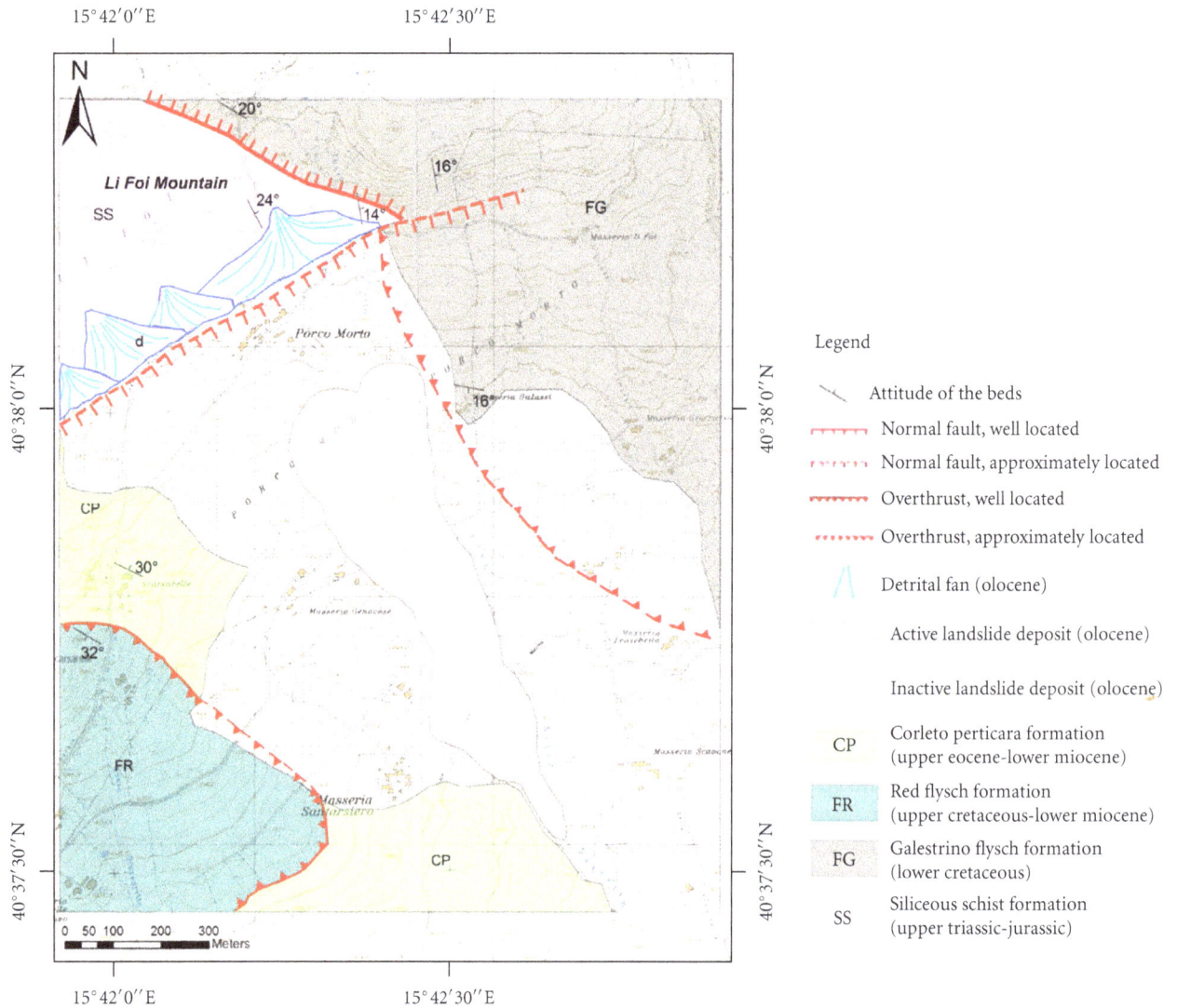

FIGURE 2: Geologic map of the area (from [1]).

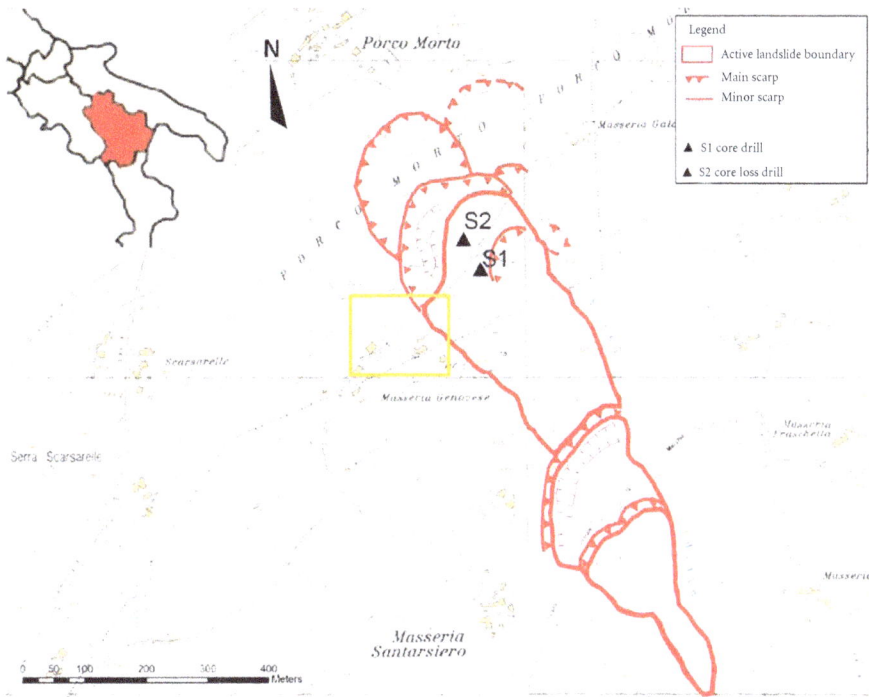

FIGURE 3: Geomorphologic map of the landslide (from [1] modified), yellow square includes the area where the system was installed.

FIGURE 4: Location of the acquisition profile (from Google Earth, modified).

Basilicata region (southern Italy), within the Apennine chain (Figure 1). The area has been frequently involved in reactivation phenomena, the most recent ones occurred on March 2006 after continuous and intense snowfalls.

Terrains affected by landslidesbelong to the Pignola, Abriola facies of the Lagonegro Unit. In particular, there are four main lithological formations outcropping on the slope that in order of sedimentation [16, 17] are (Figure 2):

(i) Siliceous Schist (upper triassic-jurassic), outcropping along the slope of Mount LiFoi;

(ii) Flysch Galestrino (lower cretaceous);

(iii) Flysch Rosso (upper cretaceous-lower miocene);

(iv) Corleto Perticara Formation (upper eocene-lower miocene).

The most recent terrains are characterized by the debris outcropping at the toe of Mount LiFoi.

The Siliceous Schist formation has a thick of 240 meters and is constituted by red and greenish shales with a typical rupture cleavage called "pencil cleavage." In addition, there are red and green jaspers embedded with layers of radiolarites and flint. Lastly, manganese jaspers until they reach the Galestri formation.

The Flysch Galestrino formation is less thick than the Shist formation (about 200 m) and consists of alternating black claystones and siliceous marls, calcilutites and marly limestones, marls, and leaf clay.

The Flysch Rosso formation made of jasper, siliceous claystones, calcarenites, red and green marls is in eteropic succession with Argille Varicolori and Corleto Perticara formations [18].

(a)   (b)

Figure 5: Syscal Pro- and Mini-Trase equipment connect to the pc (a) and the 48-channel ERT monitoring system (b).

All lithological formations outcropping along the slope are in tectonic contact with each other. Therefore, the Siliceous Schist formation is in tectonic contact with the Flysch Galestrino and Corleto Perticara formations through normal faults which lowered the young ones in respect to the ancient.

The top of the slope in the study area is characterized by recent sedimentary debris due to a certain number of mass movements occurred in the last sixty years.

The landslide classified as a complex retrogressive rototranslational slide is 600 m long and 230 m wide with an altimetry range varying between 1072 m a.s.l., at the main crown, and 978 m a.s.l. at the toe (Figure 3).

The profile of measurement, placed in the upper part of the landslide close to its right lateral boundary, is 47 m long (Figure 4).

The investigation depth reached is about 8 m according to the hydrological characteristics of the area in which the maximum piezometric level is measured at 2 m from ground level.

3. Prototype System and Acquisition Procedure

The prototype system is composed of different units and was planned with the aim to monitor in real-time the electrical resistivity and soil moisture patterns in the first layer of the subsoil.

The two fundamental units of the system are the *geoelectrical monitoring system* and the *TDR system*.

The geoelectrical monitoring system consists of 48 steel electrodes and a 48-channel cable connected to a resistivimeter Syscal Pro Switch 48 of the IRIS Instruments; the electrodes are buried in the soil at 0.5 meters depth, at the distance of 1 meter. The resistivimeter is linked to a pc used to store data and to manage the time when the acquisition starts, and the time interval between two consequential ones (Figure 5). The software used to control time-lapse acquisitions is Comsys Pro of the IRIS Instruments, operating by a script.

The TDR system is composed by 4 probes 20 cm length, buried at two different depths (1 m and 1.5 m) along the same profile of the geoelectric monitoring system. Two probes in correspondence of the 14th electrode and the other two in correspondence of 35th. All probes are connected to the Soil Moisture Equipment Corporation TRASE, which acquires and stores data. Also TDR system is connected to pc to be managed in remote control.

A weather station was installed in the area very close to the profile and linked to the pc. The station consists of a rain gauge to quantify the amount of rain falling in the area, a sensor to measure the air temperature and another one to determine speed and direction of the wind (Figure 6).

The electric current supply for the whole monitoring system is guaranteed by an uninterruptible power supply (U. P. S.).

The system was planned to be controlled in remote by an operator, who can decide day by day how to change acquisition parameters. After each acquisition the system sends an e-mail with attached the data file acquired to three different technicians involved in the check of the correct working of the system.

From September 2009 to January 2010, some ERTs were performed every week to test the system and its setting. For the acquisition three different electrode configuration were used: Wenner, Wenner-Schlumberger, and Dipole-Dipole, in order to choose the configuration that better emphasizes the features of the subsoil and the presence of the piezometric surface. At last, considering the better resolution of horizontal structures and the low noise recorded (better signal response), Wenner array was chosen. The system started to acquire and store data using "time-lapse" mode on 2 February 2010. At first, the number of acquisitions was fixed in four per day starting from 1.30 am (GMT + 1), with a time interval of 6 hours between two consecutive ones. This condition has lasted five months, till the end of June 2010. Then, the number of acquisitions was fixed in two per day, so

FIGURE 6: The weather station installed in the site.

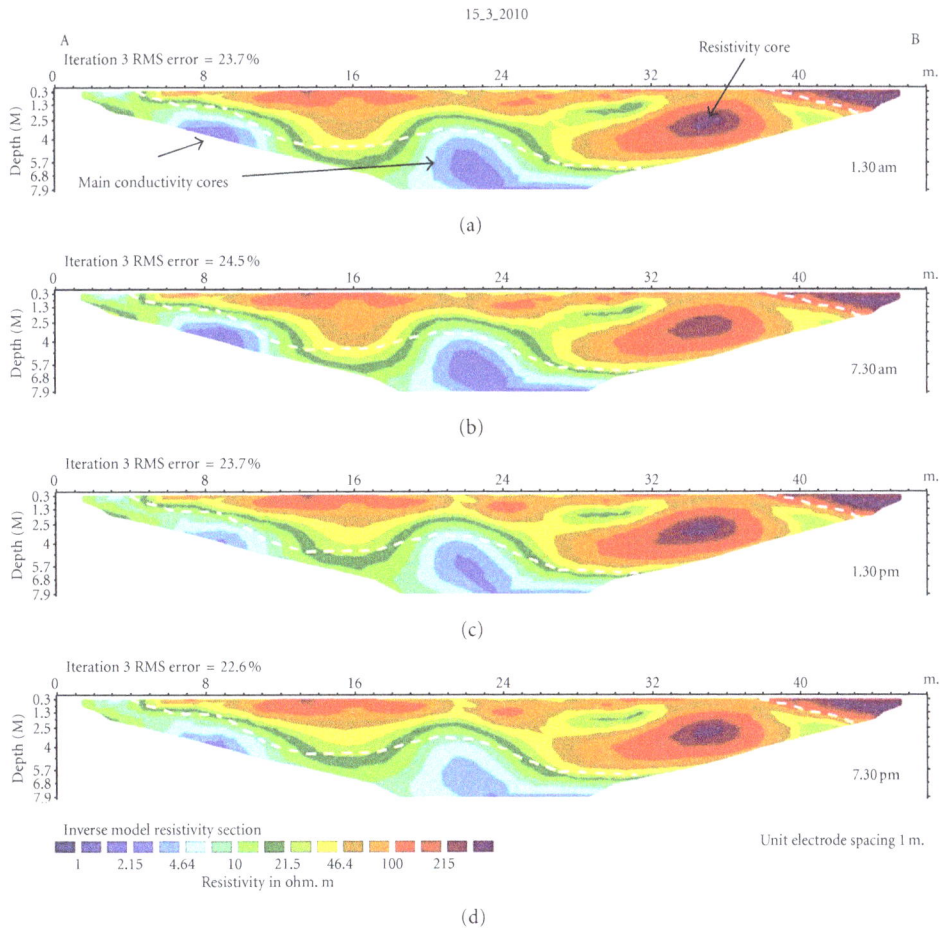

FIGURE 7: ERT related to the 15/03/2010 daily cycle of acquisition.

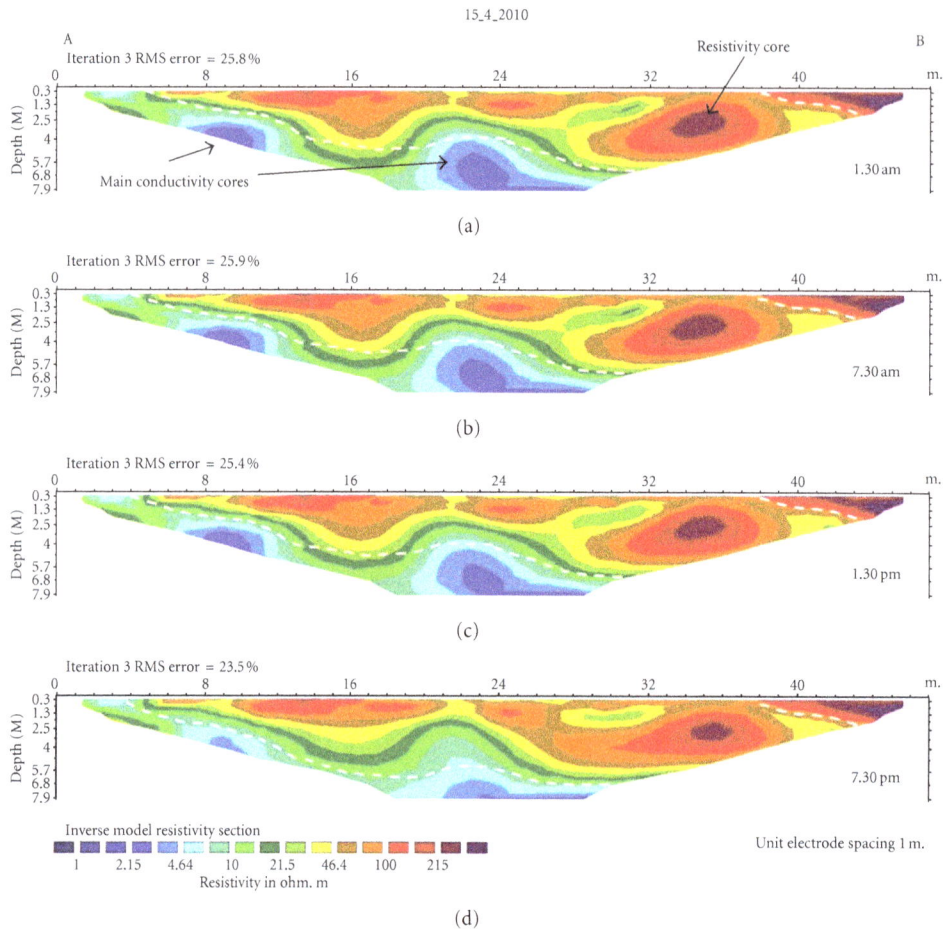

FIGURE 8: ERT related to the 15/04/2010 daily cycle of acquisition.

from 1th of July 2010 until now, system is acquiring with an interval of 12 hours, starting from 5 am (GMT + 1).

TDR acquisitions program is similar to the geoelectric one; it started with four acquisition per day, with a shift of one hour from the geoelectric survey. Then, from 1st of July 2010 it is acquiring soil moisture data at the same hour of geoelectric data, two times per day. Also for TDR survey it was decided to acquire with an interval of 12 hours, starting from 5 am (GMT + 1).

Weather station, instead, started to store data just from 2nd of February 2011, with interval of two hours. The number of acquisitions is 12 per day, so the amount of precipitations is extremely detailed. For the preliminary analysis we used the rainfall data acquired by another station located 3 kilometers far from the area of investigation.

4. Preliminary Results

The results reported in this paper concern the first acquisition period from February 2010 to June 2010.

In the first five months, from the 2nd of February to the 30th of June, time-lapse tomographies were performed every days with a frequency of four acquisition per day. Exactly, the cycle of acquisition started 1.30 am (GMT+1) and ended

at 7.30 pm (GMT+1), with a time interval of 6 hours between two consecutive ones.

At the first, the system has been tested to verify its correct working. During February 2010 it was necessary to verify the correct functioning of the monitoring system in time-lapse mode and its stability in time. Hence, the main analysis of the resistivity data started from March 2010.

A preliminary analysis of the data has shown a considerable stability of the system and its correct working. No significant variations of the resistivity values were observed during a daily cycle of acquisition. This is also highlighted by the ERTs reported in Figures 7–10 showing the resistivity distribution in the subsoil of four daily cycles; more precisely acquisitions are referred to the 15th of March (Figure 7), the 15th of April (Figure 8), the 15th of May (Figure 9), and the 14th of June (Figure 10). Time interval between each acquisition is one month. The apparent resistivity data were inverted by using the RES2DINV software that uses an algorithm based on the smoothness constrained least square inversion implemented by using a quasi-Newton optimization technique, suggested by Loke and Barker [19].

All ERTs are characterized by the same distribution of resistivity and show a large range of values ($2 < \rho < 300\,\Omega\text{m}$), picked out by the presence of two main zones with different

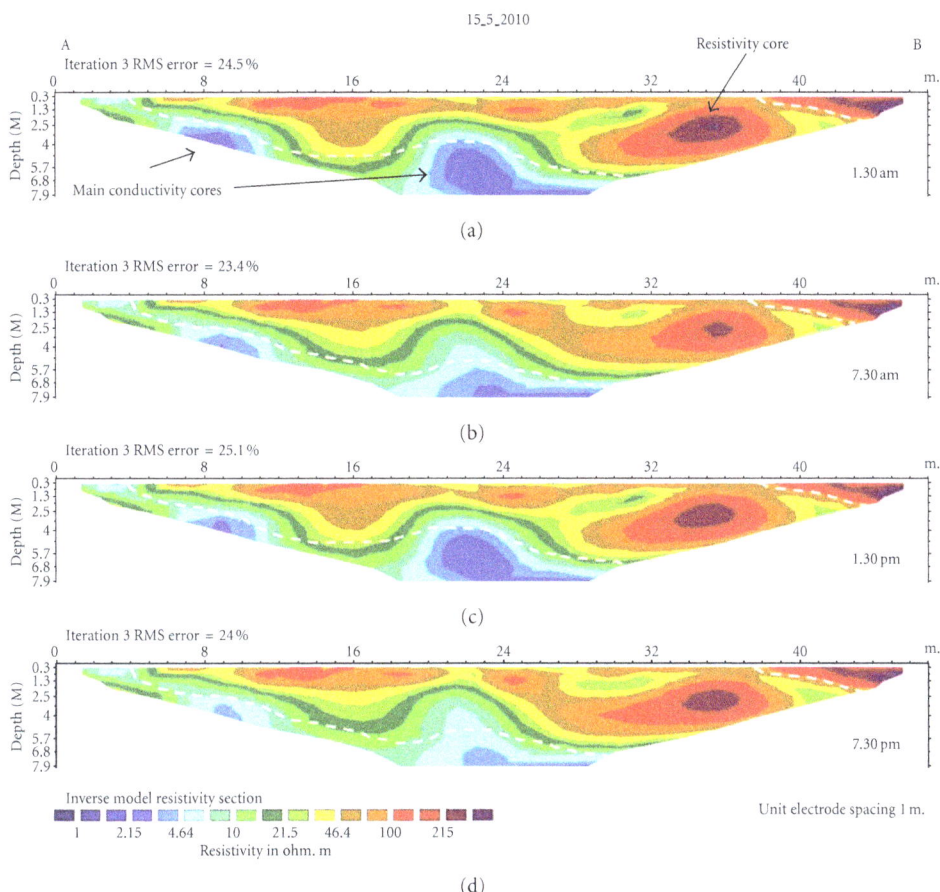

FIGURE 9: ERT related to the 15/05/2010 daily cycle of acquisition.

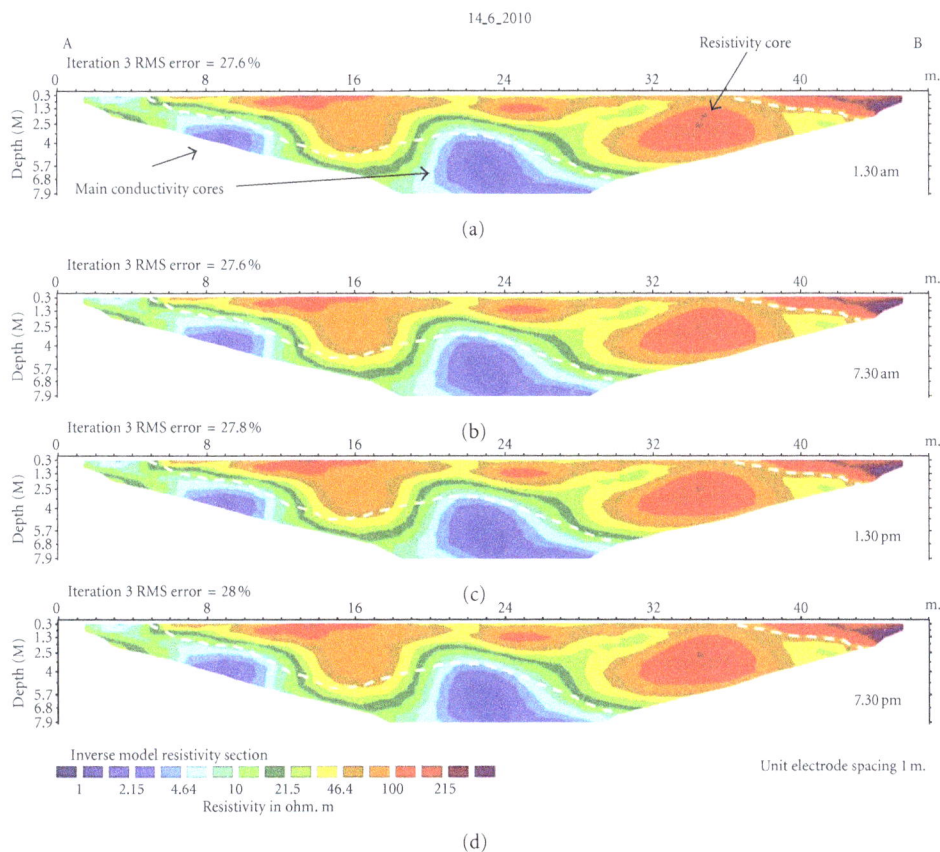

FIGURE 10: ERT related to the 14/06/2010 daily cycle of acquisition.

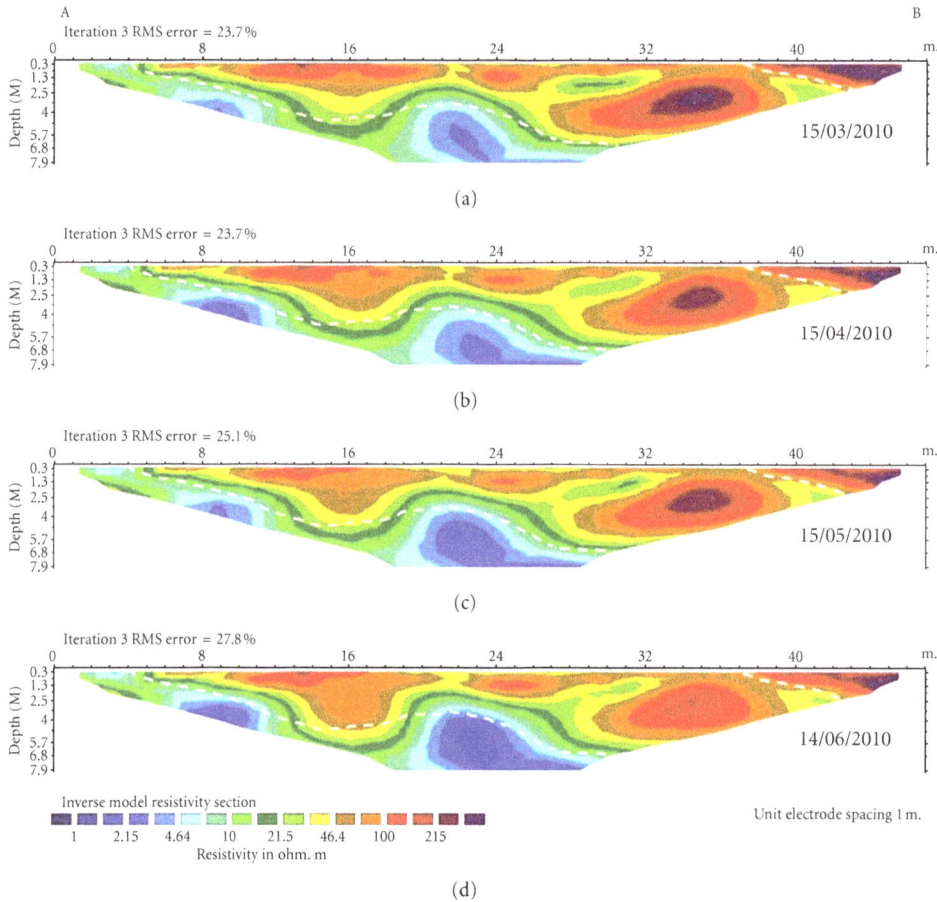

FIGURE 11: ERT performed in different days with an interval of one month; it is possible to observe the low change in resistivity during all the studied period.

resistivity values: a shallow zone, 2.5 m depth, with higher resistivity values ($50 < \rho < 300\,\Omega$m) distributed all along the profile, and a deeper one with the bulk of lower values ($2 < \rho < 10\,\Omega$m) concentrated in two main conductivity cores. A main central core with a thick of about 4 m and a smaller one in the left side of the tomographies, about 3 m deep. Moreover, by the extremity B of the tomography is evident a high conductivity layer, probably associated with the bedrock, while in the right side of the tomographies is present a large resistivity core.

Moreover, the low change in resistivity is also observed during the entire studied period (March 2010–June 2010). Figure 11 shows the comparison between the third acquisitions of each daily cycle: a general increase of conductivity in the central and right portions of the tomographies is evident, but the distribution of the resistivity values is similar for all the acquisitions.

In order to verify this resistivity distribution in the subsoil, a furthermore analysis was carried out to quantify the size of change ($\Delta\rho$) during the entire period.

In particular, the resistivity differences between the first acquisition (15th of March 2010) and the last acquisition (14th of June 2010) have been calculated, with a time interval of four months. All analyses and pseudosections have been performed by using the SURFER 8 of the Golden Software, Inc.

Variations has been calculated by (1) with respect to the acquisition of ρ_0 at time T_0 (15th of March 2010):

$$\Delta\rho = \frac{\rho_1 - \rho_0}{\rho_0}. \tag{1}$$

Figure 12 shows that the whole section is interested by negative resistivity differences, with values close to 0. This behavior seems to confirm that not important variations occurred during the entire period. Then, the same analysis was carried out in a shorter period, from the end of April 2010 ($T_0 = 28/04/2010$) to the end of May 2010, interested by intense rainfall. This analysis has demonstrated that in a short period the resistivity changes are bigger and well distributed in the whole section (Figure 13).

Pseudosection of the differences shows a variation close to the zero in the shallow layers, except some small positive variations observed in the right side. The remaining portion of the pseudosection is affected by negative variations with a range included between 0 and $-0.2\,\Omega$m, implying an unimportant increase of conductivity. Instead, resistivity differences calculated at T_2 and T_3 show an opposite behavior, with a more emphasized positive variation in the central core

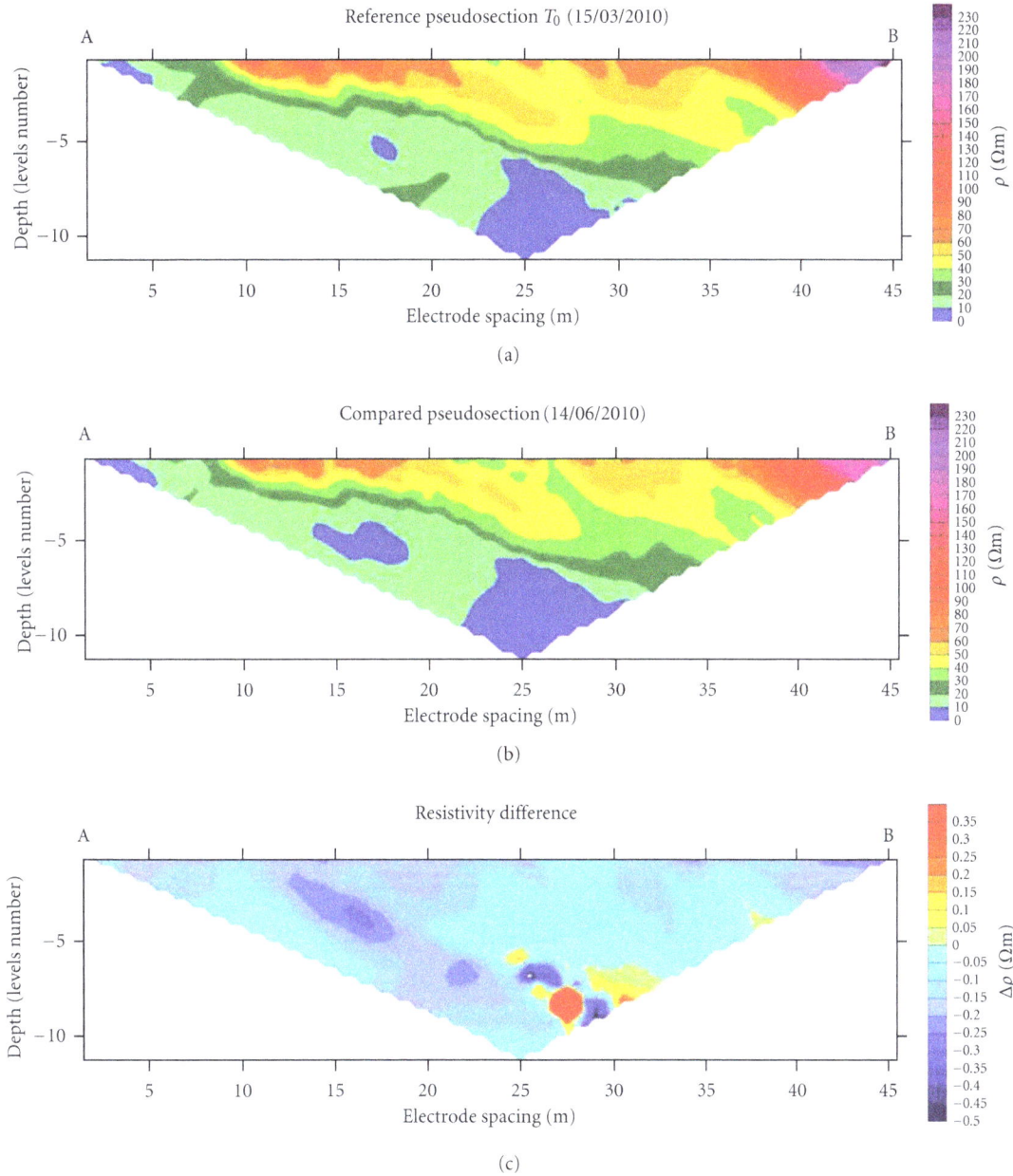

FIGURE 12: Resistivity differences calculated with respect to a fixed time T_0 (15/03/2010). No important resistivity variations are observed.

of the pseudosections and negative variations in the shallow layers (Figure 13). Probably acquisition of 16th of May 2010 (T_2) has been conditioned by the intense rainfall of two days before, explaining the shallow variation or resistivity. The electrical state of subsoil seems to remain the same also in the following day, as shown by the pseudo-section referred to the acquisition at time T_3; even if not interested by rainfall, data acquired in the 27th of May 2010 has been probably conditioned by the distribution of rainfall occurred during the week before.

Whatever, starting from July 2010, the frequency of acquisition has been changed in order to emphasize the possible resistivity variations within a day.

5. Comparison between Soil Moisture and Resistivity

To verify if the resistivity variation is related to the variations of water content in the subsoil, the resistivity trend has been also compared to TDR data. In particular, resistivity and TDR data acquired at the same depth (1 m and 1.5 m) have been considered. In some cases, soil moisture has shown a behavior opposite to resistivity, confirming that resistivity variation in the subsoil is influenced by water content. To obtain a good agreement between resistivity and TDR data, acquisition frequency of TDR has been set as similar to the resistivity one, starting from 2 am (GMT + 1).

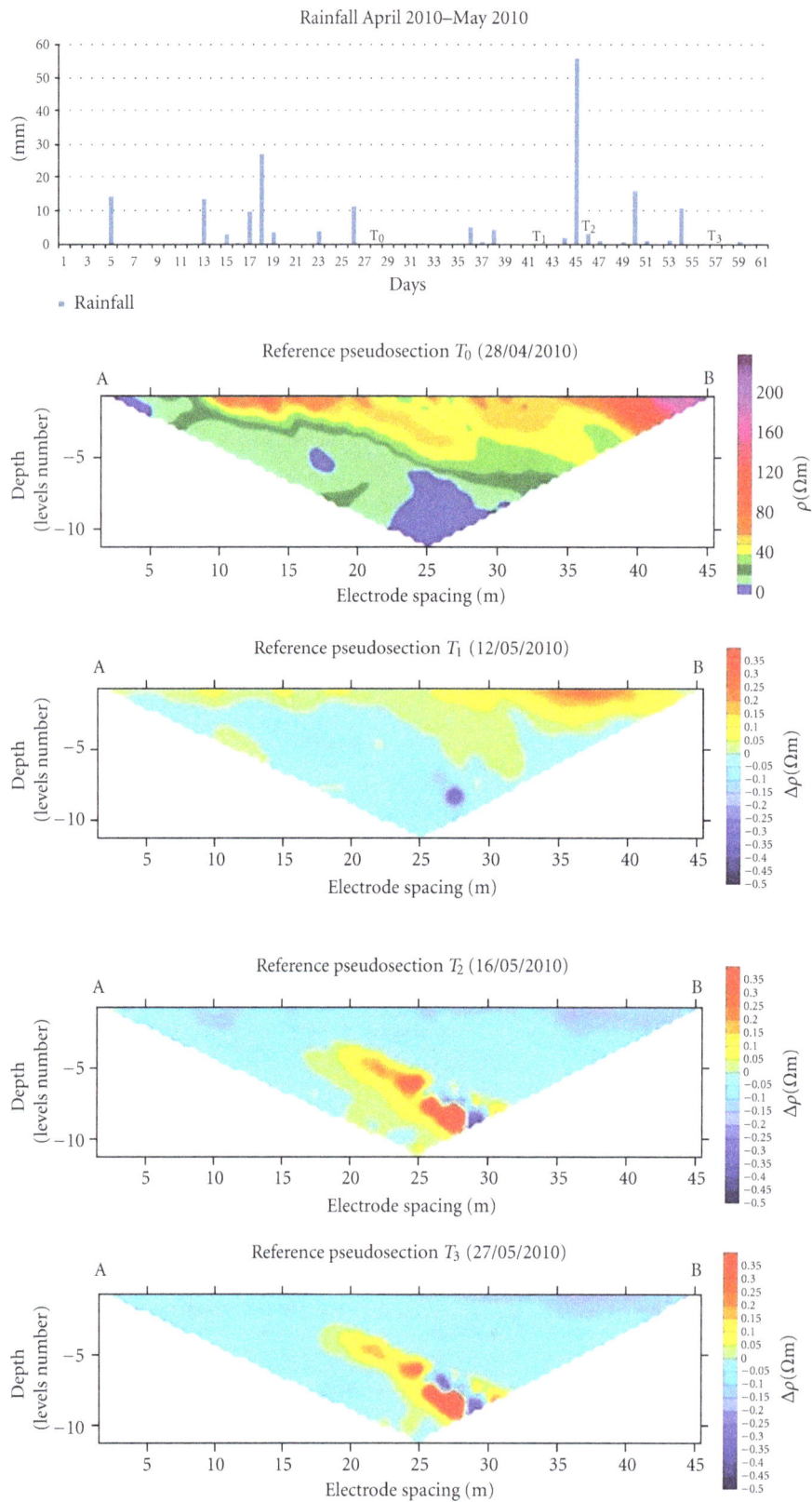

FIGURE 13: Resistivity differences calculated with respect to a fixed time T_0 (28/04/2010). Differences are referred to three days of May 2010, chosen after intense rainfall.

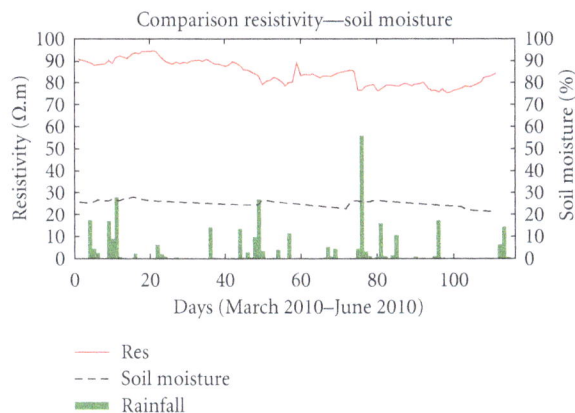

FIGURE 14: Comparison between Resistivity and TDR data acquired at 1 m depth, in correspondence of the 14 electrode.

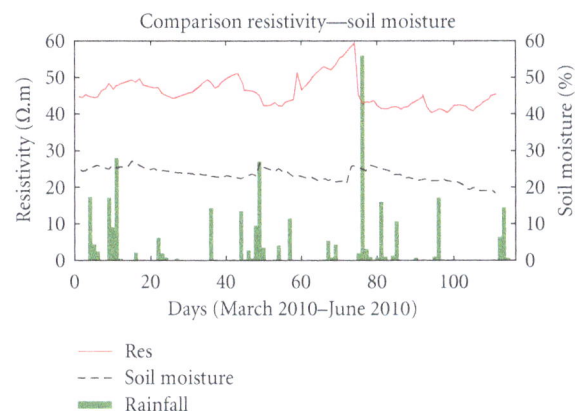

FIGURE 15: Comparison between Resistivity and TDR data acquired at 1 m depth, in correspondence of the 35 electrode.

Furthermore, also rainfall has been compared to TDR and resistivity, to understand how much the trend of the two parameters has been conditioned by it.

Figure 14 shows the comparison between resistivity and soil moisture values acquired at 1 m depth, equivalent to the 2nd level of the pseudo-section, in correspondence of the 14th electrode.

Fluctuations are present, interesting a range of values included between 75 Ωm and 91 Ωm. Instead, variations of soil moisture are smaller, included between 23% and 27%.

It is possible to observe that soil moisture and resistivity show a similar trend, even if in correspondence of some rain events an opposite behavior occurs. So it seems that the two parameters are not really conditioned by the rainfall during the four months, but just by intense spot precipitation periods.

Figure 15 is about the comparison between the two parameters acquired at 1 m depth, in correspondence of the 35th electrode. The trend of the two parameters is similar to the first example of Figure 13, but the opposite behavior of soil moisture with respect to the resistivity is more evident during the entire period.

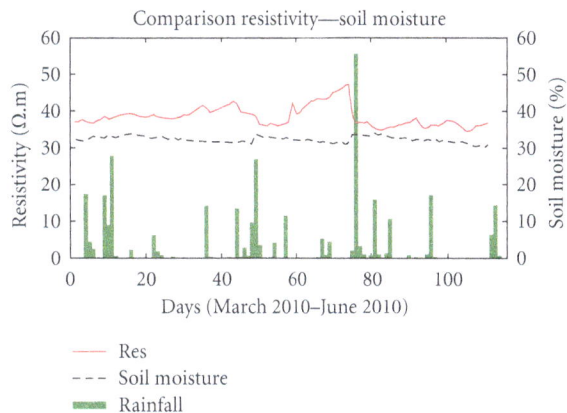

FIGURE 16: Comparison between Resistivity and TDR acquired at 1.5 m depth, in correspondence of the 35th electrode.

Also in this case the two parameters are not conditioned by rainfall during the four months but just in periods interested by intense rainfall.

The last case is about the comparison between two deeper points 1.5 m deep (3rd level of the calculated pseudosections), always in correspondence of the 35th electrode (Figure 16).

Range of resistivity is included between 38 Ωm and 50 Ωm, while soil moisture has a mean value of 30%. So, a general decrease of the resistivity values has been observed; at the same time, soil moisture percentage remains the same during the four months. A bigger number of events conditioned by the rainfall are highlighted. It has been observed that the two parameters evolve rapidly at the same time and that at this depth the variations of the two trends are more correlated than other shallower cases.

6. Conclusions

The prototype monitoring system has been developed in order to create a powerful tool to highlight the resistivity changes connected with the changes of water content in the first layers of the subsoil, within a landslide body. The real aim was to indirectly determine a preliminary estimate of the water content in an area affected by hydrogeological hazard.

Preliminary ERT have not shown great differences between the first and the last acquisition during the day, confirming the stability of the signal. Also the calculated resistivity differences between the first and the last acquisitions seem to confirm that resistivity is not subjected to big variation during the analyzed period, showing a negative distribution of values, close to 0. Instead, analysis in shorter period has demonstrated that the size of resistivity differences is bigger and well distributed in the whole subsoil section, perhaps conditioned by the intense rainfall period. Whatever, from the data trend analysis, the effect of precipitations is evident. TDR and resistivity trend comparison confirms that there is a decreasing of resistivity in correspondence of soil moisture growth, but this change seems to be stronger when intense precipitation periods occur.

Next step to confirm these preliminary results will be the statistical analysis of resistivity data along a period of 1 or 2 years, to understand if the fluctuation of values is strictly connected to the change of season. Besides, statistical results will be compared to the pluviometric data and to the soil moisture behavior, in order to obtain a more detailed description of the hydrogeological conditions of the area.

References

[1] C. De Baria, V. Lapennaa, A. Perronea, C. Puglisi, and F. Sdao, "Digital photogrammetric analysis and electrical resistivity tomography for investigating the Picerno landslide (Basilicata region, southern Italy)," *Geomorphology*, vol. 133, no. 1-2, pp. 34–46, 2011.

[2] V. A. Bogoslovsky and A. A. Ogilvy, "Geophysical methods for the investigation of landslides," *Geophysics*, vol. 42, no. 3, pp. 562–571, 1977.

[3] J. P. T. Caris and T. W. J. Van Asch, "Geophysical, geotechnical and hydrological investigations of a small landslide in the French Alps," *Engineering Geology*, vol. 31, no. 3-4, pp. 249–276, 1991.

[4] F. Ferrucci, M. Amelio, M. Sorriso-Valvo, and C. Tansi, "Seismic prospecting of a slope affected by deep-seated gravitational slope deformation: the Lago Sackung, Calabria, Italy," *Engineering Geology*, vol. 57, no. 1-2, pp. 53–64, 2000.

[5] V. Lapenna, P. Lorenzo, A. Perrone, S. Piscitelli, E. Rizzo, and F. Sdao, "2D electrical resistivity imaging of some complex landslides in the Lucanian Apennine chain, southern Italy," *Geophysics*, vol. 70, no. 3, pp. B11–B18, 2005.

[6] D. Jongmans and S. Garambois, "Geophysical investigation of landslides: a review," *Bulletin de la Societe Geologique de France*, vol. 178, no. 2, pp. 101–112, 2007.

[7] M. Pfaffhuber, S. Bastani, and M. Cornée, "Multi-method high resolutiongeophysical & geotechnical quick clay mapping A.A," *Near Surface*. In press.

[8] K. Sudha, M. Israil, S. Mittal, and J. Rai, "Soil characterization using electrical resistivity tomography and geotechnical investigations," *Journal of Applied Geophysics*, vol. 67, no. 1, pp. 74–79, 2009.

[9] D. Michot, Y. Benderitter, A. Dorigny, B. Nicoullaud, D. King, and A. Tabbagh, "Spatial and temporal monitoring of soil water content with an irrigated corn crop cover using surface electrical resistivity tomography," *Water Resources Research*, vol. 39, no. 5, pp. SBH141–SBH1420, 2003.

[10] D. H. Jayawickreme, R. L. van Dam, and D. W. Hyndman, "Subsurface imaging of vegetation, climate, and root-zone moisture interactions," *Geophysical Research Letters*, vol. 35, no. 18, Article ID L18404, 5 pages, 2008.

[11] R. De Franco, G. Biella, L. Tosi et al., "Monitoring the saltwater intrusion by time lapse electrical resistivity tomography: the Chioggia test site (Venice Lagoon, Italy)," *Journal of Applied Geophysics*, vol. 69, no. 3-4, pp. 117–130, 2009.

[12] P. Brunet, R. Clément, and C. Bouvier, "Monitoring soil water content and deficit using electrical resistivity tomography (ERT)—a case study in the Cevennes area, France," *Journal of Hydrology*, vol. 380, no. 1-2, pp. 146–153, 2010.

[13] T. Arora and S. Ahmed, "Characterization of recharge through complex vadose zone of a granitic aquifer by time-lapse electrical resistivity tomography," *Journal of Applied Geophysics*, vol. 73, no. 1, pp. 35–44, 2011.

[14] J. Travelletti, P. Sailhac, J.-P. Malet, G. Grandjean, and J. Ponton, "Hydrological response of weathered clay-shale slopes: water infiltration monitoring with time-lapse electrical resistivity tomography," *Hydrological Processes*. In press.

[15] G. Cassiani, V. Bruno, A. Villa, N. Fusi, and A. M. Binley, "A saline trace test monitored via time-lapse surface electrical resistivity tomography," *Journal of Applied Geophysics*, vol. 59, no. 3, pp. 244–259, 2006.

[16] T. Pescatore, P. Renda, and M. Tramutoli, "Rapporti tra le unità lagonegresi e le unità Sicilidi nella media valle del Basento (Appennino lucano)," *Memoriali Società Geologica Italiana*, vol. 41, pp. 353–361, 1988.

[17] S. Gallicchio, M. Marcucci, P. Pieri, I. Premoli Silva, L. Sabato, and G. Salvini, "Stratigraphical data from a Cretaceus claystones sequence of the "Argille Varicolori" in the Southern Apennines (Basilicata, Italy)," *Palaleopelagos*, vol. 6, pp. 261–272, 1996.

[18] R. Selli, "Il Paleogene nel quadro della geologia dell'Italia meridionale," *Memorie della Società Geologica Italiana*, vol. 3, pp. 737–790, 1962.

[19] M. H. Loke and R. D. Barker, "Rapid least-squares inversion of apparent resistivity pseudosections by a quasi-Newton method," *Geophysical Prospecting*, vol. 44, no. 1, pp. 131–152, 1996.

12

Imaging of Fault and Fracture Controls in the Arbuckle-Simpson Aquifer, Southern Oklahoma, USA, through Electrical Resistivity Sounding and Tomography Methods

Kumar Ramachandran,[1] Bryan Tapp,[1] Tayler Rigsby,[1, 2] and Erin Lewallen[1, 3]

[1] *Department of Geosciences, The University of Tulsa, Tulsa, OK 74104-9700, USA*
[2] *Colorado School of Mines, CO 80401-1893, USA*
[3] *Chevron Corporation, Houston, TX 77002, USA*

Correspondence should be addressed to Kumar Ramachandran, kr@utulsa.edu

Academic Editor: Michael S. Zhdanov

Arbuckle-Simpson aquifer in southern Oklahoma, USA, is a major source of water for industrial and municipal use. It is also a major source for spring-fed streams in the area. As part of an ongoing study to map and characterize the Arbuckle-Simpson aquifer, an electrical resistivity tomography (ERT) study and electrical resistivity sounding studies were conducted in Johnston County, Oklahoma, USA, to map the subsurface of a small area of the carbonate aquifer. The main aim of the study was to obtain constraints on the location of near surface faults and fractures and how they control groundwater flow in the study area. The interpreted resistivity section along an N-S profile indicates that the water table in the region is deepening to the south and probably bounded in the north by a south dipping fault. Inverse modeling of 2D electrical resistivity tomography (ERT) data acquired at two adjacent locations within the study area indicate shallow, fractured Arbuckle group rocks saturated with water adjacent to dry rocks. From electrical resistivity mapping results, it is inferred that the Mill Creek block in the Arbuckle-Simpson aquifer is an isolated system, interacting with the northern segment of a silicate-based aquifer through dissolution faults and fractures.

1. Introduction

The Arbuckle-Simpson aquifer located near Pilot Springs and the town of Mill Creek, Johnston County, Oklahoma, USA (Figure 1), is a highly fractured carbonate aquifer. This aquifer serves as a primary water source for around 39,000 residents of the surrounding areas.

Conservation of the Arbuckle-Simpson aquifer in south-central Oklahoma has been a recent concern. Citizens for the Protection of the Arbuckle-Simpson Aquifer (CPASA) have made efforts to protect against the depletion of the aquifer and its natural springs and waterways as well as protection against pollution and waste. CPASA has concerns about a quarry north of Mill Creek and its impacts to the aquifer as well as Pennington Creek and Mill Creek. The mine will quarry limestone and other materials and will cross a tributary of Mill Creek and Pennington Creek. The mine has already penetrated the water table at 15 to 44 feet beneath the surface and the mining company has proposed to produce 1.5 million gallons of water a day for mining use.

The Cambro-Ordovician units that comprise the aquifer are exposed in 500 square miles of outcrop area that serves as the recharge for the aquifer. This study is focused on the Mill Creek Block of Johnston County, Oklahoma, USA (Figure 1(a)). Two major faults (F1 and F2 in Figure 1(b)) bound the northern and southern boundaries of the Mill Creek block of the Arbuckle-Simpson aquifer.

Electrical techniques have been extensively used for mapping groundwater due to the correlation that generally exists between electrical properties of subsurface rocks and their fluid content [1–5]. Application of electrical techniques for siting wells and boreholes in crystalline basement aquifers

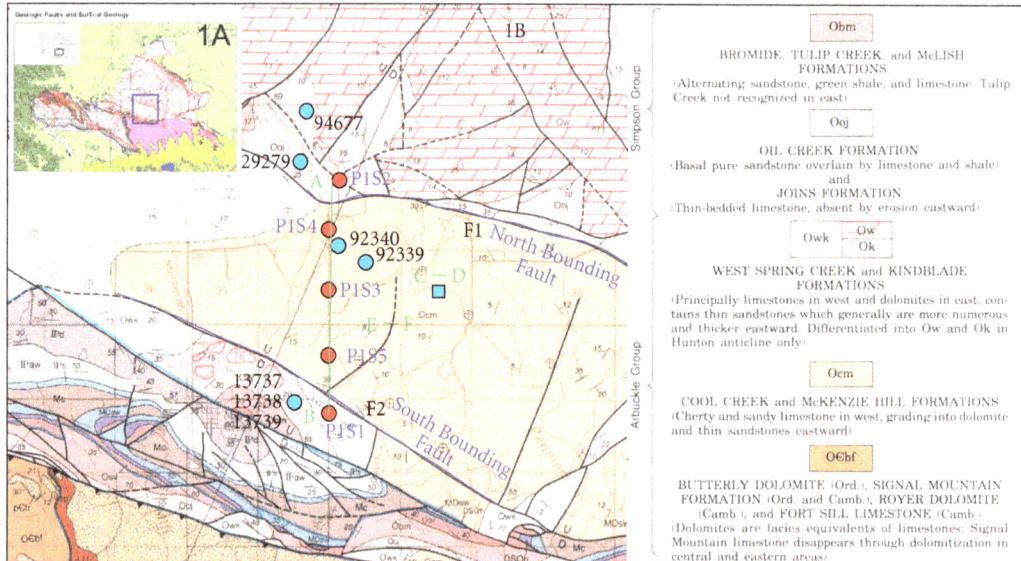

FIGURE 1: (a) Geographic location of the study area in Oklahoma. (b) Fault structure map of the Mill Creek block, Arbuckle-Simpson aquifer (From [13]). Line A-B represents the N-S sounding profile. CD and EF represent the electrical resistivity tomography profile locations. Red dots indicate the locations where electrical resistivity sounding data were collected. Faults F1 and F2 are imaged in the subsurface using electrical resistivity method to understand how they control ground water flow in the region. Well locations from which depth to groundwater level were obtained are shown by blue dots along with well numbers. Blue square indicates the location of Pilot Springs.

throughout sub-Saharan Africa has been discussed by Beeson and Jones [6], Olayinka and Barker [7], Hazell et al. [8, 9], Barker [10], and Carruthers and Smith [11]. Advances have led to developments in electrical techniques that have opened up the possibility of conducting true 2D and 3D geoelectric surveys [12].

2. Hydrogeology of the Study Area

The stratigraphy of the aquifer-bearing rocks that outcrop in the Arbuckle Mountains (after [14]) is given in Figure 2. The Arbuckle-Simpson aquifer occurs within three major rock units of Upper Cambrian to Middle Ordovician age: the Simpson, Arbuckle, and Timbered Hills Groups (Figure 3). Data from wells located in the area indicate that the aquifer contains two hydrostratigraphic units: (a) the carbonates of the Arbuckle Group and (b) sandstones of the Simpson Group. The term Arbuckle aquifer is applied to any pre-Simpson sources of water hosted in limestone or dolomitic limestone. As a result, the aquifer spans strata of the Arbuckle and Timbered Hills Groups in Cool Creek and McKenzie Hill Formations (Ocm) and West Spring Creek and Kindblade Formations (Owk and Ow). The Arbuckle Group, which comprises the major portion of the aquifer, consist of a thick sequence of carbonate rocks (limestone and dolomite) with minor layers of sandstone and shale. The Arbuckle Group ranges in thickness from 6,700 feet in the western portion of the aquifer to less than 4,000 feet in the eastern portion. Water is obtained from cavities, solution channels, fractures, and intercrystalline porosity present in the limestone and dolomite rocks. The Timbered Hills Group crops out in small areas within the study area and consists of up to 700

feet of limestone, dolomite, and sandstone. The Timbered Hills is believed to be in hydrologic connection with the Arbuckle Group and is considered part of the Arbuckle-Simpson aquifer.

The Simpson Group is the youngest, uppermost geologic unit of the aquifer and is comprised of layers of sandstone, limestone and dolomite, and some shale. Water in the Simpson Group is obtained primarily from pore spaces between the sand grains in the sandstones. The Simpson aquifer coincides with the lithostratigraphic Simpson Group and includes all water-bearing strata within it. Water-bearing sandstones of the Simpson Group occur in the Bromide, Tulip Creek, McLish (Obm), and Oil Creek formations (Ooj). About two-thirds of the Arbuckle-Simpson aquifer consists of carbonate rocks (limestones and dolomites), which are soluble. Infiltrating water slowly dissolves the rock, leading to the formation of solution channels and cavities along bedding planes, fractures, and faults. Karst (solution) features, such as sinkholes and caverns, have developed in some areas of the aquifer. The Arbuckle-Simpson aquifer is charged primarily by infiltration of water from precipitation on the outcrop area. Most of the water discharges naturally to streams, rivers, and springs. Presently, only a small portion discharges artificially through pumping and flowing wells. In this region, groundwater flows from topographically high areas to low areas, where it discharges to springs and streams.

Fairchild et al. [15] describe the Arbuckle Group specifically as having zero or only little intergranular porosity; porosity and permeability in these formations are instead due to fractures, joints, and solution features. Scheirer and Alegra [16] analyzed outcrop samples in the laboratory yielding porosity values of 5% for the Simpson Group,

Imaging of Fault and Fracture Controls in the Arbuckle-Simpson Aquifer, Southern Oklahoma, USA, through
Electrical Resistivity Sounding and Tomography Methods

131

System	Group and formation			
Ordovician	Viola Group	Fernvale Formation Viola Springs Formation		
	Simpson Group	Bromide Formation Tulip Creek Formation $\}$ Obm McLish Formation		Simpson aquifer
		Oil Creek Formation $\}$ Ooj Joins Formation		
	Arbuckle Group	West Spring Creek Formation $\}$ Owk Kindblade Formation		Arbuckle aquifer
		Cool Creek Formation $\}$ Ocm McKenie Hill Formation		
		Buttefly dolomite		
Cambrian		Signal mountain formation Royer dolomite Fort Sill limestone		
	Timbered Hills Group	Honey Creek limestone Reagan sandstone		
	Colbert rhyolite			

FIGURE 2: Stratigraphic nomenclature for the pre-Sylvan Ordovician and Cambrian rocks that outcrop in the Arbuckle Mountains. Positions of the Arbuckle and Simpson aquifers are shown on the stratigraphic column (after [14]).

based on the average porosity of six samples, and 4% for the Arbuckle Group, based on the average porosity of ten samples. Oil Creek and McLish sandstones are widely distributed in the area and represent the Simpson aquifer properties. Porosity of Oil Creek sandstone and McLish sandstone measured from well logs average about 20% [17].

3. Electrical Resistivity Surveys

To investigate the fault structure and fractures present in the Mill Creek block and to evaluate the controls exerted by them on local ground water hydrology, electrical resistivity soundings were conducted along an NS profile. The interpretation of this data has resulted in a geological cross-section for the Mill Creek block of the Arbuckle-Simpson aquifer. Electrical resistivity tomography (ERT) studies were conducted at two locations (Figure 1) near Pilot Springs in the Mill Creek block. Wenner and dipole-dipole ERT data were collected at both the locations.

3.1. Resistivity Sounding: 1D. Electrical resistivity sounding was performed and five sets of measurements (Figure 1(b)) were collected along a North-South profile to identify the presence of faults and depths to the groundwater table. The ground measurements were separated by approximately 2 km.

Soundings P1S1 and P1S2 were performed using a Wenner array configuration, while soundings P1S3, P1S4, and P1S5, used a Schlumberger array configuration. Data were analyzed by curve matching techniques [18] and inverse modeling using IPI2win public domain software, created and published by Moscow State University. Inversion models and modeled layer resistivities and thicknesses are given in Figure 4.

3.2. Electrical Resistivity Tomography Studies. Electrical resistivity tomography (ERT) studies were conducted at two locations (Figure 1) in the Mill Creek block close to Pilot Springs (Figure 5). Water emerging from Pilot Springs flows into Pennington Creek. The purpose of electrical resistivity imaging at this location was to identify the water source for Pilot Springs [19]. Water from Pilot Springs flows into Pennington Creek (Figure 5). Wenner and dipole-dipole ERT data were collected at this location and at another location south of Pilot Springs (Figure 1(b)). The data was collected using twenty-eight electrodes with five-meter electrode spacing for a total profile length of 135 m. The electrode spacing for the various survey configurations were designed prior to the survey and the data were collected using automatic switching circuitry instrument/electrode system. The dipole-dipole ERT survey resulted in 235 data points and Wenner ERT survey resulted in 135 data points.

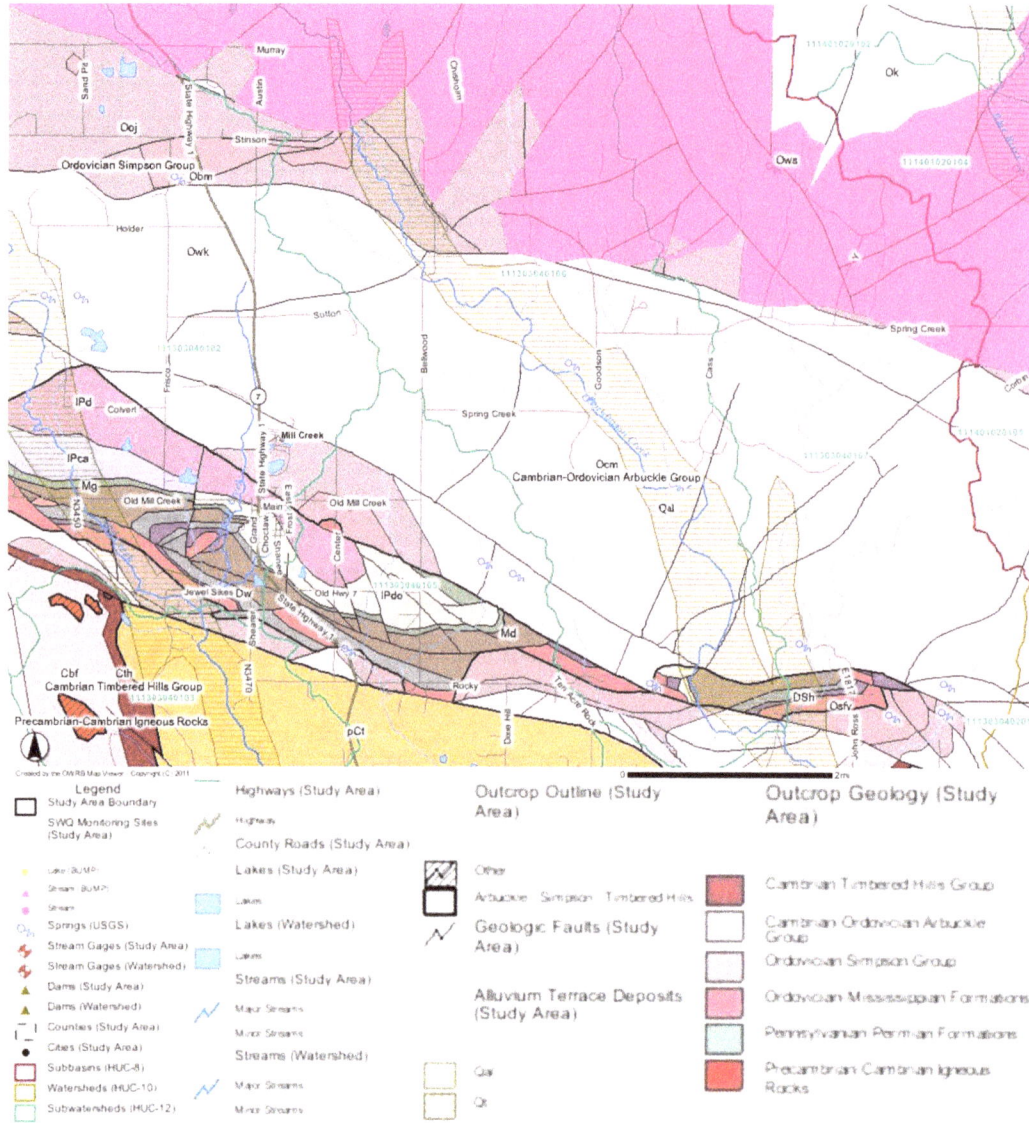

FIGURE 3: Hydrogeology map near Mill Creek Block of Arbuckle-Simpson aquifer.

4. Data Interpretation

4.1. 1D Resistivity Sounding Results. A comparison of the models for Soundings P1S2 and P1S4 are of particular interest in this study. These soundings are separated by the northern bounding fault F1 of the Mill Creek block, defining the boundary between the Oil Creek and Joins Formation (Simpson Group) to the north and the Cool Creek and McKenzie Hill Formation (Arbuckle Group) to the south (Figure 4). The relationship between the resistivity properties and depths at these two soundings aid us in understanding how this fault probably controls the aquifer in this area. The drop in resistivity between the second and third layers in both soundings is interpreted as the interface between the dry overburden rocks and the underlying aquifer. The relative position of this interface between these soundings is significant in determining how groundwater flow may be affected by the presence of the northern fault F1 between

them. For these two soundings, the depth of the inferred water-saturated layer increases toward the south, and this pattern holds true for Sounding P1S3 as well. This is significant, because it allows for an interpretation in which the northern-bounding fault itself, passing between these two soundings, is the cause of this deepening of the water-saturated layer, and may define the interface between the high-resistivity second layer and the low-resistivity third layer (Figure 6). Such an interpretation is consistent with a southward-dipping thrust model for the northern fault.

Sounding P1S5, like Sounding P1S4, is best fit to a four-layer subsurface model. The resistivity layers follow the same general pattern as those observed in the three soundings to the north; the second layer is characterized by a relatively high resistivity of 1000–1200 Ω-m, likely dry rock, overlying a low-resistivity second layer of 40–55 Ω-m, interpreted as the top of the aquifer. However, below this low-resistivity zone at a depth of at least 20 m, a fourth layer of exceptionally

Imaging of Fault and Fracture Controls in the Arbuckle-Simpson Aquifer, Southern Oklahoma, USA, through Electrical Resistivity Sounding and Tomography Methods

133

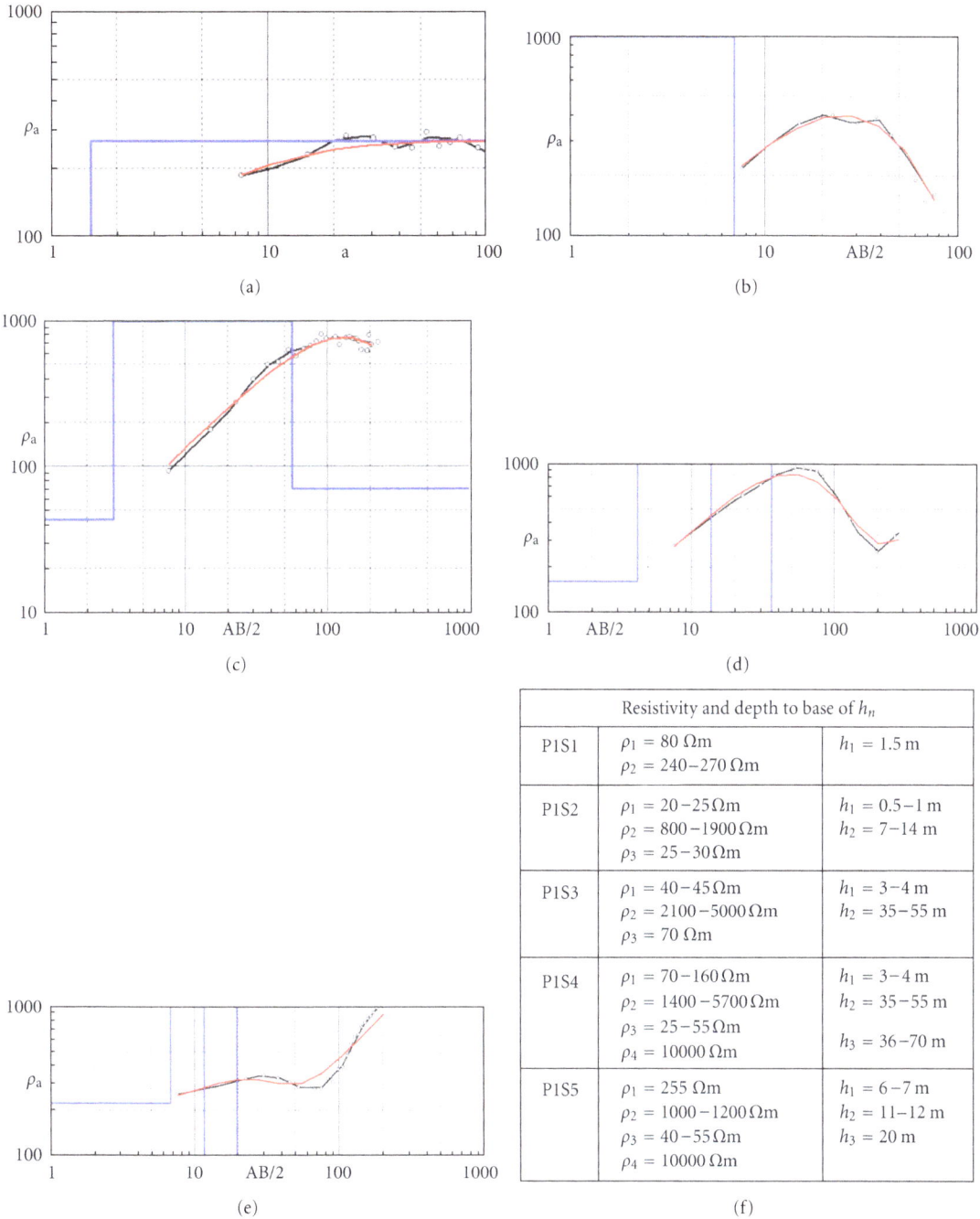

FIGURE 4: Inversion models for soundings 1 to 5 are shown in (a) to (e). (a) Sounding P1S1, (b) Sounding P1S2, (c) Sounding P1S3, (d) Sounding P1S4, and (e) Sounding P1S5. X-axis represents distance in meters and Y-axis represents resistivity in Ohm-m. In the graphs, the black line represents the observed data, blue line represents the inverse model, and the red line represents the calculated response from the inverse model. (f) Interpreted resistivities and layer thicknesses for the five electrical soundings.

	Resistivity and depth to base of h_n		
P1S1	$\rho_1 = 80\ \Omega m$ $\rho_2 = 240-270\ \Omega m$	$h_1 = 1.5\ m$
P1S2	$\rho_1 = 20-25\,\Omega m$ $\rho_2 = 800-1900\,\Omega m$ $\rho_3 = 25-30\,\Omega m$	$h_1 = 0.5-1\ m$ $h_2 = 7-14\ m$
P1S3	$\rho_1 = 40-45\,\Omega m$ $\rho_2 = 2100-5000\,\Omega m$ $\rho_3 = 70\ \Omega m$	$h_1 = 3-4\ m$ $h_2 = 35-55\ m$
P1S4	$\rho_1 = 70-160\,\Omega m$ $\rho_2 = 1400-5700\,\Omega m$ $\rho_3 = 25-55\,\Omega m$ $\rho_4 = 10000\ \Omega m$	$h_1 = 3-4\ m$ $h_2 = 35-55\ m$ $h_3 = 36-70\ m$
P1S5	$\rho_1 = 255\ \Omega m$ $\rho_2 = 1000-1200\,\Omega m$ $\rho_3 = 40-55\,\Omega m$ $\rho_4 = 10000\ \Omega m$	$h_1 = 6-7\ m$ $h_2 = 11-12\ m$ $h_3 = 20\ m$

high resistivity (as high as $10^5\ \Omega$-m) occurs. Such a high-resistivity layer below a low-resistivity layer is also observed at Sounding P1S4; however, there it occurs at a much greater depth. The relative shallowness of this dry fourth layer at Sounding P1S5 is especially noteworthy juxtaposed with Sounding P1S3, which represents the deepest water-saturated layer of all five soundings. Even the shallowest depth estimate for this layer in Sounding P1S3 is considerably

deeper than the high-resistivity layer in Sounding P1S5. An interpretation for this difference in models is interpreted to be the presence of a possible fault between Soundings P1S3 and P1S5 that may account for the relative displacement of rocks with varying properties (Figure 6). If such a fault acted as a barrier to groundwater flow, this interpretation could also offer an explanation for a high water level inferred in Sounding P1S5 relative to that at Sounding P1S3. A fault

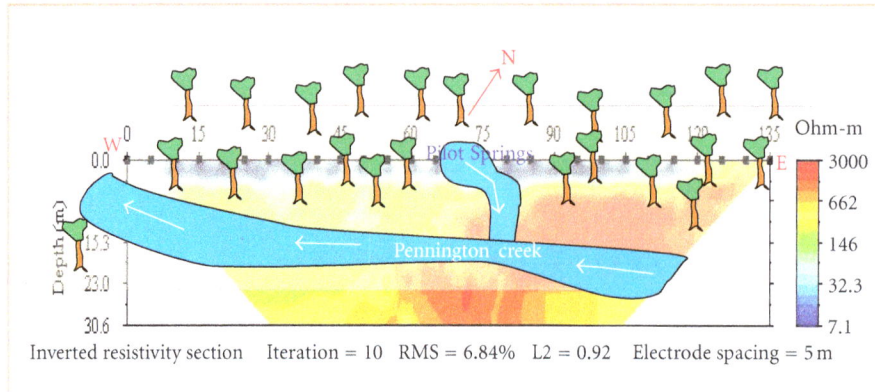

FIGURE 5: Location of Pilot Springs and Pennington Creek with respect to ERT profile CD. The orientation of the profile is W-E. Arrow marks shown in Pennington Creek and Pilot Springs indicate the direction of water flow.

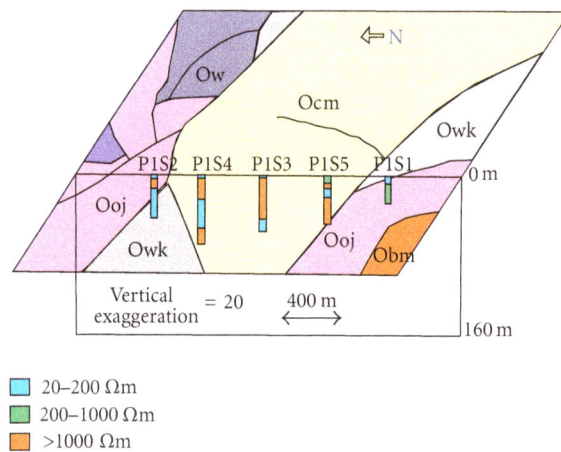

FIGURE 6: Diagram of subsurface resistivity model relative to surface geology. Cool Creek and McKenzie Hill Formations (Ocm), West Spring Creek and Kindblade Formations (Owk and Ow), Oil Creek Formation and Joins Formation (Ooj), and Bromide, Tulip Creek, and McLish Formations (Obm).

at this location is not indicated on the geologic map, but the resistivity data in this study raises the possibility that one may be present. Alternatively, this model difference may represent the presence of another type of discontinuity, such as a lithostratigraphic facies change.

Examining Soundings P1S5 and P1S1 to the south, data limitations prevent making a conclusive assessment of the interaction of groundwater with the southern bounding fault lying between them. While the penetration depth for Sounding P1S5 is estimated to be at least 46 m, for Sounding P1S1 it is only 22 m. In Sounding P1S5, the top of the low-resistivity layer interpreted as the saturation zone occurs at a depth of 11-12 m below the surface; a resistivity layer of 240–270 Ω-m is observed in Sounding P1S1, which likely represents water saturated layer. However, whether this layer can be correlated with the low resistivity layer of Sounding P1S5 as the top of the aquifer would require further data at this sounding point.

Analysis of the 1D sounding data resulted in a preliminary model of the subsurface fault geometry in depth as indicated in Figure 6. The northern-bounding fault (F1

in Figures 1(b) and 7) in this region is hypothesized to be a southward-dipping thrust fault; this northern fault and the fault north of that appear to define the boundary between the high-resistivity second layer and the water-saturated rocks below it. In this model, the northern faults in the study area are interpreted as south-dipping thrust faults. This interpretation has significant implications for the groundwater hydrology in this area. The south-dipping thrust fault may serve to isolate the deeper Simpson Group aquifer from the overlying Mill Creek block of the Arbuckle group, acting to confine this part of the aquifer.

4.2. Comparison of Sounding Results with Data from Groundwater Wells. To independently confirm this interpretation of groundwater level based on resistivity, water levels from monitored groundwater wells in the study area were examined; data for these wells were obtained from the Oklahoma Water Resources Board. Figure 1(b) displays the locations of groundwater wells.

The two northernmost wells, No. 94677 and No. 29279, are located in the Ooj Formation (Simpson Group). Recent

Imaging of Fault and Fracture Controls in the Arbuckle-Simpson Aquifer, Southern Oklahoma, USA, through
Electrical Resistivity Sounding and Tomography Methods

135

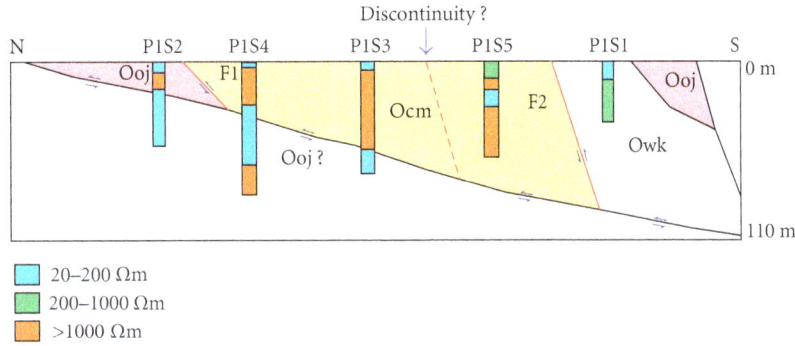

FIGURE 7: Interpreted fault geometry and subsurface geology along profile A-B relative to resistivity soundings.

water level information for well No. 29279 is not available, but the depth at which the driller first encountered water in the year 1991 was approximately 9.1 m. Water depth at well No. 94677 was measured at 7.5 m in June 2005 and 8.7 m in June 2006. Although farther north than Sounding P1S2, these wells show a relatively shallow groundwater table in this area, consistent with the depth (7–14 m) to the low-resistivity layer at this sounding.

To the south, Sounding P1S4 is located close to well No. 92340. The water level information available for this well is from February 2005; no water was encountered at a depth of approximately 13.1 m. This is consistent with the low-resistivity layer of Sounding 4 occurring deeper than 12 m. Well No. 92339 is located between Soundings P1S4 and P1S3 approximately a mile to the east. Here, water depth was measured at 25.8 m in June 2005 and 24.9 m in June 2006. Water depth at this well is shallower than the low-resistivity layer range offered for Sounding 3 (35–55 m), but it lies within the range of that of Sounding P1S4 (12–34 m). Data from these two wells is thus consistent with the resistivity interpretations and confirms the trend found in the resistivity soundings of increasing aquifer depth toward the south.

Finally, the southernmost three wells lie in the Ooj Formation, west of Sounding P1S1. No recent water level data could be found for these wells; however, the water table measured at these wells in the 1970s was between 3 and 6 m. Although this data is relatively old and must be used with caution, it is consistent with the interpretation at Sounding P1S1 of a very shallow low-resistivity layer.

Overall, the water table depths available from well data in the study area consistent with resistivity data [20]; this independent corroboration allows considerable confidence to be placed in the resistivity groundwater table interpretation.

4.3. ERT Data Analysis. In general, data collected using Wenner array provides resistivity control to intermediate depth range, has an intermediate resolution, and shows moderate sensitivity to near-surface lateral effects; dipole-dipole array has the smallest depth of investigation range, highest resolution, and greatest sensitivity to near-surface lateral effects [21]. In addition, inverse modeling of ERT

data from different electrode configurations leads to models that are dissimilar to some extent owing to the resolution, geometry, and noise tolerance characteristics of the electrode arrays. To see the difference in the final resistivity sections obtained from inversion of Wenner array, dipole-dipole array and combined Wenner and dipole-dipole array data are shown along with sensitivity sections in Figure 8. The sensitivity values provides a measure of the amount of information offered by the resistivities of the model blocks [22]. Higher sensitivity values indicate reliable model resistivity. In general, the cells near the surface usually have higher sensitivity values because the sensitivity function has very large values near the electrodes. Also, inversion results of Wenner and dipole-dipole data results in dissimilar resistivity cross-sections due to their differing subsurface imaging capabilities [23]. Inversion of combined datasets of Wenner and dipole-dipole yield a more robust solution to subsurface resistivity distribution (Figure 8).

To address some of the issues mentioned above, data from both electrode arrays were inverted jointly to provide a smooth model. The resulting ERT images provide significant information about lateral resistivity variability close to Pilot Springs and Pennington Creek. ERT image of the northern profile located close to Pilot Springs (Profile C-D) (Figure 9(a)) shows low-resistivity Arbuckle group rocks below Pilot springs. The low resistivity is inferred to be water-saturated rocks. ERT image acquired along a southern profile (Profile E-F) (Figure 9(b)) also indicates the presence of a fault contact and fluid filled fractures, similar to the resistivity cross-section obtained at Pilot Springs. It is inferred from the above ERT images that faults and fractures in the Arbuckle group rocks provide conduits for ground water to rise up from below.

In the area around Pilot Springs (Figure 5), Pennington Creek is observed with gas bubbles ascending to the surface in many places along the stream's course. Analysis of water samples taken at Pilot Springs and Pennington Creek in this area reveals a relatively low pH, indicating that the water emanating from them is not in equilibrium with the greater carbonate reservoir, and instead implies a silicate source for this water. A possible hypothesis is that the source of this water is the silicate aquifer system to the north. Groundwater from this northern aquifer may flow south, encounter a fault

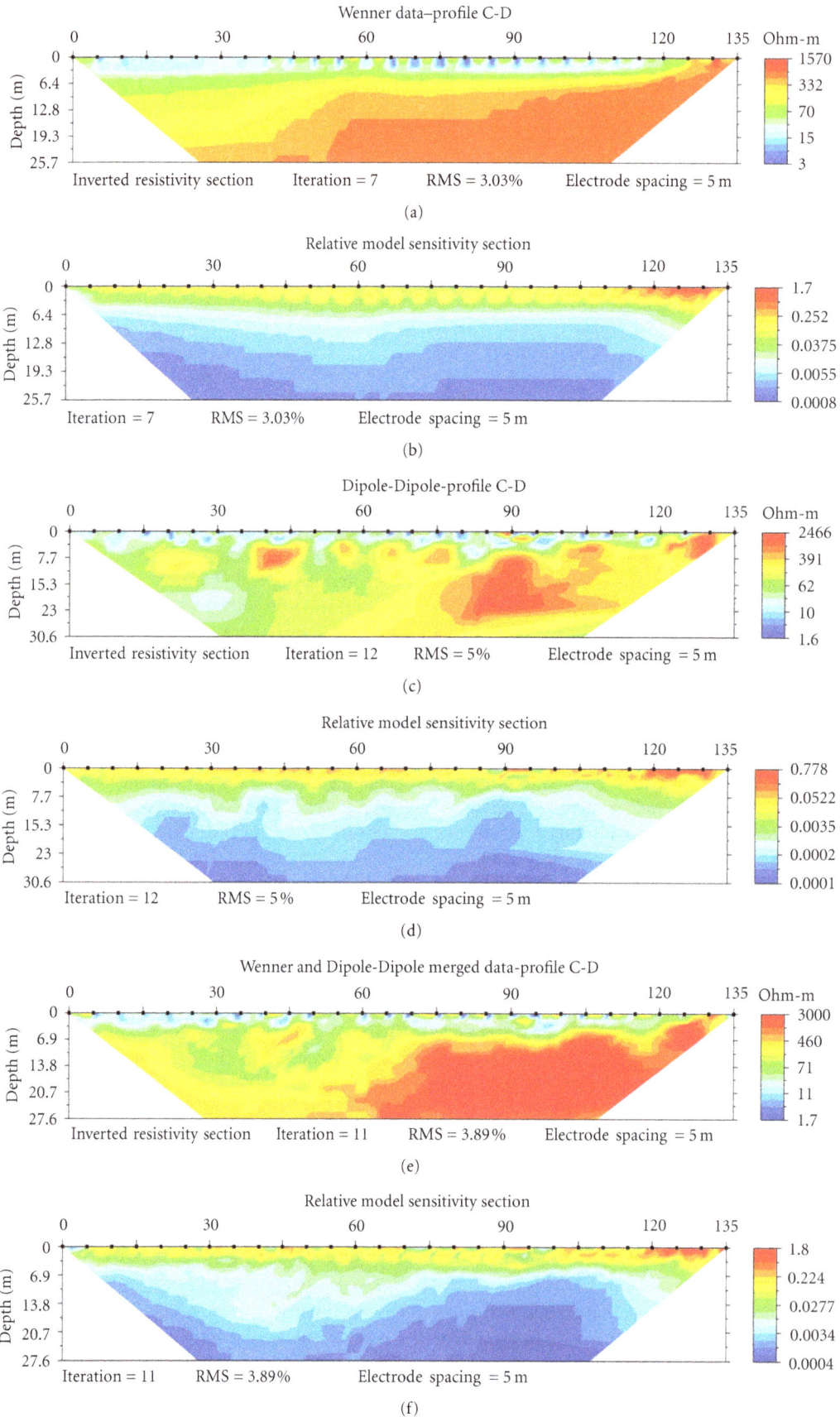

FIGURE 8: Resistivity sections constructed from Wenner, dipole-dipole, and merged Wenner and dipole-dipole data along with their corresponding sensitivity sections.

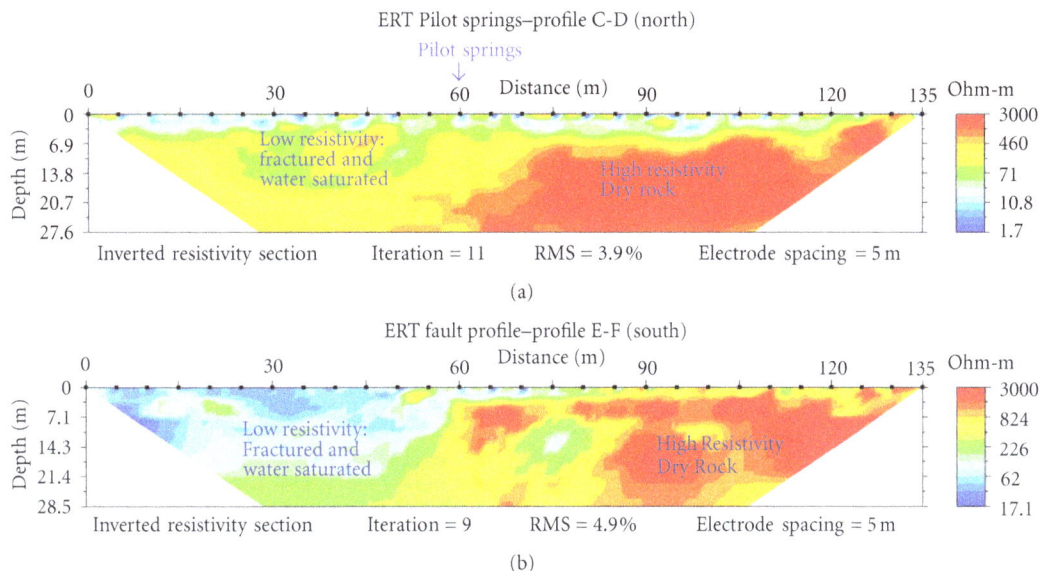

FIGURE 9: (a) ER tomography resistivity section close to Pilot Springs using Wenner and dipole-dipole arrays. (b) ER tomography resistivity section across mapped fault using Wenner and dipole-dipole arrays.

that acts as a baffle and redirects water upward, and emerge at the surface to create Pilot Springs and feed Pennington Creek.

5. Conclusions

Electrical resistivity surveys were carried out in the Arbuckle-Simpson aquifer in Mill Creek Block to (1) understand how faults and fractures control groundwater flow in this area, and (2) delineate the geometry of the northern fault bounding the Mill Creek block. Five electrical resistivity soundings were carried out in the study area to achieve these goals. The resistivity data provides insight into both the hydrogeologic and the structural environment present in the study area of the aquifer. The trend of a deepening low-resistivity third layer from Sounding P1S2 in the north to Sounding P1S3 farther south likely reflects a deepening groundwater table, which is consistent with a south-dipping northern fault delineating the overlying Arbuckle Group formations from an underlying confined Simpson Group aquifer. Within this interpretation, the Mill Creek block is likely an isolated system, interacting with the northern silicate-based aquifer only through dissolution faults and fractures. Such features might allow water to ascend directly from the underlying aquifer to the surface, which could account for the presence of Pilot Springs and Pennington Creek and the properties of the water emanating from them, which imply a silicate rather than a carbonate source.

It is inferred from ERT image obtained at Pilot Springs (Figure 9(a)) that water saturated rock section lies in the shallow subsurface indicating a fractured rock strata through which deeper water is ascending and feeding Pilot Springs, which in turn flows into Pennington Creek. About half a kilometer south of Pilot Springs, the southern ERT profile

(Figure 9(b)) shows a low resistivity section that signifies fractured and water-filled rock section that is juxtaposed to a dry rock section. This section of low resistivities seems to correlate spatially with the low resistivity section observed at Pilot Springs to the north. If this interpretation is accepted, then this trend indicates a fracture system in extending the near surface from the northern profile to the southern profile.

From the analysis of 1D sounding data and 2D ERT data, we interpret the Mill Creek block of the Arbuckle-Simpson aquifer as an isolated system, interacting with the northern silicate-based aquifer only through dissolution faults and fractures. Such features might allow water to ascend directly from the underlying aquifer to the surface, which could account for the presence of Pilot Springs. Hence, if large amounts of water were to be pumped out of this aquifer block, it might lead to drying up of the streams in the area.

References

[1] H. Flathe, "Possibilities and limitations in applying geoelectrical methods to hydrogeological problems in coastal areas of north west Germany," *Geophysical Prospecting*, vol. 3, pp. 95–110, 1955.

[2] A. A. R. Zohdy, "The use of schlumberger and equatorial soundings in groundwater investigations near el paso, texas," *Geophysics*, vol. 34, no. 5, pp. 713–728, 1969.

[3] H. Flathe, "Interpretation of geoelectircal rsistivity measuremnts for solving hydrogeological problems," in *Mining and Groundwater Geophysics: Geological Survey of Canada Economic Geologica Report, no. 26*, E. W. Morely, Ed., pp. 580–597, 1970.

[4] A. A. Ogilvy, "Geophysical prospecting for groundwater in the Soviet Union," in *Mining and Groundwater Geophysics: Geological Survey of Canada Economic Geological Report, no. 26*, E. W. Morely, Ed., pp. 536–543, 1970.

[5] A. A. R. Zohdy, G. P. Eaton, and D. R. Mabey, *Application of Surface Geophysics to Ground Water Investigation*, U. S. G. S. Techniques of Water-Resource Investigation, 1974.

[6] S. Beeson and C. R. C. Jones, "The combined EMT/VES geophysical method of siting boreholes," *Ground Water*, vol. 26, no. 1, pp. 54–63, 1988.

[7] A. Olayinka and R. Barker, "Borehole siting in crystalline basement areas of Nigeria with a microprocessor-controlled resistivity traversing system," *Ground Water*, vol. 28, no. 2, pp. 178–183, 1990.

[8] J. R. T. Hazell, C. R. Cratchley, and A. M. Preston, "The location of aquifers in crystalline rocks and alluvium in Northern Nigeria using combined electromagnetic and resistivity techniques," *Quarterly Journal of Engineering Geology*, vol. 21, no. 2, pp. 159–175, 1988.

[9] J. R. T. Hazell, C. R. Cratchley, and C. R. C. Jones, "The hydrogeology of crystalline aquifers in northern. Nigeria and geophysical techniques used in their exploration," in *The Hydrogeology of Crystalline Basement Aquifers in Africa. Geological Society Special Publication, no. 66*, E. P. Wight and W. G. Burgess, Eds., pp. 155–182, 1992.

[10] R. D. Barker, "The offset system of electrical resistivity sounding and its use with a multicore cable," *Geophysical Prospecting*, vol. 29, no. 1, pp. 128–143, 1992.

[11] R. M. Carruthers and I. F. Smith, "The use of ground electrical methods for siting water supply boreholes in shallow crystalline basement terrains," in *The Hydrogeology of Crystalline Basement Aquifers in Africa. Geological Society Special Publication, no. 66*, E. P. Wight and W. G. Burgess, Eds., pp. 203–220, 1992.

[12] R. D. Barker, "Application of electrical tomography in groundwater contamination studies: 61st Mtg. Eur. Assoc. Expl Geophys," Extended Abstracts, European Association of Geophysical Exploration, Session:P082, 1996.

[13] W. E. Ham, M. E. McKinley et al., "Geologic map and sections of the Arbuckle Mountains, Oklahoma, Plate 1," in *Hydrology of the Arbuckle Mountains Area, South-Central Oklahoma: Oklahoma Geological Survey Circular 91*, K. S. Johnson, Ed., 1990.

[14] R. O. Fay, Geology of the Arbuckle Mountains along Interstate 35, Carter and Murray Counties, Oklahoma: Oklahoma Geological Survey Guidebook 26, 1989.

[15] R. W. Fairchild, R. L. Hanson, and R. E. Davis, Hydrology of the Arbuckle Mountains area, south-central Oklahoma: Oklahoma Geological Survey Circular 91, 1990.

[16] D. S. Scheirer and H. S. Alegra, "Gravity investigations of the Chickasaw National Recreation Area, south-central Oklahoma," USGS Open-File Report 2006-1083, 2006.

[17] J. Puckette, T. Halihan, and F. Faith,, "Characterization of the Arbuckle-Simpson Aquifer," Report for the Arbuckle-Simpson Hydrology Study, The Oklahoma Water Resources Board, 2009.

[18] W. M. Telford, L. P. Geldart, R. E. Sheriff, and D. A. Keys, *Applied Geophysics*, Cambridge University Press, 1976.

[19] K. Ramachandran, T. Bryan, R. Tayler, and L. Erin, "Characterizing the Arbuckle-Simpson aquifer through electrical methods," *SEG Technical Program Expanded Abstracts*, vol. 29, no. 1, pp. 2014–2018, 2010.

[20] E. Lewallen, *Geophysical investigation of the arbuckle-simpson aquifer, Southern Oklahoma*, M.S. thesis, The University of Tulsa, 2008.

[21] R. Barker and J. Moore, "The application of time-lapse electrical tomography in groundwater studies," *The Leading Edge*, vol. 17, no. 10, pp. 1454–1458, 1998.

[22] M. H. Loke and R. D. Barker, "Rapid least-squares inversion of apparent resistivity pseudosections by a quasi-Newton method," *Geophysical Prospecting*, vol. 44, no. 1, pp. 131–152, 1996.

[23] Loke M.H., "Tutorial: 2-D and 3-D electrical imaging surveys," course notes, http://www.geoelectrical.com/coursenotes.zip.

13

Analysis of Petroleum System for Exploration and Risk Reduction in Abu Madi/Elqar'a Gas Field, Nile Delta, Egypt

Said Keshta,[1] Farouk J. Metwalli,[2] and H. S. Al Arabi[3]

[1] *Geology Department, Faculty of Education, Suez Canal University, Arish, Egypt*
[2] *Geology Department, Faculty of Science, Helwan University, Cairo, Egypt*
[3] *Geology Department, Faculty of Science, Suez Canal University, Ismailia, Egypt*

Correspondence should be addressed to Said Keshta, said974@yahoo.com

Academic Editor: Jean-Pierre Burg

Abu Madi/El Qar'a is a giant field located in the north eastern part of Nile Delta and is an important hydrocarbon province in Egypt, but the origin of hydrocarbons and their migration are not fully understood. In this paper, organic matter content, type, and maturity of source rocks have been evaluated and integrated with the results of basin modeling to improve our understanding of burial history and timing of hydrocarbon generation. Modeling of the empirical data of source rock suggests that the Abu Madi formation entered the oil in the middle to upper Miocene, while the Sidi Salem formation entered the oil window in the lower Miocene. Charge risks increase in the deeper basin megasequences in which migration hydrocarbons must traverse the basin updip. The migration pathways were principally lateral ramps and faults which enabled migration into the shallower middle to upper Miocene reservoirs. Basin modeling that incorporated an analysis of the petroleum system in the Abu Madi/El Qar'a field can help guide the next exploration phase, while oil exploration is now focused along post-late Miocene migration paths. These results suggest that deeper sections may have reservoirs charged with significant unrealized gas potential.

1. Introduction

The Nile Delta basin contains a thick sequence of potential hydrocarbon source rocks that generate essentially gas and condensate.

The Nile Delta is generally known as a natural gas-prone (essentially methane/gas condensate) region with production from Miocene and Pliocene fields. However, temperature and pressure data from these fields suggest that the Nile Delta basin should be oil rather than gas prone [1].

The purpose of this paper is to evaluate potential source rocks in north eastern part of Nile Delta from available geochemical data from two wells. An additional objective is to study the regional petroleum systems by using numerical models which provide information about burial and temperature history, maturation of source rocks and timing of expulsion of hydrocarbons. Maturity information was used for calibration of numerical models.

Sharaf [2] showed from organic geochemical and petrographic analyses that the kerogen in the early Pliocene mudstones of the Kafr El-Sheikh formation and the Tortonian Wakar formation in the NE of the Nile Delta has poor capability to generate gas and minor oil. These formations are immature in all of the wells he studied. However, the early Miocene Qantara and the Middle Miocene Sidi Salem formations have a poor potential to generate gas and minor oil in the southern part of the area, further north Sharaf [2]. They have improved capability to generate oil and minor gas. These formations are immature in the southern parts of the area but are within the oil zone in the northern part [2].

In general, the kerogen in the Pliocene and Middle-Late Miocene samples from the NE Nile Delta is mainly of Continental origin. Terrestrial woody and herbaceous fragments [2] are the main components with a minor content of amorphous organic matter (AOM) and marine phytoplankton. In the Early Miocene and Oligocene samples

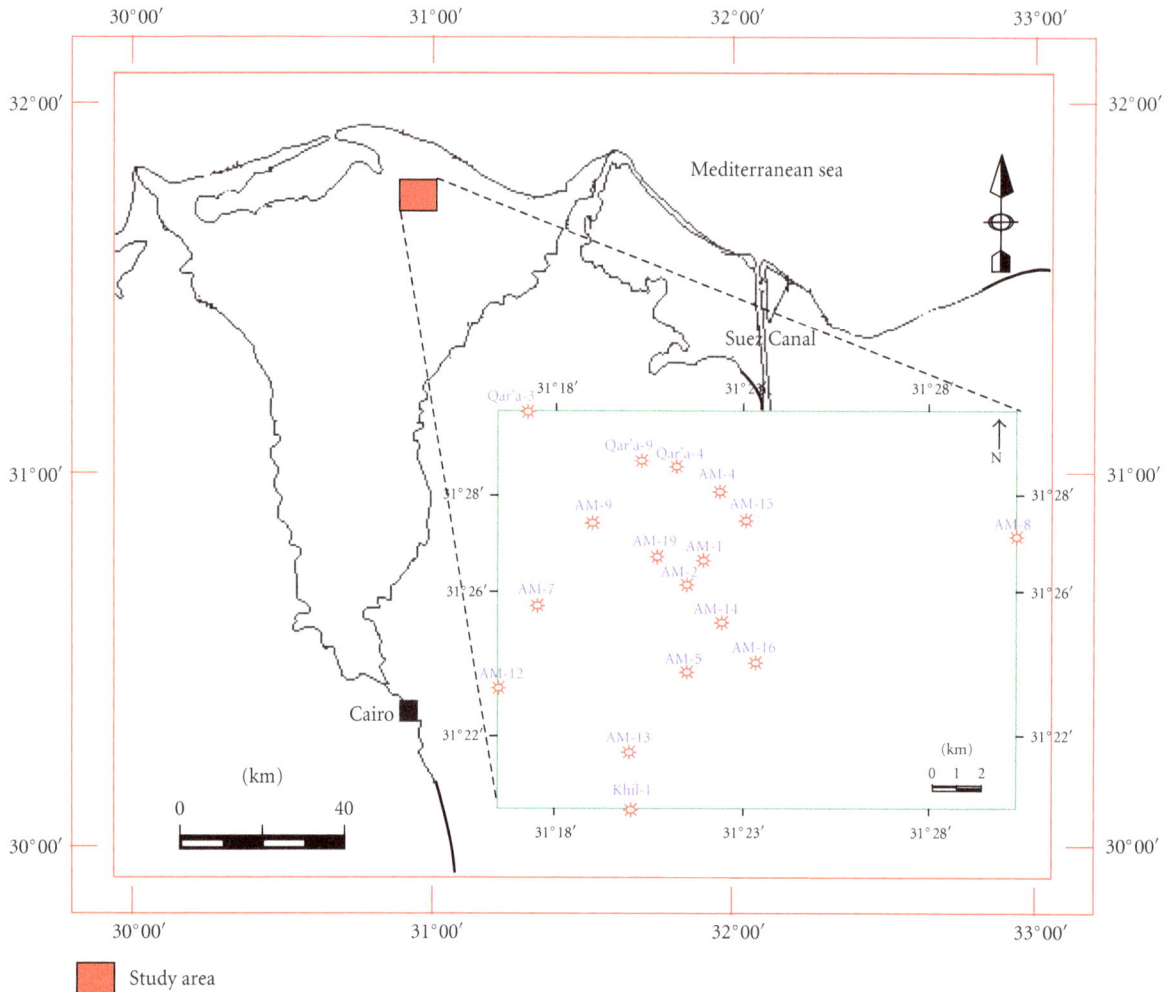

FIGURE 1: Location map of the study area.

from above 3500 m, the kerogen is mainly of Terrestrial origin. Below this depth, the kerogen quality improves with increasing depth (the content of AOM increases to 75% and the Terrestrial fragments decrease to 20%). The Eocene to late Cretaceous samples are characterized by moderate to high AOM and marine phytoplankton contents typical of kerogen with a good to fair petroleum generation potential. In the early-middle Cretaceous and late Jurassic samples, the kerogen is an amorphous-woody-algal assemblage accompanied by significant proportion of inertinitic debris [2].

2. The Area of Study

The area of study covers the onshore concession within Nile Delta. It lies between latitudes 31°20′ and 31°35′N and longitudes 31°15′ and 31°30′E (Figure 1). A total of 30 wells were drilled, of which 21 are in Abu Madi lease and 9 wells are in the El Qar'a lease. Twenty-four wells gave positive results and six wells were plugged and abandoned as dry holes, (five wells in Abu Madi lease and one well in El Qar'a lease (Figure 1). As a matter of fact, Abu Madi field is the first commercial discovery in the Nile Delta, where the IEOC

achieved the first gas discovery from the early pliocene Abu Madi formation in the northeastern part of the onshore Delta by drilling the Abu madi-1 well, encounteringtwo pay zones with more than 50 m thick gas and condensate bearing sands in Abu Madi formation [4]. The field was put on production in 1975 and was producing 355 MMSCF per day in 1995 [4].

3. Geological Setting

The sedimentary section in the Nile Delta area with gas potential seems to be limited to the Neogene formations trapped against listric faults or draped over tilted fault blocks. However pre-Miocene formations of the base of this Neogene sequence may also be considered as future exploration plays. Mesozoic reservoirs are present at greater depth and have been only penetrated by a few wells which are mostly located in the south delta block.

The sedimentary rocks penetrated in Abu Madi/El Qar'a field consist of thick clastics representing Miocene-Holocene time interval. These rocks were described by El Heiny and Enani [3], Figure 2, Alfy et al. [5], and Sarhan and Hemdan [6].

FIGURE 2: Generalized lithostratigraphic column of Abu Madi gas field modified after [3].

TABLE 1: Summary of pyrolysis data of Abu Madi-9 well.

Depth (M)	TOC	S1	S2	S3	T_{max}	HI	OI	S2/S3	%R_o	TTI
3067	0.68	0.15	0.51	0.85	427	75	125	0.6	0.6	9
3154	1.21	0.17	2.74	1.1	429	226	91	2.49	0.62	12
3184	1.12	0.14	2.1	0.95	429	187	85	2.21	0.6	9
3224	1.37	0.1	1.06	1.01	431	77	74	1.05	0.5	3
3256	1.54	0.05	1.1	1.06	434	72	69	1.04	0.7	20
3310	1.85	0.1	1.74	0.7	434	94	38	2.49	1.1	98
3344	0.81	0.09	0.71	0.25	436	87	31	2.84	1.3	165
3358	0.58	0.13	0.63	0.24	437	108	42	2.63	0.9	48
3390	0.51	0.25	0.25	0.17	439	49	34	1.47	1.35	179

The studied section involves lithostratigraphic units ranging in age from Miocene to Holocene. The studied section is differentiated into the rock units: Qantara, Sidi Salem, Qawasim, and Abu Madi formations of the Miocene age; Kafr El Sheikh and El Wastani formations of the Pliocene age; Mit Ghamer formation of late Pliocene-Pleistocene age

and the Bilqas formation of recent age. All these Formations consist essentially of clastic sediments (shale, sand, and silt).

4. Materials and Methods

One directional modeling of burial history and thermal maturity was preformed on five wells using basin mod Platter

TABLE 2: Summary of pyrolysis data of Abu Madi-11 well.

Depth (M)	TOC	S1	S2	S3	T_{max}	HI	OI	S2/S3	$\%R_o$	TTI
3108	1.5	0.07	1.38	1.95	429	89	125	0.71	0.49	0.8
3186	1.26	0.1	1.27	1.41	430	101	112	0.9	0.55	7
3210	1.65	0.06	1.42	1.7	431	86	103	0.84	0.71	21
3231	1.57	0.09	1.54	1.56	432	98	99	0.99	0.67	18
3250	1.62	0.11	1.41	1.07	432	87	66	1.32	0.88	44
3293	1.27	0.2	1.42	1.04	436	112	82	1.37	0.7	20
3318	0.82	0.05	0.75	0.77	435	91	94	0.97	0.68	18
3400	1.66	0.08	1.61	0.86	435	97	52	1.87	1.1	99
3425	2.03	0.1	0.91	1.02	436	45	50	0.89	1.15	112
3460	1.74	0.14	1.53	0.84	436	88	48	1.82	1.3	165

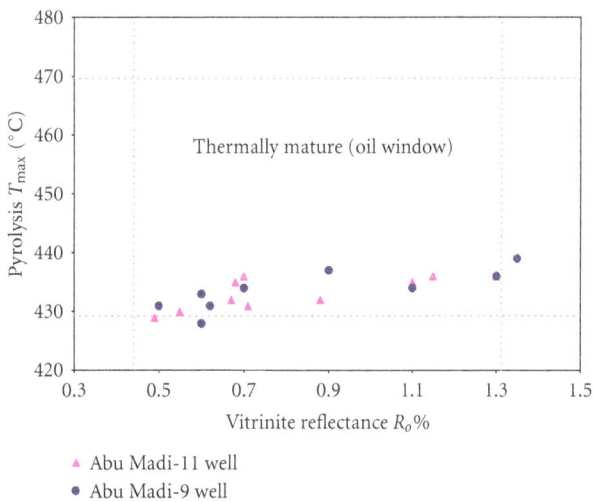

▲ Abu Madi-11 well

● Abu Madi-9 well

FIGURE 3: Plot of T_{max} versus Vitrinite reflectance values (R_o), showing good agreement between T_{max} and Vitrinite reflectance data.

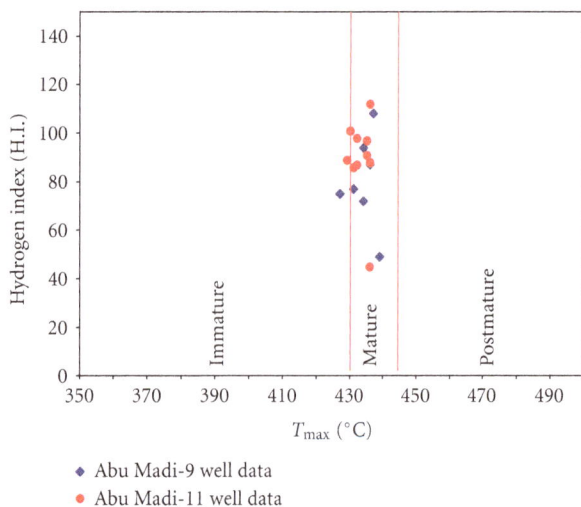

Abo Madi-9
Nile delta, Egypt

Abo Madi-9 well. mod

$Y = 1.61X + 23.22$
Goodness of fit = 0.42

FIGURE 5: Plot of Oxygen index versus hydrogen index for the analyzed samples from Abu Madi-9 well.

River (2003) software in addition to kerogen data measured by Khaled et al. [7], these samples were taken from Abu Madi formation and they represent a depth interval from 3067 m to 3390 m in well Abu Madi-9 and, 3108 m to 3460 m in Abu Madi-11 well (Tables 1 and 2). Parameters measured include TOC, S1, S2, S3 and temperature of maximum pyrolysis yield (T_{max}). Hydrogen index and oxygen index were calculated as described by Espitalie et al. [8] and peters and Cassa [9].

5. Modeling Procedures

To Construct the burial history, the essential input data included formation tops from ground level, absolute time

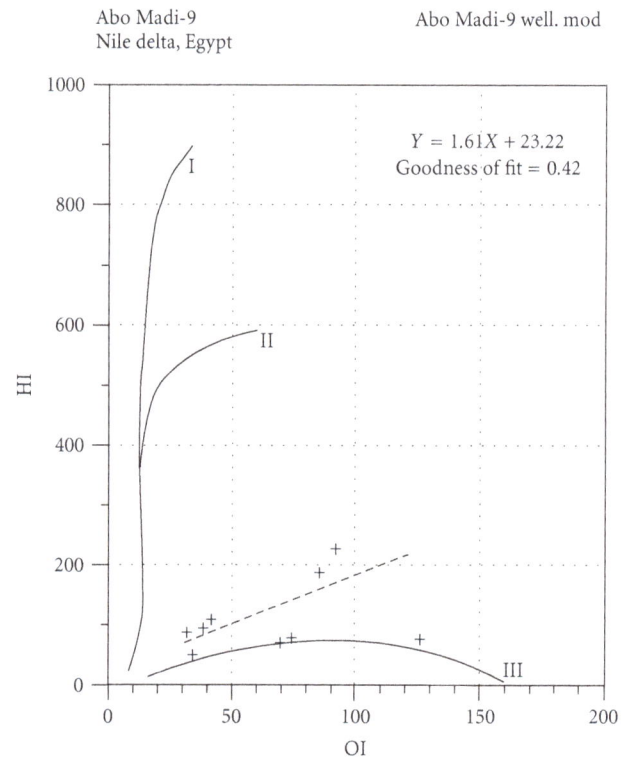

◆ Abu Madi-9 well data

● Abu Madi-11 well data

FIGURE 4: Plot of T_{max} versus H.I. for the analyzed samples from Abu Madi-9 and Abu Madi-11 wells well.

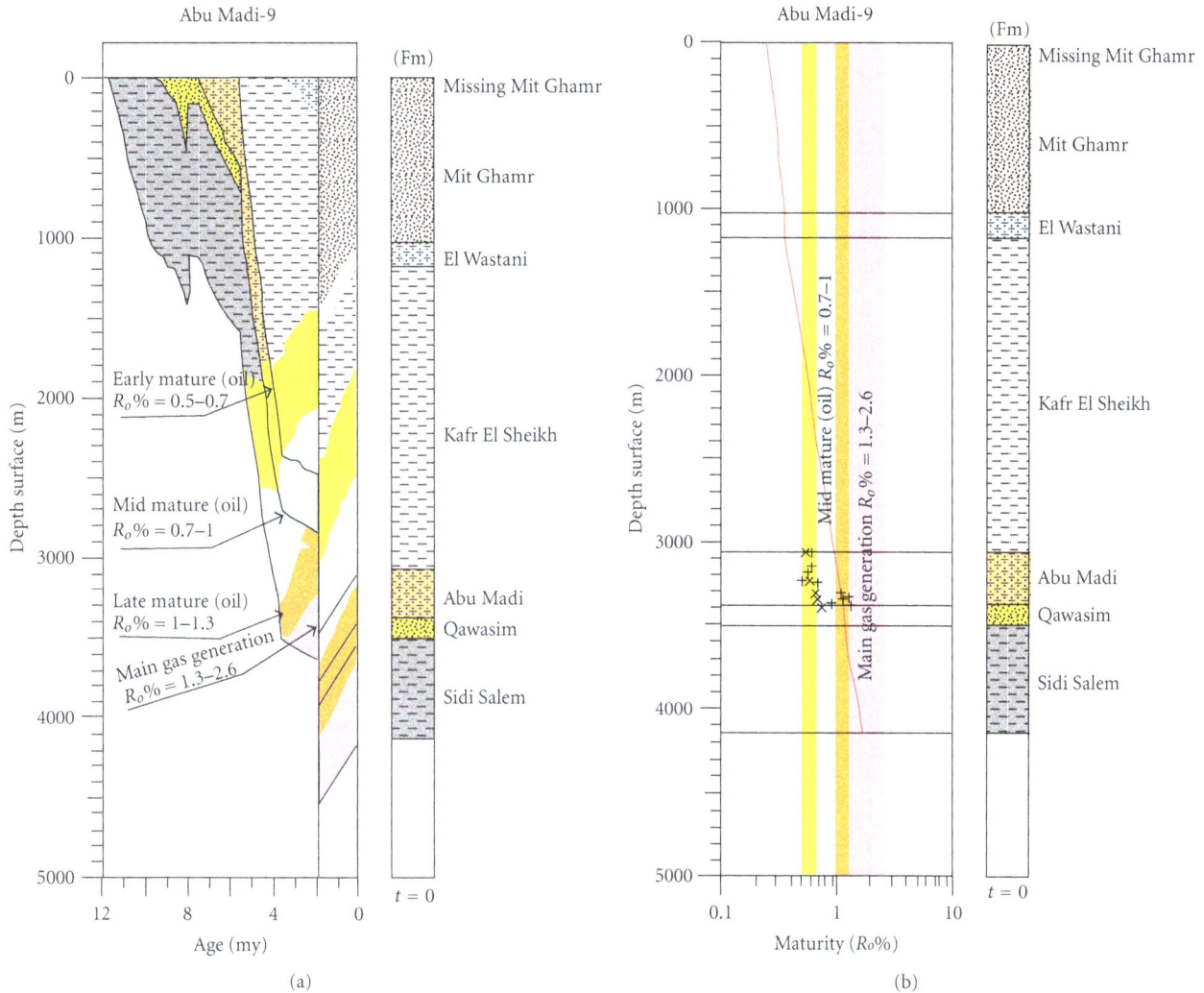

FIGURE 6: (a) Burial history curve for Abu Madi-9 well showing the calculated maturity (b) the measured maturity for samples taken from Abu Madi formation. Both figures agree the level of maturity for Abu Madi formation (early to mid mature).

of deposition in Ma (millions of years), lithological composition, hiatus age, thickness and age of eroded intervals, and the heat flow data calculated from observed geothermal gradients. The study area had sufficient information for modeling. Absolute age in many of the different stratigraphic units was defined using the global stratigraphic chart complied by Cowie and Bassett [10]. The lithological composition of the stratigraphic units was obtained from composite logs, whereas the average porosities and densities of these reservoir units were determined from petrophysical analysis.

In the basin modeling, the sedimentary sequence at the wells is subdivided into layers from the Tortonian (12 Ma) to the present and have vertical continuity in lithology and lateral continuity in time. The vertical continuity is essential to correctly compute pressure and temperature histories. The lateral continuity in time is needed to accurately define chronostratigraphy and to plot the result at any time during the geologic history [11].

6. Source Rock Evaluation

6.1. TOC and Pyrolysis Data. Total organic carbon (TOC) of samples from Abu Madi formation (Tables 1 and 2) ranges from 0.51 to 2.03 Wt% which is considered good potential source rock. The migration index (S1/TOC) ranges from 0.03 to 0.49 mg HC/g TOC with an average value of 0.11 mg HC/g TOC. This value lies in the range of 01–0.2 mg HC/g TOC suggested by Hunt [13] to characterize the oil expulsion zone. In the Rock-Eval pyrolysis analysis, free hydrocarbons (S1) in the rock and the amount of hydrocarbons (S2) and CO_2 (S3) expelled from pyrolysis of kerogen are measured (Tables 1 and 2). In addition, the T_{max} value, which represents the temperature at the point where the S2 peak is maximum, is also determined [14]. Pyrolysis data from 19 samples from Abu Madi Formation presents low values of S1 (average 0.12 mg HC/g rock), S2 value ranges from 0.25 to 2.74 mg HC/g rock (average is 1.27 mg HC/g rock), the (S1 + S2) range from 0.50 to 2.91 mg HC/g rock

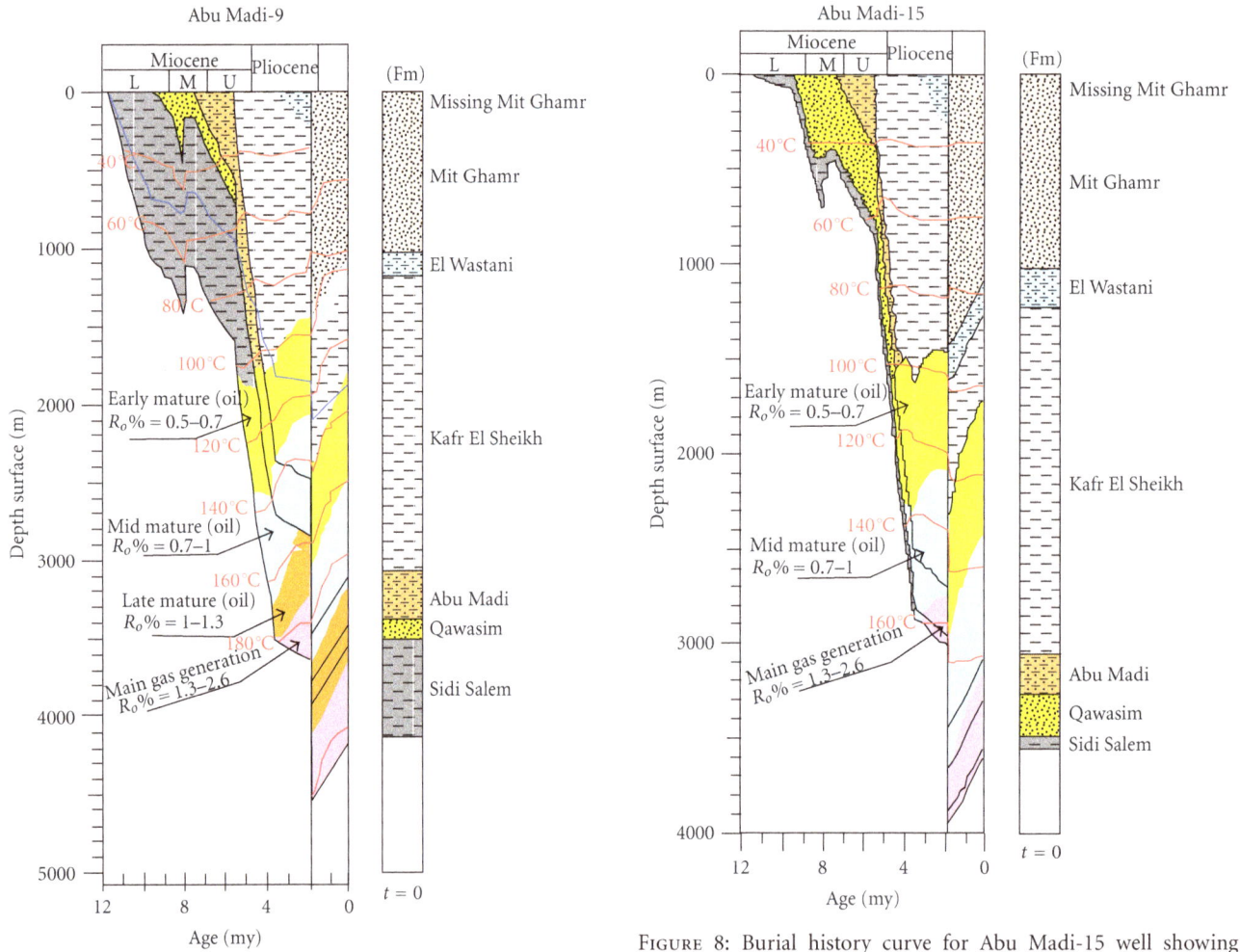

FIGURE 7: Burial history curve for Abu Madi-9 well showing kinetic window of stratigraphic units.

FIGURE 8: Burial history curve for Abu Madi-15 well showing kinetic window of stratigraphic units.

(average 1.38 mg HC/g rock). The calculated S2/S3 equals 1.3 (less than 2).These values indicate gas-prone organic matter and poor to fair hydrocarbon generation potential. The values of Production Index (PI) expressed by $\{S1/(S1 + S2)\}$ range from 0.04 to 0.5 (average 0.11) lie in the range of oil window [13]. Accordingly, organic matter in mudstone bed within Abu Madi formation is suggested to be fairly mature and gas prone; they have reached the early oil generation zone very close to the roof of the oil window.

6.2. Vitrinite Reflectance. The vitrinite reflectance R_o values range from 0.5 to 1% indicating that the samples are thermally mature and have entered the mature to late mature stage of hydrocarbon generation. This is supported by pyrolysis T_{max} values as shown in Figure 3.

6.3. Type of Organic Matter. The stage of maturity can be estimated using the temperature of maximum pyrolysis yield (T_{max}), although this is partly dependent on other factors such as the type of organic matter or mineral matrix effects [9]. In general, T_{max} values less than 435°C indicate

immature organic matter, whereas values of about 445°C indicate the end of the oil window and the beginning of the wet gas zone [8]. A plot of T_{max} versus hydrogen index from Abu Madi-9 and Abu Madi-11 wells is shown in Figure 4. The maturity stages determined from this plot (Figure 4) indicate that the kerogen of mature type agrees with that determined from vitrinite reflectance.

The hydrogen index (HI = S2/TOC) versus oxygen index (OI = S3/TOC) plot on the Van Krevelen diagram [15, 16], for the data from Abu Madi-9 well is shown in Figure 5. The plot indicates the predominance of organic matter of type III.

6.4. Thermal Maturity of Organic Matter. The heat flow history of a basin is proposed by establishing an agreement between a calculated (or modeled) maturity parameter and the equivalent observed maturity parameter (such as vitrinite reflectance or Rock-Eval T_{max}). The maturity profile (Figure 6) reveals a good match between the measured and calculated vitrinite reflectance values. Windows boundary of oil and gas varies with type of organic matter, ranging from 0.5% to 1.0% R_o and 1.3–3.5 R_o, respectively [14, 16]. Generally, vitrinite reflectance values increase with depth due

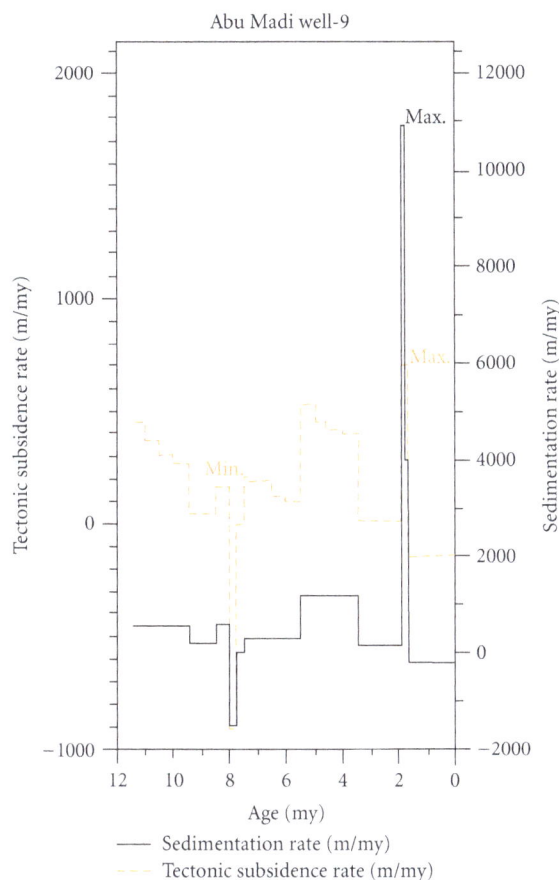

FIGURE 9: Tectonic subsidence rate for Abu Madi-9 well, showing period of nondeposition and period of increasing rate of subsidence/sedimentation.

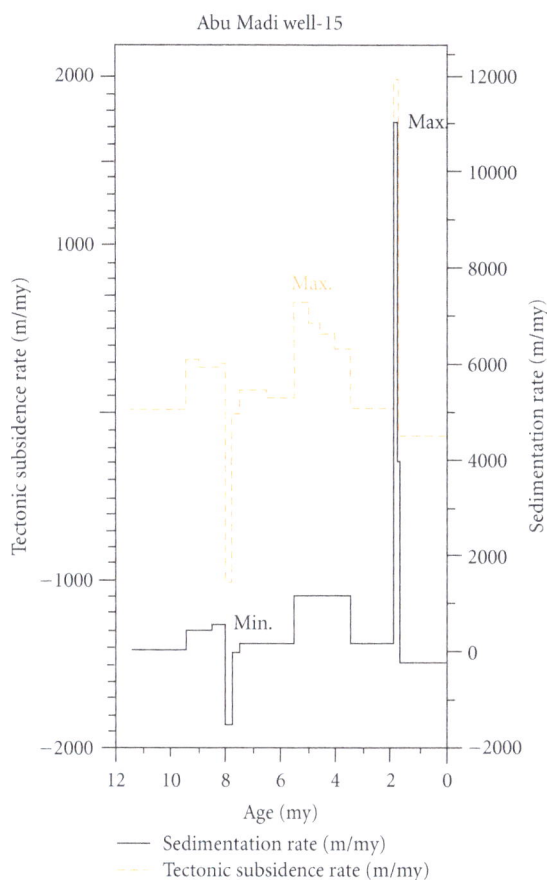

FIGURE 10: Tectonic subsidence rate for Abu Madi-15 well, showing period of nondeposition and period of increasing rate of subsidence/sedimentation.

to an increase in temperature and age of the rock with depth. The available data generally suggest that the majority of the kerogen of Abu Madi formation belong to mature type III in the principal zone of oil generation (oil window), where R_o values range from 0.5 to 1%, with small amounts of kerogen immature type III, where R_o is less than 0.5%. Figure 3 shows distribution of vitrinite reflectance data indicating that the samples taken from late Miocene Abu Madi formation are sufficiently thermally mature (oil window) for hydrocarbon generation (Figure 3).

7. Numerical Modeling

7.1. Burial History. Supporting evidence for the existence of prolific petroleum systems in the Abu Madi/El Qar'a gas field area comes from basin modeling in the area, which incorporated an analysis of the petroleum system criticals [17]. For effective exploration, a better understanding of the processes that led to the generation, migration, and accumulation of hydrocarbons is necessary. The purpose of the modeling is to evaluate the maturity of the potential source rocks and to estimate their timing of generation and expulsion. The calculated maturities of the potential source rocks were calibrated against the available measured maturity

parameters, which included mainly vitrinite reflectance data ($\%R_o$).

The burial history curves of two wells penetrating the study area are shown in Figures 7 and 8. Theses curves show the subsidence history of Abu Madi-09 and Abu Madi-15 wells including both steady state and rifting where the Basin mod software uses two basic assumptions for heat flow histories. The steady state uses a constant heat flow over time, whereas the nonsteady state (rifting) uses a variable heat flow over time. These curves (Figures 7 and 8) reveal three distinct pluses of subsidence and uplift. These pluses are from 12 to 5.5 Ma, from 5.5 to 3.5 Ma and from 3.5 Ma to the present.

The burial history in Abu Madi/El Qar'a field is characterized by a relatively low rate of sediment accumulation from 12 to 5.5 Ma in the middle to upper Miocene (Tortonian-Messinian), during the deposition of Sidi Salem, Qawasim, and Abu Madi formations. A relatively brief period of uplift and erosion occurred between 8.5 and 7.5 Ma and it was ended with deposition of Abu Madi formation.

The rate of deposition increased substantially between 5.5 and 3.5 Ma in the lower Pliocene during deposition of Kafr El Sheikh formation, when most of the sediments (between 1810 ft to 2003 ft) were deposited. A slowing rate of sedimentation in the upper Pliocene occurred during

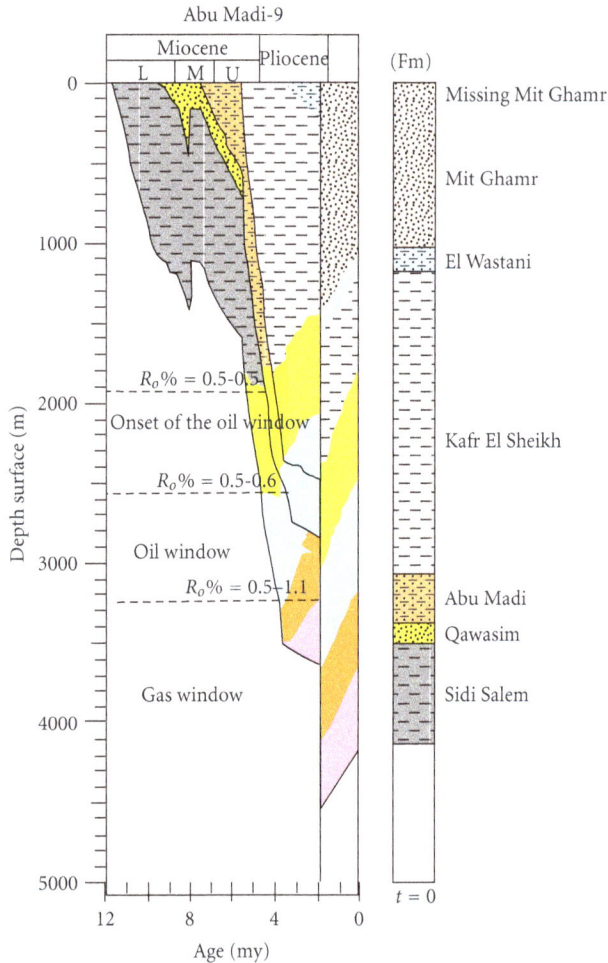

FIGURE 11: Burial history curve with hydrocarbon zones for the Abu Madi and Sidi Salem formations.

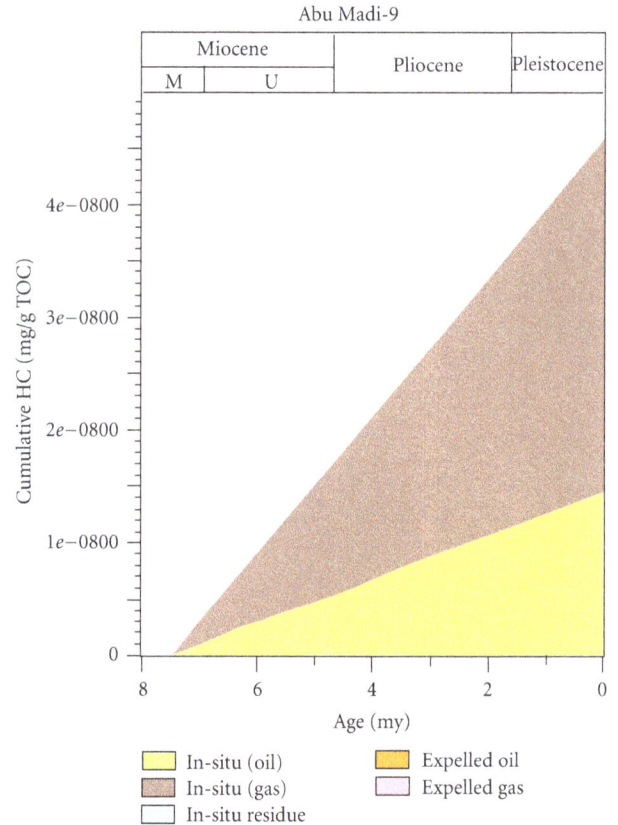

FIGURE 12: Calculated (cumulative model) of hydrocarbon generation from organic matter in the Abu Madi formation from Abu Madi-9 well, Abu Madi/El Qar'a Gas field.

deposition of El Wastani formation. This is followed by more rapid, post upper Pliocene, burial from 1.9 Ma to recent during deposition of Mit Ghamer formation.

7.2. *Tectonic Subsidence Curves.* The tectonic subsidence curves of the two selected wells are shown in Figures 9 and 10. These illustrate the tectonic subsidence and sedimentation rate in the Abu Madi/El Qar'a field area. The curves clearly show the period of non deposition that occurred in the Middle Miocene, prior to deposition of the Abu Madi Formation (from 8.5 to 7.5 Ma). The curves also reveal the existence of two periods marked by an increase of subsidence/sedimentation rate, during deposition of Kafr El Sheikh and Mit Ghamer Formations.

7.3. *Timing of Hydrocarbon Generation of Miocene Source Rocks.* The timing of hydrocarbon generation from the Miocene Abu Madi and Sidi Salem formations in the studied wells was analyzed and determined based on temperature and maturation history. Petroleum generation stages were calculated assuming mainly type III kerogen and using a

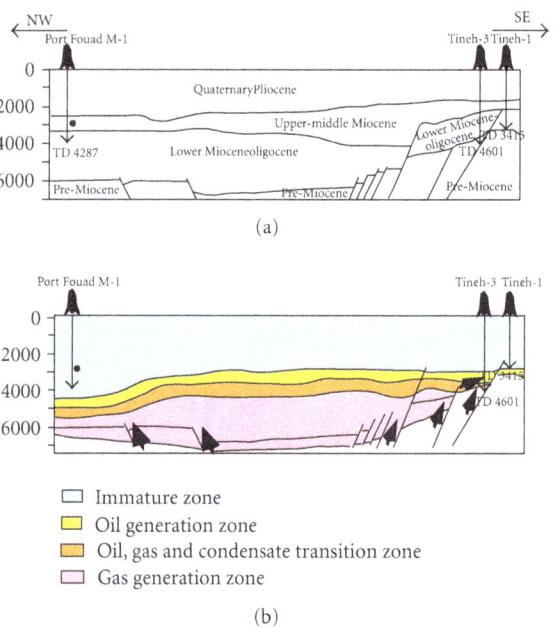

FIGURE 13: NW-SE cross sections showing the migration paths through the Nile Delta (modified after [12]).

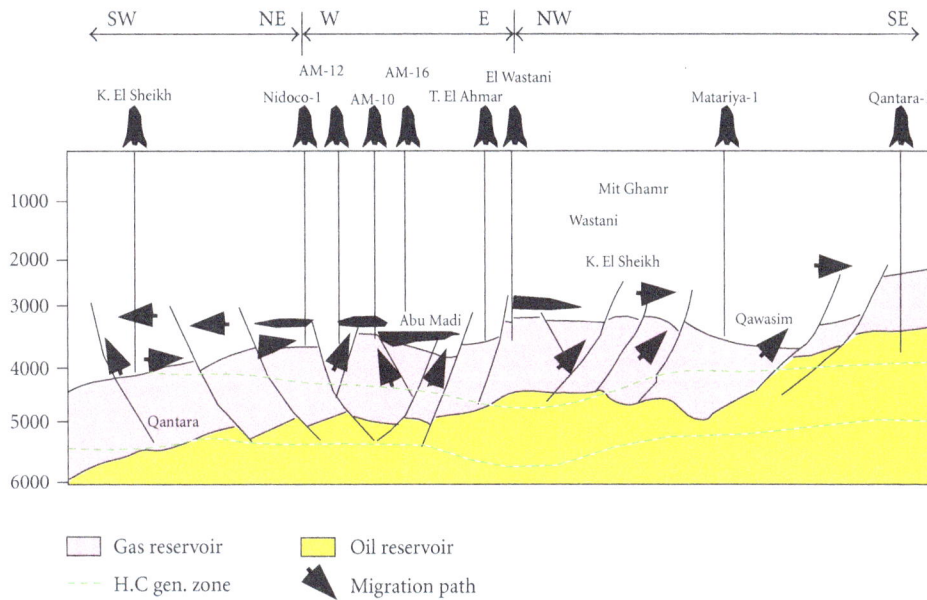

FIGURE 14: Hydrocarbon reservoirs and migration paths through the Nile Delta (modified after [12]).

reaction kinetic data set based on Burnham [18]. The modeled hydrocarbon generation from Abu Madi-9 well is shown in Figure 11. This model shows that the corresponding to onset of the oil window (0.5-0.6 R_o) of the Miocene source rocks was during early Miocene at depths greater than 2000 m (Figure 11). The Miocene source rocks reached late mature stage at middle to upper Miocene age and the gas window in upper Miocene to lower Pliocene. Accordingly the hydrocarbon generation (oil and gas) started in middle to upper Miocene and peak hydrocarbon generation occurred during lower Pliocene (Figures 11 and 12).

In the area of study geochemical analysis from two wells (Abu Madi-9 and Abu Madi-11) and burial history curves were taken to identify and characterize rich source intervals that probably the source of hydrocarbons, where organic matter (OM) in mudstone beds within Abu Madi Formation and shale beds of Sidi Salem formation is considered as effective hydrocarbon source rocks in study area.

8. Migration and Entrapment

Understanding the process of hydrocarbon generation and migration coupled with good geological data will assist to predict the ultimate hydrocarbon accumulations. The generation and migration of the hydrocarbons are thought to have reached their peak at the end of Miocene. This occurred after the main structural features had been imposed on the area and the main reservoirs had been deposited [12].

Migration along faults is responsible for vertical migration pathways from mature source rocks to shallower reservoirs (Figures 13 and 14). The presence of deep-cut channels, old valleys during Messinian, and the unconformities allows a good path for lateral updip migration and further entrapment in the shallow closures [12].

In the study area, where the shallow Pliocene growth faults are not well developed, only the normal fault pattern plays the main role in the vertical migration for the generated hydrocarbons and its ultimate entrapment and accumulation in the Miocene and older reservoirs. The possibilities of finding out commercial Pliocene gas accumulation are relatively low and depending on the presence of syndepositional growth faults in the Pliocene.

9. Conclusions

Majority of Abu Madi kerogen belongs to mature type III in the principal zone of oil generation (oil window), R_o values range from 0.5 to 1%. With small amounts of kerogen immature type III, where R_o is less than 0.5%. Plot of hydrogen index versus oxygen index also indicates the predominance of organic matter of type III.

Subsidence history of Nile Delta basin classified into two phases, the first phase from early Cretaceous to middle Miocene is a mechanical (faulted-controlled) subsidence phase prevailed which continued from subsidence initiation (9 Ma) to (3 Ma). The second phase is a nonmechanical (thermal-controlled) subsidence from (3 Ma) and continued through Tertiary.

Numerical modeling of various wells indicates that onset of the oil window (0.5-0.6 R_o) of the Miocene source rocks was during early Miocene at depths greater than 2000 m. The Miocene source rocks reached late mature stage at middle to upper Miocene age and the gas window in upper Miocene to lower Pliocene. Accordingly the hydrocarbon generation (oil and gas) started in middle to upper Miocene and peak hydrocarbon generation occurred during lower Pliocene.

The maturation modeling of the study area revealed that the hydrocarbon compositions (gas and condensate) of Abu

Madi formation are sourced from both mudstone beds in Abu Madi Formation and shale beds of Sidi Salem formation, where vitrinite reflectance estimated from models Maturity degree of Abu Madi Formation is early mature according to Vitrinite Reflectance determined from models (R_o range from 0.5 to 0.76) and Sidi Salem formation to main gas generation zone (R_o range from 1.3 to 2.6).

References

[1] Z. M. Zaghloul, A. A. Taha, and A. Gheith, *Microfacies studies and paleo-environmental trends on the subsurface sediments of Kafr el sheikh well no. 1*, vol. 5 of *Nile Delta Area, Bulletin*, Faculty of Science, Mansoura University, 1977.

[2] L. M. Sharaf, "Source rock evaluation and geochemistry of condensates and natural gases, offshore Nile Delta, Egypt," *Journal of Petroleum Geology*, vol. 26, no. 2, pp. 189–209, 2003.

[3] I. El-Heiny and N. Enani, "Regional stratigraphic interpretation pattern of Neogene's sediments, Northern Nile Delta, Egypt," in *Proceedings of the 13th EGPC, Exploration and Production Conference*, Cairo, Egypt, 1996.

[4] EGPC (Egyptian General Petroleum Cooperation), *Nile Delta and North Sinai: A Field, Discoveries and Hydrocarbon Potentials (A Comprehensive Overview)*, Egyptian General Petroleum, Cairo, Egypt, 1994.

[5] M. Alfy, F. Polo, and M. Shash, "The geology of Abu Madi Gas Field," in *Proceedings of the 11th Petroleum Exploration and Production Conference, EPGC, Cairo*, vol. 2, pp. 485–513, 1992.

[6] M. Sarhan and K. Hemdan, "North Nile Delta structural setting and trapping mechanism, Egypt," in *Proceedings of the 12th Petroleum Conference of EGPC*, vol. 12, no. 1, pp. 1–17, Cairo, Egypt, 1994.

[7] K. A. Khaled et al., "Maturation and source-rock evaluation of mudstone beds in Abu Madi and Kafr el Sheikh Formations," in *Proceedings of the Abu Madi Gas Field, Nile Delta, Egypt, Annals of the Egyptian Geological Survey*, vol. 26, pp. 449–474, 2004.

[8] J. Espitalie, M. Madec, B. Tissot, J. J. Mennig, and P. Leplat, "Source rock characterization, methods of petroleum exploration," in *Proceedings of the Offshore Technology Conference*, vol. 3, no. 9, pp. 439–444, May 1977.

[9] K. E. Peters and M. R. Cassa, "Applied source rock geochemistry," in *The Petroleum System, from Source to Trap. AAPG Mem*, B. Magoon Leslie and W. G. Dow, Eds., pp. 93–120, 1994.

[10] J. W. Cowie and M. G. Basset, "Global stratigraphic chart with geochronometric and magneto-stratigraphic calibration," in *International Union of Geological Sciences*, vol. 12, no. 2, 1989.

[11] M. A. Yukler and Welte, "A three—dimensional deterministic dynamic model to determined geologic history and hydrocarbon generationmigration and accumulation in a sedimentary basin: Fossil Fuels," 1980.

[12] H. Kamel, T. Eita, and M. Sarhan, "Nile Delta hydrocarbon potentiality," in *Proceeding of the 14th Petroleum Conference*, vol. l2, pp. 485–503, EGPC, Cairo, Egypt, October 1998.

[13] J. M. Hunt, *Petroleum Geochemistry and Geology*, W. H. Freeman, New York, NY, USA, 2nd edition, 1996.

[14] J. Espitalie, "Use of Tmax as a maturation index for different types of organic matter-comparison with vitrinite reflectance," in *Thermal modeling in Sedimentary Basins*, J. Burrus, Ed., pp. 475–496, Editions Technip, Paris, France, 1985.

[15] D. W. Van Krevelen, "Organic geochemistry-old and new," *Organic Geochemistry*, vol. 6, pp. 1–10, 1984.

[16] B. P. Tissot and D. H. Welte, *Petroleum Formation and Occurrence*, Springer, 2nd edition, 1984.

[17] F. I. Metwalli and J. D. Pigott, "Analysis of petroleum system criticals of the Matruh-Shushan Basin, Western Desert, Egypt," *Petroleum Geoscience*, vol. 11, no. 2, pp. 157–178, 2005.

[18] A. K. Burnham, A simple kinetic model of petroleum formation and cracking Lawrence Livermore National Laboratory Report UCID-21665, 1989.

14

The Study of Westward Drift in the Main Geomagnetic Field

G. Bayanjargal

Research Center for Astronomy and Geophysics, Mongolian Academy of Sciences (MAS), Ulaanbaatar, Mongolia

Correspondence should be addressed to G. Bayanjargal; readersman@yahoo.com

Academic Editor: Rudolf A. Treumann

We have obtained a solution for the velocity of westward drift from the induction equation in which an approach for main geomagnetic field was built. Distribution functions $B(r, t)$ entered into the induction equation have been built by the observatories' data in North America and the Europe from 1991 to 2006. The longitudinal −0.123 degree/year and latitudinal 0.068 degree/year drifts were defined in North America. And the longitudinal −0.257 degree/year drift was defined in Europe from 1991 to 2006. These drifts are similar to results of other studies.

1. Introduction

Westward drift in the main geomagnetic field (MGF) has been studied since discovered by Halley [1]. We could write it briefly from the previous results of investigations by other people. Bauer [2] studied the drift of the zero line of declination tracing. He detected that the mean velocity was −0.22 deg/yr. Bullard et al. [3] also studied the nondipole field at epoch 1907.5 and 1945. They determined that the drift velocity was −0.26 deg/yr. Yukutake [4] estimated that the mean value of westward drift was −0.20 deg/yr in 1850–1950. Wei and Xu [5] estimated that the mean value of westward drift was −0.18 deg/yr in 1900–2000. These researches were based mainly on the spherical harmonic potential's formula for main geomagnetic field.

If we have got the geomagnetic field's distribution function of space and time, can we define the westward drift on the Earth's little area such as a continent or any part of continents by other method? Is there any theoretical possibility to decide this problem? When I read Yukutake's paper of 1962, the spirit of solving the previous problem arose to me. Our task is to answer these questions or to find a new method. Yukutake [4] hinted an idea to write a similar equation with the induction equation written in liquid core for Earth's surface. But Yukutake did not estimate the westward in MGF from the induction equation. He studied the westward in MGF by scalar potential's method.

The westward drift in MGF is the result of interaction between fluid motions on the core mantle boundary CMB and the strong toroidal field. And the westward drift is observed in steady MGF in a large interval of time scale. The effects of fluid motions with small scales of space and time on CMB are not observed on the Earth's surface because these effects could be decreased by mantle.

Thus, we have built an approach for the induction equation on CMB by mean field's dynamic theory. And we have obtained the form of the induction equation for Earth's surface.

I think that the westward drift in MGF was not obtained from the induction equation in the past. The reason is that many satellites have been blasted in orbital paths around the Earth since 1962. And scalar potential's method for MGF has been rapidly developed since 1962. Thus, the idea of studying the westward drift in MGF by the induction equation has been forgotten since 1962.

In this paper, we have studied the westward drift in MGF by the induction equation. And we have used the geomagnetic field's distribution functions of space and time in North America and Europe, because internet network had been built since around 1980. And http://intermagnet.org/ had been built since 1990. Therefore, we could build $B(r, t)$ distribution functions of space and time by data of http://intermagnet.org/ in Europe and North America. These distribution functions have been defined by the data of observatories in Figures 1 and 2.

2. The Approach in the Induction Equation for Main Geomagnetic Field

Strong toroidal field is generated by Herndon's postulate [6] in subshell of georeactor in the Earth's inner core. The strong toroidal field is steady field. In other words, the energy of the strong toroidal field does not change in a few years. The strong toroidal field generated in the center of inner core spreads through the inner core, liquid core, and mantle. And it is observed as the main geomagnetic field (MGF) on the Earth's surface.

Most effectual movements of electrical conduction substances, which are interacted with the strong toroidal field, exist only in the Earth's liquid core.

The Earth's liquid core is about 23 percent denser at the bottom than at top. This difference of density blocks the possibility that radial flows are generated by the thermal convection in Earth's liquid. Thus, thermal equilibrium does not establish for whole Earth's liquid core. And the flows in Earth's liquid move in a horizontal direction. In other words, the radial component of the flow is very small in the Earth's liquid core. And pressure increases in the direction of depth of the liquid core. The fluidity of liquid core also decreases in the direction of depth.

Thus, the following representation exists in the geodynamic theory. The most effectual flows or fluid motions, which interact with the strong toroidal field, exist on the core mantle boundary (CMB). The interaction between the fluid motions on the CMB and the strong toroidal field produces irregular magnetic fields with small scale. The irregular magnetic fields are called non-dipole part of the MGF. And variation with the long time period exists in the non-dipole part of the MGF. The variation of MGF is called secular variation (SV). On the Earth's surface, the westward drift in the MGF is also observed by the influence of the fluid motions on the CMB. The interactions between the fluid motions on the CMB and the strong toroidal field are studied by the governing differential equation.

The governing differential equation is written in the form

$$\frac{d\vec{B}}{dt} = \vec{\nabla} \times \left(\vec{V} \times \vec{B} \right) + \eta \nabla^2 \vec{B}. \tag{1}$$

Equation (1) is called the induction equation. Here, η is the coefficient of magnetic diffusivity and $\vec{V}$ is the local velocity of the conductive fluid motions on the CMB.

It is possible that the fluid motions on the CMB are divided into two components such as mean motions and turbulent motions. Velocity of the mean motions is steady and its L-scale is long. l-scale of the turbulent motions is small as $L \gg l$. The velocity of the turbulent motions rapidly changes and it has fluctuations. They are written in the form

$$\vec{V} = \vec{V}_0 + \vec{v} = \text{mean} + \text{fluctuations},$$

$$\vec{B} = \vec{B}_0 + \vec{b} = \text{mean} + \text{fluctuations}. \tag{2}$$

These effects of fluctuations on the mean field on CMB are very small. They are also decreased by mantle. Thus, the effects of fluctuations are mostly not observed in the main geomagnetic field on the Earth's surface.

Therefore, (1) must be averaged when (2) is substituted into it. Equation (1) is averaged by (3), called Reynold's rules for averaging, which are

$$\left\langle \vec{V}_0 \right\rangle = \vec{V}_0, \qquad \left\langle \vec{v} \right\rangle = 0,$$

$$\left\langle \vec{V} \right\rangle = \left\langle \vec{V}_0 + \vec{v} \right\rangle = \left\langle \vec{V}_0 \right\rangle + \left\langle \vec{v} \right\rangle = \vec{V}_0,$$

$$\left\langle \vec{B}_0 \right\rangle = \vec{B}_0, \qquad \left\langle \vec{b} \right\rangle = 0, \tag{3}$$

$$\left\langle \vec{B} \right\rangle = \left\langle \vec{B}_0 + \vec{b} \right\rangle = \left\langle \vec{B}_0 \right\rangle + \left\langle \vec{b} \right\rangle = \vec{B}_0.$$

where $\langle \cdots \rangle$ shows statistical averaging, ($\approx$ volume or time averaging). The substitution and averaging of (1), (2), and (3) are shown as follows:

$$\left\langle \frac{d\vec{B}_0}{dt} + \frac{d\vec{b}}{dt} \right\rangle = \vec{\nabla} \times \left\langle \left(\vec{V}_0 + \vec{v} \right) \times \left(\vec{B}_0 + \vec{b} \right) \right\rangle$$

$$+ \eta \nabla^2 \left\langle \vec{B}_0 + \vec{b} \right\rangle. \tag{4}$$

Equation (4) can be written in the following form:

$$\frac{d\vec{B}_0}{dt} = \vec{\nabla} \times \left\langle \left(\vec{V}_0 + \vec{v} \right) \times \left(\vec{B}_0 + \vec{b} \right) \right\rangle + \eta \nabla^2 \vec{B}_0. \tag{5}$$

The term $\vec{\nabla} \times \langle (\vec{V}_0 + \vec{v}) \times (\vec{B}_0 + \vec{b}) \rangle$ in (5) can be written in the expanded form

$$\vec{\nabla} \times \left\langle \left(\vec{V}_0 + \vec{v} \right) \times \left(\vec{B}_0 + \vec{b} \right) \right\rangle$$

$$= \vec{\nabla} \times \left(\left\langle \vec{V}_0 \times \vec{B}_0 \right\rangle + \left\langle \vec{V}_0 \times \vec{b} \right\rangle \right.$$

$$\left. + \left\langle \vec{v} \times \vec{B}_0 \right\rangle + \left\langle \vec{v} \times \vec{b} \right\rangle \right) \tag{6}$$

$$= \vec{\nabla} \times \left(\vec{V}_0 \times \vec{B}_0 \right) + \vec{\nabla} \times \left\langle \vec{v} \times \vec{b} \right\rangle.$$

The terms $\langle \vec{V}_0 \times \vec{b} \rangle = \vec{V}_0 \times \langle \vec{b} \rangle = 0$ and $\langle \vec{v} \times \vec{B}_0 \rangle = \langle \vec{v} \rangle \times \vec{B}_0 = 0$ in (6) are vanished by Reynold's rules. When (6) is substituted into (5), it will be like

$$\frac{d\vec{B}_0}{dt} = \vec{\nabla} \times \left(\vec{V}_0 \times \vec{B}_0 \right)$$

$$+ \vec{\nabla} \times \left\langle \vec{v} \times \vec{b} \right\rangle + \eta \nabla^2 \vec{B}_0. \tag{7}$$

Equation (7) is called the averaged induction equation. The term $\vec{\nabla} \times \langle \vec{v} \times \vec{b} \rangle$ in (7) expresses the effects of turbulence on the mean magnetic field. The Equation

$$\vec{\varepsilon} = \left\langle \vec{v} \times \vec{b} \right\rangle \tag{8}$$

is called the mean electromotive force. There is a linear relation between the mean electromotive force $\vec{\varepsilon}$ and

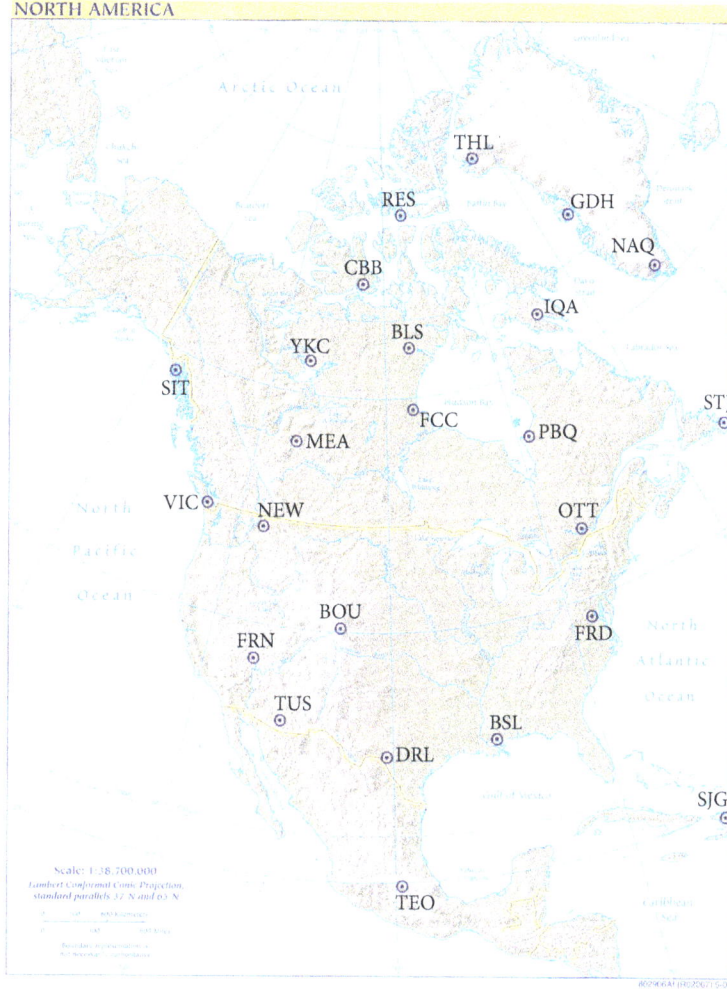

FIGURE 1: The locations of the observatories in North America.

the mean magnetic field $\vec{B}_0$. The relation is expressed in the following equation:

$$\vec{\varepsilon} = \alpha \vec{B}_0 - \beta \vec{\nabla} \times \vec{B}_0 + \cdots,$$

$$\alpha = -\frac{1}{3} \left\langle \vec{v} \cdot \left(\vec{\nabla} \times \vec{v} \right) \right\rangle \tau, \tag{9}$$

$$\beta = \frac{1}{3} v^2 \tau,$$

called Taylor's expansion. Here, α is the coefficient of the helical effects of fluid motion on the CMB, β is the coefficient for the turbulent diffusivity, and τ is lifetime of the helixes.

When (9) is substituted into (7), it becomes

$$\frac{d\vec{B}_0}{dt} = \vec{\nabla} \times \left(\vec{V}_0 \times \vec{B}_0 \right)$$

$$+ \vec{\nabla} \times \alpha \vec{B}_0 + (\eta + \beta) \nabla^2 \vec{B}_0. \tag{10}$$

The resulting equation (10) is the governing equation of the mean field on the CMB.

After expanding the term as $\vec{\nabla} \times (\vec{V}_0 \times \vec{B}_0) = -(\vec{V}_0 \cdot \vec{\nabla}) \vec{B}_0 + (\vec{B}_0 \cdot \vec{\nabla}) \vec{V}_0 - \vec{B}_0 (\vec{\nabla} \cdot \vec{V}_0)$ in (10), it becomes

$$\frac{d\vec{B}_0}{dt} = -\left(\vec{V}_0 \cdot \vec{\nabla} \right) \vec{B}_0 + \left(\vec{B}_0 \cdot \vec{\nabla} \right) \vec{V}_0$$

$$- \vec{B}_0 \left(\vec{\nabla} \cdot \vec{V}_0 \right) + \vec{\nabla} \times \alpha \vec{B}_0 \tag{11}$$

$$+ (\eta + \beta) \nabla^2 \vec{B}_0,$$

where $-(\vec{V}_0 \cdot \vec{\nabla}) \vec{B}_0$ is advection term, $(\vec{B}_0 \cdot \vec{\nabla}) \vec{V}_0$ is stretching term, $-\vec{B}_0 (\vec{\nabla} \cdot \vec{V}_0)$ is compression term, and $(\eta + \beta) \nabla^2 \vec{B}_0$ is diffusion and decay term. $\vec{V}_0$ is the mean velocity of the conductivity of the fluid motion on the CMB.

Equation (11) can be written in the form

$$\frac{d\vec{B}_0}{dt} + \left(\vec{V}_0 \cdot \vec{\nabla} \right) \vec{B}_0$$

$$= \left(\vec{B}_0 \cdot \vec{\nabla} \right) \vec{V}_0 - \vec{B}_0 \left(\vec{\nabla} \cdot \vec{V}_0 \right) \tag{12}$$

$$+ \vec{\nabla} \times \alpha \vec{B}_0 + (\eta + \beta) \nabla^2 \vec{B}_0.$$

FIGURE 2: The locations of observatories in Europe.

Generally, $\vec{V}_0$ is steady on the CMB and its local variation is very small. Thus, the effects of the terms $(\vec{B}_0 \cdot \vec{\nabla})\vec{V}_0$ and $-\vec{B}_0(\vec{\nabla} \cdot \vec{V}_0)$ on (12) for the CMB are very small. In other words, there are almost no stretching and the compression or we can say that they do not exist in the conductivity of the fluid motion on the CMB.

The terms $\vec{\nabla} \times \alpha\vec{B}_0$ and $\beta\nabla^2\vec{B}_0$ in (12) show turbulent effects in the fluid motion on the CMB. The scales of α and β are small. The life-time τ in α and β in (9) is also small. Therefore, the effects of the terms may be observed around one point or one observatory on the Earth's surface during short time. But the effects of the terms with α and β are small in the large interval of space and time. The reason is that the effects of these terms are decreased when they are integrated by the large interval of space and time. For example, the distribution functions of Bx, By, and Bz had covered continents such as Europe and North America in 1991–2006. Thus, the influence of the terms with α and β is very small in the same equation as (12) that will be written for the surface of the Earth or in the mantle.

As aftermentioned, (12) is only valid for the liquid core. But it can be seen from (12) that the advection term is the main factor in the secular variation of MGF on the Earth's surface.

We can write an equation of the same kind with (12) on the Earth's surface. The equation is

$$\frac{d\vec{B}}{dt} + \left(\vec{v}_0' \cdot \vec{\nabla}\right)\vec{B} = \vec{R}\left(\vec{r}, t\right), \tag{13}$$

where $\vec{r}$ denotes the positional vector and $\vec{v}_0'$ is the drift velocity of the field on the Earth's surface or in the mantle. $\vec{R}(\vec{r}, t)$ is the residual field after the westward drift has been subtracted from the observed time variation. $\vec{R}(\vec{r}, t)$ can also be described as the right-hand side of (12). Because the influence of these terms is decreased by mantle. The effects of these terms on the residual field in (13) are observed as weaker than those in (12) on the Earth's surface.

Therefore, it is seen from (12)-(13) that the variation of MGF on the Earth's surface depends on the mean velocity of the conductivity of the fluid motion on the CMB.

Now, we can rewrite the total energy of the residual field in the form

$$E = \int \left[\vec{R}\left(\vec{r}, t\right)\right]^2 ds, \tag{14}$$

where ds is an interval of space and time, $\vec{R}(\vec{r}, t)$ is a residual field, $[\vec{R}(\vec{r}, t)]^2$ is a variance of the energy density of the residual field in the small time interval, and E is the energy of the residual field.

The energy of the residual field weakly depends on the same terms as the right-hand side of (12). The energy variation of the strong toroidal field sourced in subshell in the Earth's inner core is also very small and it equals to zero in a time of 20 years.

Therefore, the energy variation of the residual field equals to zero in a time of 20 years as

$$\delta E = 0. \tag{15}$$

Once we know the field distribution $\vec{B}$ and its time derivative $d\vec{B}/dt$, we would be able to find the velocity $\vec{v}'_0$ from the condition that the energy variation of the residual field equals to zero on (15).

In this paper, we have solved the drift velocity $\vec{v}'_0$ of field from the condition in (15).

3. Construction of the Distribution Functions of Main Geomagnetic Field

In this paper, we have studied the westward drift in the MGF on the territories of North America and Europe. We have the geomagnetic field's data of 1991–2006 in the observatories located in North America and Europe. The locations and the territories of the observatories are shown in Figures 1 and 2.

One-minute data of X, Y, and Z in observatories in Figures 1 and 2 are used in this study. The components of the geomagnetic field can be displayed as Figure 3. Figure 3 is called the magnetogram.

The data of geomagnetic field noted in the observatories are the summary field that can be expressed as

$$B = M + m + D + N + S + L, \tag{16}$$

where M is the MGF, m is the secular variation of MGF, and D is a regular and irregular part of the disturbance field that is sourced outer from the Earth. N is a long period field and it has no frequency that is sourced outer from the Earth. N is also observed on the geomagnetic quiet days and it is not frequently observed. S, sourced by the solar effects, is the periodic field. The solar magnetic variation in the quiet days is denoted by Sq. L, sourced by the lunar effects, is periodic field. The term $M + m$ or MGF is about 90 percent of the summary intensity of magnetic field in (16). And the term $D + N + S + L$ is about 10 percent of the summary intensity.

Our aim is to subtract the term $M + m$ from the summary magnetic field in (16). We use a relatively simple method. The reason is that the maximum and minimum values of amplitude of any variation locate in a periodic time interval. We have averaged the geomagnetic field noted in observatories by (17). When the values of the amplitude in one period have been averaged, the maximum and minimum values are compensated. And the variations are decreased.

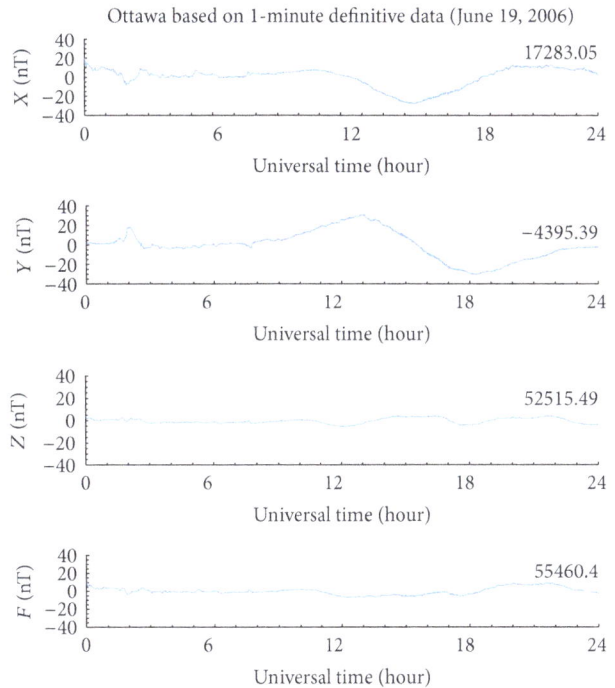

FIGURE 3: The magnetogram of a day in OTT.

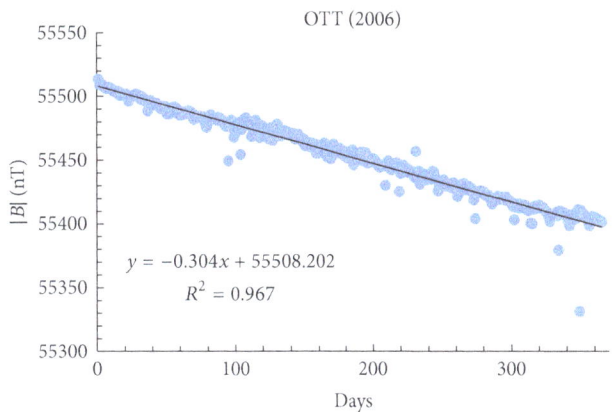

FIGURE 4: The dependence of the averaged value of MGF on the day order for OTT in 2006.

Thus, variations, period which is shorter than 24 hours, are decreased by the equation

$$\overline{B}_i = \frac{\sum_{j=1}^{1440} B_{ij}}{1440}, \tag{17}$$

where i is the number of days, j is the minutes, B_{ij} is the value of geomagnetic field noted in observatory in one minute, and $\overline{B}_i$ is the mean value of geomagnetic field for ith day.

But variation D is still there in the averaged values. Generally, the figure of the secular variation m is a monotonic and regular function. We can depict the general tendency of the secular variation by plotting the averaged values in (17). The depiction is in Figure 4.

We have also managed to subtract some days with great fluctuation by controlling them in the equation

$$\sigma_i = \left(B(t) - \overline{B}_i \right)^2, \tag{18}$$

where σ_i is the square of deviation of ith day from the general tendency for one year. $B(t)$ is the general tendency for one year detected by the mean values of the days. $\overline{B}_i$ is the mean value for ith day obtained using (17).

Generally, the effects of the magnetic storm are in the subtracted days. The effects of the magnetic storm have been eliminated by (18).

We have defined the mean geomagnetic field for a year by the equation

$$\overline{B} = \frac{\sum_{j=1}^{N} \overline{B}_i}{N}, \tag{19}$$

where $\overline{B}$ is the mean geomagnetic field for a year, $\overline{B}_i$ is the mean value of the magnetic field for the remaining days, and N is the number of the remaining days.

Variations with short period, such as minutes, days, months, and seasons, are decreased and eliminated by the previous averaging method. Thus, the MGF and its secular variation are in the mean value of the geomagnetic field for one year. The reason is that the previous averaging method cannot decrease or eliminate the secular variation with long period. Therefore, the mean values in (19) can represent the MGF and its secular variation.

We have defined the general tendency of the secular variation of MGF in the observatories in Figures 1 and 2.

Generally, it is seen that the secular variation of MGF in North America linearly decreases from 1991 to 2006. But in Europe, it linearly increases in the same time. Therefore, it can be considered that they are governed by one general dynamics.

We have defined the distribution functions of the MGF for each component in North America and Europe for each year. They are expressed by the polynomial regression which is

$$Bx = C_{00} + C_{01}\varphi + C_{02}\varphi^2 + C_{03}\varphi^3$$
$$+ C_{10}\theta + C_{11}\theta\varphi + C_{12}\theta\varphi^2$$
$$+ C_{20}\theta^2 + C_{21}\varphi\theta^2 + C_{30}\varphi^3,$$
$$By = A_{00} + A_{01}\varphi + A_{02}\varphi^2 + A_{03}\varphi^3$$
$$+ A_{10}\theta + A_{11}\theta\varphi + A_{12}\theta\varphi^2 \tag{20}$$
$$+ A_{20}\theta^2 + A_{21}\varphi\theta^2 + A_{30}\theta^3,$$
$$Bz = K_{00} + K_{01}\varphi + K_{02}\varphi^2 + K_{03}\varphi^3$$
$$+ K_{10}\theta + K_{11}\theta\varphi + K_{12}\theta\varphi^2$$
$$+ K_{20}\theta^2 + K_{21}\varphi\theta^2 + K_{30}\theta^2,$$

where Bx, By, and Bz are the components of MGF in the coordinate system NED, φ is longitude, θ is latitude, and

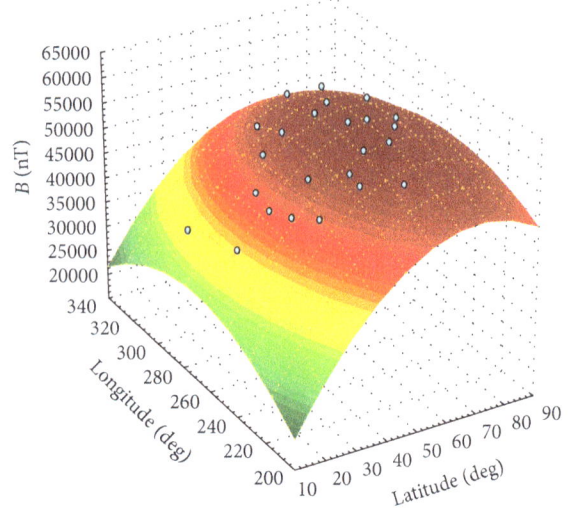

FIGURE 5: The MGF distribution functions in North America in 2006.

$C_{00}, \dots, C_{30}$, $A_{00}, \dots, A_{30}$, and $K_{00}, \dots, K_{30}$ are polynomial coefficients, and

$$Bx = D_{00} + D_{01}\varphi + D_{02}\varphi^2 + D_{03}\varphi^3$$
$$+ D_{10}\theta + D_{11}\theta\varphi + D_{12}\theta\varphi^2$$
$$+ D_{20}\theta^2 + D_{21}\varphi\theta^2 + D_{30}\theta^3,$$
$$By = Q_{00} + Q_{01}\varphi + Q_{02}\varphi^2 + Q_{03}\varphi^3$$
$$+ Q_{10}\theta + Q_{11}\theta\varphi + Q_{12}\theta\varphi^2$$
$$+ Q_{20}\theta^2 + Q_{21}\varphi\theta^2 + Q_{30}\theta^3,$$
$$Bz = H_{00} + H_{01}\varphi + H_{02}\varphi^2 + H_{03}\varphi^3$$
$$+ H_{10}\theta + H_{11}\theta\varphi + H_{12}\theta\varphi^2$$
$$+ H_{20}\theta^2 + H_{21}\varphi\theta^2 + H_{30}\theta^3, \tag{21}$$

where $D_{00}, \dots, D_{30}$, $Q_{00}, \dots, Q_{30}$, and $H_{00}, \dots, H_{30}$ are polynomial coefficients.

For example, we have displayed the distribution functions of the MGF in Figures 5 and 6.

The form of the distribution functions of space and time is not important. And the confidence level of the formula must be of high percentage. In case the relative errors of the extracted values of MGF are smaller than 1%–5%, P values of the distribution functions in (20) and (21) are 95%–98% by the reduced chi-square test.

The polynomial coefficients in (20)-(21) for each year were defined using the least square method.

Afterwards, we have defined the dependence of the polynomial coefficients on time. They are written in (B.1)–(B.3) and (B.4)–(B.6) in Appendix B.

We now have the distribution functions of MGF that depend on the space and time for North America and

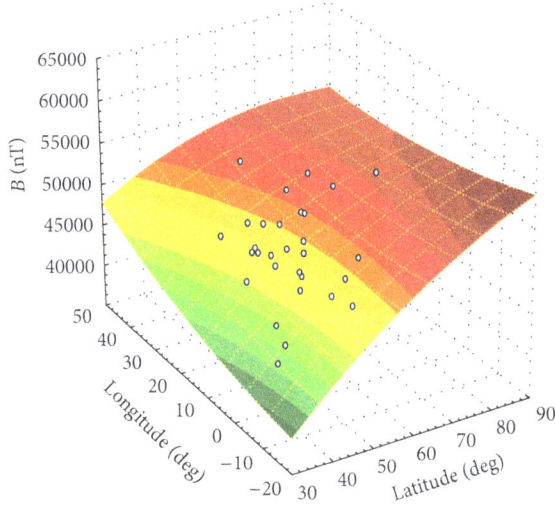

FIGURE 6: The MGF distribution functions in Europe in 2006.

Europe for the time between 1991 and 2006. Equations (20)-(21) are also written in the coordinate system NED. The coordinate system NED (x-north, y-east, z-down) is used in observatories of http://intermagnet.org/. The velocity of the westward drift is defined for the coordinate system Earth-Centered Earth-Fixed (ECEF). The ECEF or conventional terrestrial coordinate system rotates together with the Earth and has its origin at the center of the Earth. The x-axis passes through the equator at the prime meridian. The z-axis passes through the North Pole. The y-axis can be determined by the right-hand rule to be passing through the equator at 90 degrees of longitude.

Thus, the distribution functions in (20)-(21) for NED must be transformed to ECEF by (22) on the Earth's surface. Equation (22),

$$\vec{B}_e = \widehat{I}\vec{B},\qquad(22)$$

is expanded to (A.2) which is in Appendix A. Here, $\widehat{I}$ is an inversion tensor that transforms it from NED to ECEF. The inversion tensor is determined as

$$\left|\widehat{I}\right| = 1,\qquad(23)$$

and it is expanded to (A.1) in Appendix A.

4. Solution of the Drift Velocity

Now, the residual field in (13) is rewritten in the form of (24). And (22) is substituted into (24) which is

$$\vec{R}_e\left(\vec{r},t\right) = \frac{d\vec{B}_e}{dt} + \left(\vec{v}_0' \cdot \vec{\nabla}_e\right)\vec{B}_e$$
$$= \frac{d\left(\widehat{I}\vec{B}\right)}{dt} + \left(\vec{v}_0' \cdot \vec{\nabla}_e\right)\left(\widehat{I}\vec{B}\right).\qquad(24)$$

We must define the quadrate of the residual field. It is written in (25) which is

$$\left[\vec{R}_e\left(\vec{r},t\right)\right]^2 = \left[\widehat{I}\frac{d\vec{B}}{dt} + \widehat{I}\left(\vec{v}_0' \cdot \vec{\nabla}_e\right)\vec{B}\right]^2$$
$$= \widehat{I}^2\left[\frac{d\vec{B}}{dt} + \left(\vec{v}_0' \cdot \vec{\nabla}_e\right)\vec{B}\right]^2,\qquad(25)$$

where $\widehat{I}^2$ is the square of the inversion tensor. $\widehat{I}^2 = 1$. $\vec{\nabla}_e$ is Nable's operator in ECEF.

The reason is that quadrate of a vector is not varied by the inversion operation or the rotation of coordinate system on the Earth's surface. Equation (25) can be written as

$$\left[\vec{R}_e\left(\vec{r},t\right)\right]^2 = \left[\frac{d\vec{B}}{dt} + \left(\vec{v}_0' \cdot \vec{\nabla}_e\right)\vec{B}\right]^2.\qquad(26)$$

Equation (26) is called the variance of the energy density of the residual field in the small interval of time. Thus, we can define the energy of the residual field via

$$E = \int\left[\vec{R}_e\left(\vec{r},t\right)\right]^2 ds,\qquad(27)$$

where ds is an interval of space and time, $\vec{R}_e(\vec{r},t)$ is the residual field, $\left[\vec{R}_e(\vec{r},t)\right]^2$ is the variance of the energy density of the residual field in the small interval of time, and E is the energy of the residual field.

Components of the residual field in (27) are

$$R_{ex}\left(\vec{r},t\right) = \vec{i}\frac{dBx}{dt} + \vec{i}\left[v_{0x}'\frac{dBx}{dx} + v_{0y}'\frac{dBx}{dy} + v_{0z}'\frac{dBx}{dz}\right],$$
$$R_{ey}\left(\vec{r},t\right) = \vec{j}\frac{dBy}{dt} + \vec{j}\left[v_{0x}'\frac{dBy}{dx} + v_{0y}'\frac{dBy}{dy} + v_{0z}'\frac{dBy}{dz}\right],$$
$$R_{ez}\left(\vec{r},t\right) = \vec{k}\frac{dBz}{dt} + \vec{k}\left[v_{0x}'\frac{dBz}{dx} + v_{0y}'\frac{dBz}{dy} + v_{0z}'\frac{dBz}{dz}\right].\qquad(28)$$

When the study of westward drift is done in North America, (20) is substituted into (28).

In the coordinate system ECEF, the components of the drift velocity in (28) are

$$v_{0x}' = -r\sin\theta\cos\varphi \cdot \dot{\theta} - r\cos\theta\sin\varphi \cdot \omega,$$
$$v_{0y}' = -r\sin\theta\sin\varphi \cdot \dot{\theta} + r\cos\theta\cos\varphi \cdot \omega,\qquad(29)$$
$$v_{0z}' = r\cos\theta \cdot \dot{\theta}.$$

In case (28) and (29) are substituted into the functional in (27), it becomes

$$E_{am} = \int_{1991}^{2006}\left[\int_0^{R_0}\int_{\pi/9}^{4\pi/9}\int_{10\pi/9}^{16\pi/9}\left(R_{ex}^2 + R_{ey}^2 + R_{ez}^2\right)r^2\cos\theta\,d\varphi\,d\theta\,dr\right]dt,\qquad(30)$$

where R_{ex}, R_{ey}, and R_{ez} are

$$R_{ex} = \vec{i}\left(a_1 + b_1 \cdot \omega + c_1 \cdot \dot{\theta}\right),$$

$$R_{ey} = \vec{j}\left(a_2 + b_2 \cdot \omega + c_2 \cdot \dot{\theta}\right), \qquad (31)$$

$$R_{ez} = \vec{k}\left(a_3 + b_3 \cdot \omega + c_3 \cdot \dot{\theta}\right),$$

and where a_1, a_2, a_3, b_1, b_2, b_3, c_1, c_2, and c_3 are expanded in (A.3) in Appendix A.

We have defined the latitudinal and longitudinal drift velocities from the condition that the energy variation of the residual field is zero in (15).

5. Results and Discussion

When the functional in (30) is integrated by the intervals of space and time, it becomes

$$E_{am}\left(\omega, \dot{\theta}\right) = R_0^3 \left(m_1\dot{\theta} + m_2\omega\right.$$
$$\left. + m_3\dot{\theta}\omega + m_4\dot{\theta}^2 + m_5\omega^2\right), \qquad (32)$$

where m_1, m_2, m_3, m_4, and m_5 are the coefficients that are calculated in (B.7) in Appendix B.

Equation (15) can be rewritten in the form

$$\delta E_{am}\left(\omega, \dot{\theta}\right) = \frac{\partial E_{am}\left(\omega, \dot{\theta}\right)}{\partial \omega}\delta\omega + \frac{\partial E_{am}\left(\omega, \dot{\theta}\right)}{\partial \dot{\theta}}\delta\dot{\theta} = 0. \quad (33)$$

From (33), we can write the following condition:

$$\frac{\partial E_{am}\left(\omega, \dot{\theta}\right)}{\partial \omega} = 0,$$

$$\frac{\partial E_{am}\left(\omega, \dot{\theta}\right)}{\partial \dot{\theta}} = 0. \qquad (34)$$

Then, we can also write the following linear equation system:

$$m_2 + m_3\dot{\theta} + 2m_5\omega = 0,$$

$$m_1 + m_3\dot{\theta} + 2m_4\omega = 0. \qquad (35)$$

We can solve the previous linear equation system, and its solutions will be

$$\omega = -0.123 \text{ degree/year},$$

$$\dot{\theta} = 0.068 \text{ degree/year}. \qquad (36)$$

From the results of this investigation, we see that the longitudinal and latitudinal drifts existed in North America in 1991–2006.

When the study of westward drift is done for Europe, (21) is substituted into the functional in (30). Moreover, intervals of the integral in (30) are changed as

$$E_{eu} = \int_{1991}^{2006}\left[\int_0^{R_0}\int_{\pi/6}^{4\pi/9}\int_{-\pi/18}^{2\pi/9}\left(R_{ex}^2 + R_{ey}^2 + R_{ez}^2\right)r^2\cos\theta\,d\varphi\,d\theta\,dr\right]dt. \qquad (37)$$

After the functional in (37) is integrated by the previous intervals, it becomes

$$E_{eu}\left(\omega, \dot{\theta}\right) = R_0^3\left(n_1\dot{\theta} + n_2\omega + n_3\dot{\theta}\omega + n_4\dot{\theta}^2 + n_5\omega^2\right), \quad (38)$$

where n_1, n_2, n_3, n_4 and n_5 are coefficients, values of which are calculated in (B.8) in the Appendix B.

The linear equation system

$$n_2 + n_3\dot{\theta} + 2n_5\omega = 0,$$

$$n_1 + n_3\dot{\theta} + 2n_4\omega = 0 \qquad (39)$$

can be written from the condition like in (33). The solutions of (39) are

$$\omega = -0.257 \text{ degree/year},$$

$$\dot{\theta} = 7.356 \cdot 10^{-4} \text{ degree/year} \approx 0. \qquad (40)$$

Wei and Xu [5, 7] defined the longitudinal and the global mean velocity of westward drift as −0.18 deg/yr in MGF for the period 1900–2000. Bullard et al. [3] also defined the mean westward drift as −0.26 deg/yr in secular variation by the study for the period 1907–1945.

Their method is based on the spherical harmonic formula for the scalar potential of MGF. And the spherical harmonic formula is the global properties.

MGF distribution functions of space and time can be clearly defined on the little area of Earth's surface. On the little area, MGF distribution functions are more local properties than the spherical harmonic formula. Thus, our method is more accurate than the method of spherical harmonic analyze on the local little areas.

The westward drift in MGF on the Earth's surface is observed by the influence of fluid motions on CMB. Therefore, the properties of fluid motions on CMB can also be studied, in the opposite way, using the local properties of westward drift on the Earth's surface.

The probability of a dependence of the velocity of fluid motions on the latitude is high.

If we could build MGF distribution functions of space and time on the little areas differed with the latitudes as in Figure 7, we could be able to study the velocity of the fluid motions on the CMB using the velocity of westward drift. If the velocity of westward drift on the Earth's surface depends on the latitude, a similar dependence on the latitude must exist for the velocity of the fluid motions on the CMB. This information is important to the modeling of MGF. These are the advantages of our method.

6. Conclusions

The advection term $-(\vec{V}_0 \cdot \vec{\nabla})\vec{B}_0$ in (11) is the main factor in the time variation or the secular variation of MGF. In other words, (12) for the CMB very weakly depends on the terms in the right-hand side. And the energy variation does not exist in the strong toroidal field in a few years. They are the theoretical basis of this method to define the westward drift.

The westward drift can also be estimated using this method for the little area of Earth's surface, the MGF

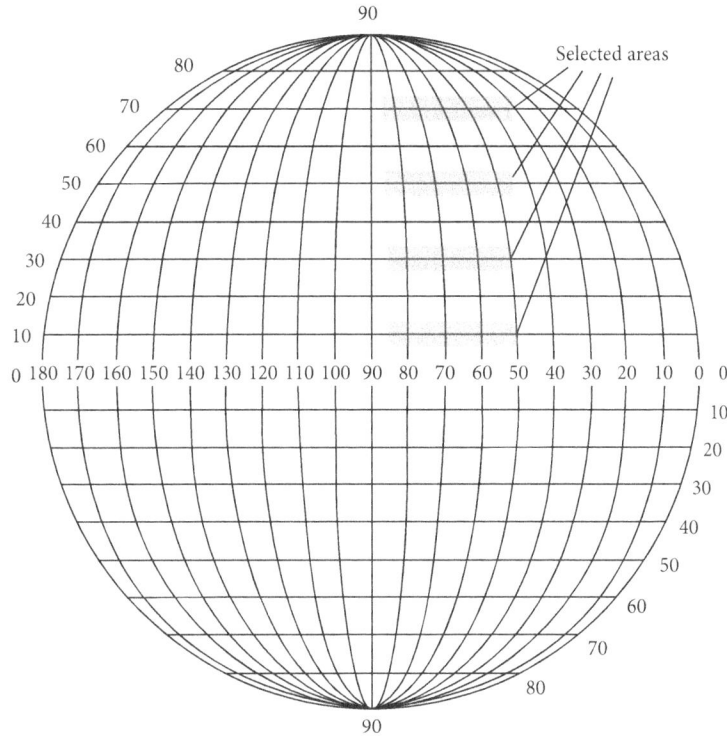

FIGURE 7: Samples of the selected areas.

distribution functions of space and time of which are well known.

In this paper, we have also defined the MGF distribution functions of space and time using the data of observatories located in North America and Europe. And the westward drifts have been estimated in the following forms. The longitudinal −0.123 degree/year and latitudinal 0.068 degree/year drifts existed in North America in 1991–2006. And the longitudinal −0.257 degree/year drift has been defined for Europe in 1991–2006. A latitudinal drift was not Europe in 1991–2006.

The results of these drifts are similar to the results by other studies which used the method of spherical harmonic formula.

Appendices

A. Mathematics

Inversion tensor that transforms NED to ECEF is written in the equation

$$
|\hat{I}| = \begin{vmatrix} -\sin\theta \cdot \cos\varphi & -\sin\varphi & -\cos\theta \cdot \cos\varphi \\ -\sin\theta \cdot \sin\varphi & \cos\varphi & -\cos\theta \cdot \sin\varphi \\ \cos\theta & 0 & -\sin\theta \end{vmatrix} = 1, \quad \text{(A.1)}
$$

where θ and φ are the latitude and longitude in the geodetic coordinates.

Equation (22) is rewritten in the following form:

$$
\begin{bmatrix} B_{xe} \\ B_{ye} \\ B_{ze} \end{bmatrix} = \begin{vmatrix} -\sin\theta \cdot \cos\varphi & -\sin\varphi & -\cos\theta \cdot \cos\varphi \\ -\sin\theta \cdot \sin\varphi & \cos\varphi & -\cos\theta \cdot \sin\varphi \\ \cos\theta & 0 & -\sin\theta \end{vmatrix} \begin{bmatrix} B_x \\ B_y \\ B_z \end{bmatrix},
$$
$$
\text{(A.2)}
$$

where B_{xe} is the parallel component with x axis of ECEF of geomagnetic fields on the points of the Earth's surface, B_{ye} is the parallel component with y axis of ECEF of geomagnetic fields, and B_{ze} is also the parallel component with z axis of ECEF of geomagnetic field.

The coefficients in (31) are rewritten in the following forms:

$$
a_1 = \frac{dB_x}{dt}, \qquad a_2 = \frac{dB_y}{dt}, \qquad a_3 = \frac{dB_z}{dt},
$$

$$
b_1 = \frac{dB_x}{dy} r \cos\theta \cos\varphi - \frac{dB_x}{dx} r \cos\theta \sin\varphi,
$$

$$
b_2 = \frac{dB_y}{dy} r \cos\theta \cos\varphi - \frac{dB_y}{dx} r \cos\theta \sin\varphi,
$$

$$
b_3 = \frac{dB_z}{dy} r \cos\theta \cos\varphi - \frac{dB_z}{dx} r \cos\theta \sin\varphi,
$$

$$
c_1 = \frac{dB_x}{dz} r \cos\theta - \frac{dB_x}{dy} r \sin\theta \sin\varphi - \frac{dB_x}{dx} r \sin\theta \cos\varphi,
$$

$$c_2 = \frac{dB_y}{dz} r \cos\theta - \frac{dB_y}{dy} r \sin\theta \sin\varphi - \frac{dB_y}{dx} r \sin\theta \cos\varphi,$$

$$c_3 = \frac{dB_z}{dz} r \cos\theta - \frac{dB_z}{dy} r \sin\theta \sin\varphi - \frac{dB_z}{dx} r \sin\theta \cos\varphi.$$

$$(A.3)$$

And the derivatives in (A.3) could be written in the following forms:

$$\frac{dB_x}{dx} = \frac{dB_x}{d\varphi}\frac{d\varphi}{dx} + \frac{dB_x}{d\theta}\frac{d\theta}{dx},$$

$$\frac{dB_y}{dx} = \frac{dB_y}{d\varphi}\frac{d\varphi}{dx} + \frac{dB_y}{d\theta}\frac{d\theta}{dx},$$

$$\frac{dB_z}{dx} = \frac{dB_z}{d\varphi}\frac{d\varphi}{dx} + \frac{dB_z}{d\theta}\frac{d\theta}{dx},$$

$$\frac{dB_x}{dy} = \frac{dB_x}{d\varphi}\frac{d\varphi}{dy} + \frac{dB_x}{d\theta}\frac{d\theta}{dy},$$

$$\frac{dB_y}{dy} = \frac{dB_y}{d\varphi}\frac{d\varphi}{dy} + \frac{dB_y}{d\theta}\frac{d\theta}{dy}, \qquad (A.4)$$

$$\frac{dB_z}{dy} = \frac{dB_z}{d\varphi}\frac{d\varphi}{dy} + \frac{dB_z}{d\theta}\frac{d\theta}{dy},$$

$$\frac{dB_x}{dz} = \frac{dB_x}{d\varphi}\frac{d\varphi}{dz} + \frac{dB_x}{d\theta}\frac{d\theta}{dz},$$

$$\frac{dB_y}{dz} = \frac{dB_y}{d\varphi}\frac{d\varphi}{dz} + \frac{dB_y}{d\theta}\frac{d\theta}{dz},$$

$$\frac{dB_z}{dz} = \frac{dB_z}{d\varphi}\frac{d\varphi}{dz} + \frac{dB_z}{d\theta}\frac{d\theta}{dz}.$$

Now, it is easy to define the derivatives dB_x/dx, dB_y/dx, dB_z/dx, dB_x/dy, dB_y/dy, dB_z/dy, dB_x/dz, dB_y/dz, and dB_z/dz using the equations

$$\frac{d\varphi}{dx} = \frac{d}{dx}\arccos\left(\frac{x}{\sqrt{x^2+y^2}}\right) = -\frac{\sin\varphi}{r\cos\theta},$$

$$\frac{d\varphi}{dy} = \frac{\cos\varphi}{r\cos\theta},$$

$$\frac{d\varphi}{dz} = 0,$$

$$\frac{d\theta}{dx} = \frac{d}{dx}\arccos\left(\frac{\sqrt{x^2+y^2}}{\sqrt{x^2+y^2+z^2}}\right) = -\frac{\cos\varphi\sin\theta}{r},$$

$$\frac{d\theta}{dy} = -\frac{\sin\varphi\sin\theta}{r},$$

$$\frac{d\theta}{dz} = \frac{\cos\theta}{r}.$$

$$(A.5)$$

For example, dB_x/dx can be written like

$$\begin{aligned}
\frac{dB_x}{dx} = &-\frac{\sin\varphi}{r\cos\theta}\left(C_{01} + 2C_{02}\varphi + 3C_{03}\varphi^2 \right. \\
&\left. + C_{11}\theta + 2C_{12}\theta\varphi + C_{21}\theta^2\right) \\
&-\frac{\cos\varphi\sin\theta}{r}\left(C_{10} + 2C_{20}\theta + 3C_{30}\theta^2 \right. \\
&\left. + C_{11}\varphi + 2C_{21}\varphi\theta + C_{12}\varphi^2\right).
\end{aligned}$$

$$(A.6)$$

B. Calculations

Our calculation should be as clear as possible. Thus, we have done some normalization in the scales of time and coordinate. When we define the coefficients in (20)-(21), we have changed the scale of time for 1991–2006 years by the 1–16 years. And the φ longitude is changed by $\varphi - 180$ only in North America. The values of westward drift must not be changed in the previous normalization. The normalization has only reduced the order of the polynomial coefficients in (20)-(21). And it helps to make the estimation clear.

In North America, the following polynomial coefficients have been defined for the longitudinal scale $\varphi - 180$ and the time scale of 1–16 years. And in Europe, the polynomial coefficients are also defined for the time interval of 1–16 years. Thus, we have integrated (30) and (37) by the previous intervals of time and coordinate in North America and Europe. But these are the technical problems in the calculation. Therefore we do not change the intervals of time and longitude in (30) and (37), because (30) and (37) should be understandable for researchers.

The dependences of the coefficients in (20) and (21) on time are defined as

$$C_{00} = -19.375t^2 + 898.247t + 32074.697,$$

$$C_{01} = 27.286t^2 - 1319.913t - 9026.928,$$

$$C_{02} = -10.194t^2 + 595.632t + 1691.978,$$

$$C_{03} = 0.652t^2 - 78.083t - 91.66,$$

$$C_{10} = -1065.704t + 52252.804,$$

$$C_{11} = -17.426t^2 + 962.015t - 14691.562,$$

$$C_{12} = 5.645t^2 - 222.753t + 5722.045,$$

$$C_{20} = -3.849t^3 + 108.95t^2 + 115.72t - 89763.49,$$

$C_{21} = -186.744t + 410.167,$

$C_{30} = 1.637t^3 - 46.433t^2 + 68.74t + 33421.397;$

$$(B.1)$$

$A_{00} = 16.855t^2 - 0.913t - 16101.254,$

$A_{01} = 1.395t^3 - 60.858t^2 + 204.93t + 54636.953,$

$A_{02} = -0.664t^3 + 29.015t^2 - 121.327t - 39358.43,$

$A_{03} = 0.125t^3 - 5.309t^2 + 30.47t + 7480.44,$

$A_{10} = -14.693t^2 - 220.23t + 25874.22,$

$A_{11} = 15.659t^2 + 168.957t - 26226.375,$

$A_{12} = -1.58t^2 - 25.624t + 5391.304,$

$A_{20} = 170.34t - 16502.614,$

$A_{21} = -5.76t^2 - 32.61t + 9350.964,$

$A_{30} = -0.443t^3 + 14.337t^2 - 126.78t + 1100.061;$

$K_{00} = -3.697t^3 + 145.03t^2 - 878.612t - 109655.43,$

$K_{01} = 3.29t^3 - 142.08t^2 + 1220.517t + 132037.83,$

$K_{02} = -0.882t^3 + 44.47t^2 - 579.29t - 37867.043,$

$K_{03} = -1.98t^2 + 51.055t + 401.07,$

$K_{10} = 6.912t^3 - 243.205t^2 + 955.721t + 206508.148,$

$K_{11} = -3.075t^3 + 120.14t^2 - 769.72t - 113743.611,$

$K_{12} = 0.99t^3 - 36.184t^2 + 418.863t + 25710.492,$

$K_{20} = -5.42t^3 + 174.353t^2 - 638.703t - 55351.172,$

$K_{21} = -299.78t + 14775.592,$

$K_{30} = 2.29t^3 - 68.197t^2 + 490.495t - 5461.943;$

$D_{00} = -0.79t^3 + 28.22t^2 - 253.338t + 49320.938,$

$D_{01} = -290.1645t + 6472.407,$

$D_{02} = 212.34t + 5017.157,$

$D_{03} = -0.0484t^3 - 1.4166t^2 + 34.979t - 681.798,$

$D_{10} = 2.612t^3 - 91.474t^2 + 922.318t - 38396.7694,$

$D_{11} = 453.385t - 16664.086,$

$D_{12} = -191.376t - 6551.478,$

$D_{20} = -2.751t^3 + 95.569t^2 - 1006.358t + 5209.597,$

$D_{21} = -200.728t + 10926.704,$

$D_{30} = 0.945t^3 - 32.644t^2 + 349.745t + 73.998;$

$$(B.2)$$

$$(B.3)$$

$$(B.4)$$

$Q_{00} = -0.601t^3 + 24.324t^2 - 269.764t + 1814.066,$

$Q_{01} = -1.122t^3 + 32.228t^2 - 162.816t + 4613.38,$

$Q_{02} = -6.53t^2 + 315.947t - 16129.512,$

$Q_{03} = -2.482t^2 + 96.678t + 655.054,$

$Q_{10} = 2.222t^3 - 85.493t^2 + 1037.27t - 7733.423,$

$Q_{11} = -446.872t + 11705.439,$

$Q_{12} = -228.862t + 12840.865,$

$Q_{20} = -2.72t^3 + 98.783t^2 - 1096.699t + 5678.082,$

$Q_{21} = 304.279t - 8005.954,$

$Q_{30} = 1.059t^3 - 36.481t^2 + 366.018t - 1303.48;$

$$(B.5)$$

$H_{00} = -2.811t^3 + 86.553t^2 - 730.614t - 42252.381,$

$H_{01} = -4.596t^3 + 140.1074t^2 - 833.826t - 1043.21,$

$H_{02} = 33.781t^2 - 1224.008t + 15682.71,$

$H_{03} = -5.79t^2 + 148.61t - 3773.1026,$

$H_{10} = 9.595t^3 - 308.561t^2 + 2816.664t + 192757.05,$

$H_{11} = 9.245t^3 - 298.329t^2 + 2267.262t + 8822.885,$

$H_{12} = -1.882t^3 + 16.501t^2 + 777.267t - 7568.86,$

$H_{20} = -10.759t^3 + 357.114t^2 - 3361.87t - 142531.088,$

$H_{21} = -4.553t^3 + 156.063t^2 - 1390.52t - 7658.873,$

$H_{30} = 3.912t^3 - 132.968t^2 + 1283.618t + 38877.982.$

$$(B.6)$$

The values of the coefficients in (35) and (38) are calculated as follows:

$$m_1 = -32357481.6,$$

$$m_2 = 7677960.7,$$

$$m_3 = 2726671352.51,$$

$$m_4 = 16031093418.8,$$

$$m_5 = 2597509055.6;$$

$$(B.7)$$

$$n_1 = 2751704.38,$$

$$n_2 = 1117572.227,$$

$$n_3 = 645280739.23,$$

$$n_4 = 5626256620.658,$$

$$n_5 = 125423203.758.$$

$$(B.8)$$

References

[1] E. Halley, "On the cause of the change in the variation of the magnetic needle, with an hypothesis of the structure of the internal parts of the earth," *Philosophical Transactions of the Royal Society of London*, vol. 17, pp. 470–478, 1692.

[2] L. A. Bauer, "On the secular motion of a free magnetic needle," *Physical Review (Series I)*, vol. 2, pp. 455–465, 1895.

[3] E. C. Bullard, C. Freedman, H. Gellman, and J. Nexon, "The westward drift of the Earth's magnetic field," *Philosophical Transactions of the Royal Society A*, vol. 243, pp. 67–92, 1950.

[4] T. Yukutake, "The westward drift of the magnetic field of the Earth," *Bulletin of the Earthquake Research Institute*, vol. 40, pp. 1–65, 1962.

[5] Z. Wei and W. Xu, "Latitude-dependence and dispersion of the westward drift in the geomagnetic field," *Chinese Science Bulletin*, vol. 47, no. 4, pp. 330–333, 2002.

[6] J. M. Herndon, "Nature of planetary matter and magnetic field generation in the solar system," *Current Science*, vol. 96, no. 8, pp. 1033–1039, 2009.

[7] Z. Wei and W. Xu, "Differential rotation of geomagnetic field," *Chinese Science Bulletin*, vol. 48, no. 24, pp. 2739–2742, 2003.

[8] T. Yukutake and H. Tachinaka, "The westward drift of the geomagnetic secular variation," *Bulletin of the Earthquake Research Institute*, vol. 46, pp. 1075–1102, 1968.

15

Imaging the Morrow A Sandstone Using Shear Wave VSP Data, Postle Field, Oklahoma

Naser Tamimi and Thomas L. Davis

Reservoir Characterization Project, Department of Geophysics, Colorado School of Mines, Golden, CO 80401, USA

Correspondence should be addressed to Naser Tamimi, ntamimi@mines.edu

Academic Editor: Joerg Schleicher

Morrow sandstones constitute important oil-producing reservoirs in the Anadarko Basin in the Mid-Continent Region of the USA. Characterization of the Morrow A sandstone reservoir in Postle Field, Texas County, Oklahoma, is challenging due to its small thickness, low acoustic impedance contrast with the surrounding Morrow shale, and lithological heterogeneity. Shear wave data have been documented as a promising solution for imaging the Morrow A sandstone. Vertical seismic profiling (VSP) offers the potential to enhance shear wave imaging of the thin heterogeneous Morrow A sandstone at Postle Field. The zero-offset VSP results confirm the advantages of shear wave over compressional wave in imaging the Morrow A sandstone. Also, the final shear wave VSP image shows that, applying the proposed processing flow, we are able to image the Morrow A sandstone where the thickness is as thin as 8.5 m.

1. Introduction

Pennsylvanian upper Morrow sandstones (Figure 1) constitute major oil-producing reservoirs across southeastern Colorado, southwestern Kansas, western Oklahoma, and northern Texas. They consist of multiple-stacked lenticular sandstone bodies formed within valley-fill complexes. The Morrow A is an oil-producing sandstone mainly in Oklahoma [1–3]. In this study, we mainly focus on the Morrow A sandstone in Postle Field, Texas County, Oklahoma (Figure 2).

Compressional wave studies have been mostly used in the past for characterizing the Morrow sandstones [4–7]. Briefly, there are two main challenges with characterizing the Morrow A sandstone based on compressional data: first, the thickness of the Morrow A sandstone is below the tuning thickness [6, 8] and secondly, the acoustic impedance contrast between the Morrow A and the surrounding Morrow shale layer is extremely low to the point that these sandstones are acoustically transparent [9, 10]. Dipole sonic well logs reveal the low acoustic impedance contrast between the Morrow A and overlying shale (Table 1 and Figure 3).

Singh and Davis [10] showed that the low acoustic contrast between the Morrow A sandstone and surrounding Morrow shale layer and the presence of strong multiples from shallow anhydrite layers are important challenges associated with the detection of the Morrow A sandstone in compressional data (Figure 4). Full waveform modeling [10, 11] suggested that shear wave data can improve the Morrow A imaging and characterization significantly in Postle Field. Hardage et al. [12] drew the same conclusion after their study on thicker section of the Morrow sandstone using three zero-offset 9C VSP surveys. Also, a 3D VSP study using shear data addressed the advantages of shear wave in imaging of sandstone layers which have low acoustic impedance is contrast with surrounding layers [13].

Shear VSP data have characteristics which make this method a unique source of information. The advantages of VSP methods used in this study involve higher frequency, better wavefield separation, and higher signal-to-noise ratio. In this study, we expect to map the Morrow A sandstone and its variations around the VSP well. To achieve this purpose, certain processing steps and enhancements were applied that will be discussed.

TABLE 1: Physical and elastic properties of the Morrow shale and the Morrow A sandstone from well log data (well HMU 24-4).

Properties	Morrow shale	Morrow A sandstone	Change from Sh to Sst
V_P (m/sec)	3335	3875	+16.21%
V_S (m/sec)	1518	2279	+50.13%
V_P/V_S	2.20	1.70	−22.59%
ρ (gm/cm³)	2.57	2.44	−5.01%
Z_P	8571	9462	+10.38%
Z_S	3901	5563	+42.60%

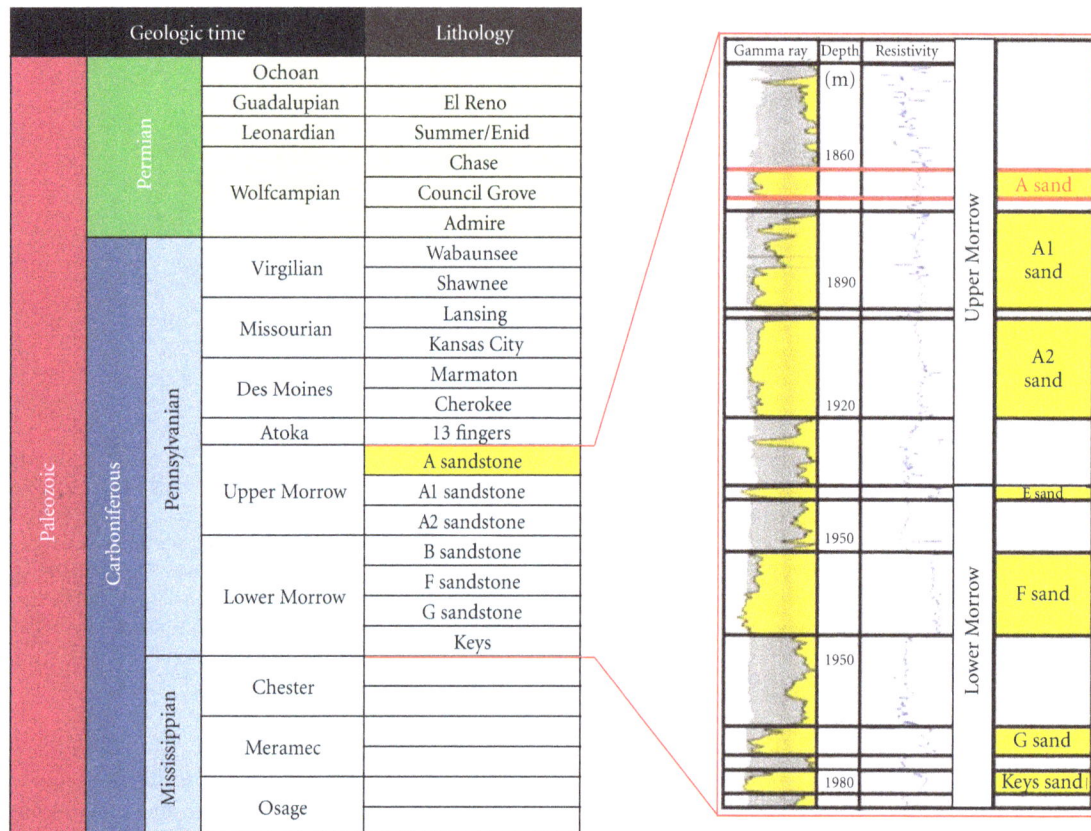

FIGURE 1: General stratigraphic column (left) and detailed stratigraphy column (right) show the main formations at Postle Field, Oklahoma. The Morrow A sandstone is the main oil-producing reservoir in this field. Shale layers between the Morrow sandstones are called the Morrow shales which are of important significance for characterizing heterogeneity in the reservoir.

2. Field VSP Data

As a part of Reservoir Characterization Project Phase XII at Colorado School of Mines, two 3D multicomponent VSP surveys were acquired from well HMU 13-2 in March and December 2008. Figures 5(a) and 5(b) demonstrate source locations and 3C geophone arrays in the well, respectively. For imaging purposes, we used only the March 2008 VSP survey in this paper.

The March 2008 survey was mainly acquired with horizontal vibrators, with the exception of two vertical shots: one 91 m to the north of the VSP well (called zero-offset data) and one 1829 m to the south of the VSP well (for the purpose of receiver reorientation). The horizontal vibrator truck generated two mutually orthogonal horizontal (E-W

and N-S) ground motions. The geophones in the borehole were located from 933 m to 1818 m MD (their depth spacing is 15 m) from well head (Figure 5(b)). The last geophone was 60 m above the top of the Morrow A sandstone. Table 2 summarizes source and receiver parameters of the March 2008 VSP survey.

In addition to VSP data, well log and surface seismic data were used in this study. The locations of studied wells and surface seismic area are depicted in Figure 6.

3. VSP Data Processing

VSP processing was applied to three datasets from the March 2008 survey: zero-offset shear dataset, zero-offset

TABLE 2: Source and receiver parameters of the March 2008 VSP survey.

Date	Parameter	Horizontal Vib.	Vertical Vib.
March 2008	Start frequency	4 Hz	6 Hz
	End frequency	60 Hz	100 Hz
	Start taper	500 ms	250 ms
	End taper	500 ms	250 ms
	Sweep type	Linear	Linear
	Listen time	6 sec	5 sec
	Sweep length	8 sec	8 sec
	Sampling rate	2 ms	2 ms

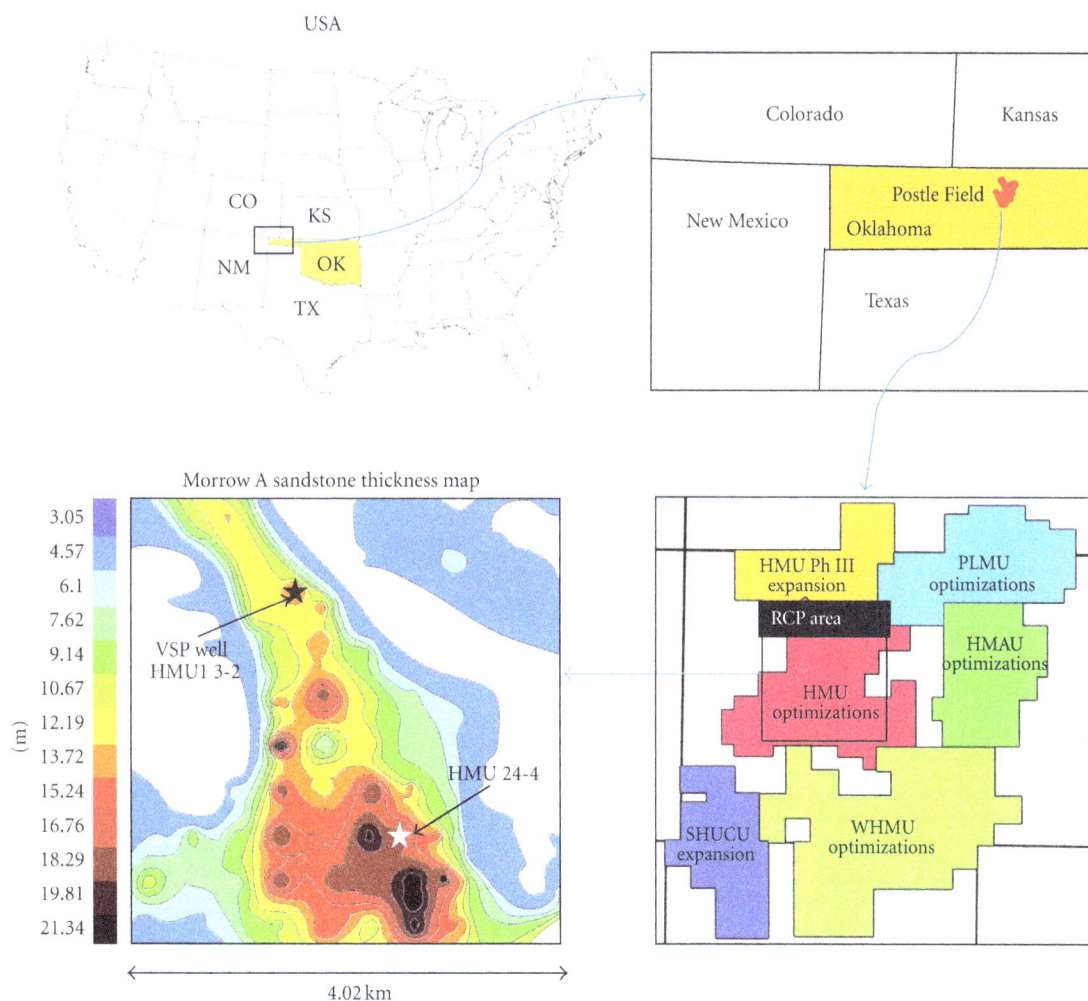

FIGURE 2: Postle Field is located in Texas County, Oklahoma. The RCP study area, 4.02 km × 4.02 km, is in the Hovey Morrow Unit (HMU). The Morrow A sandstone thickness, according to the well data, varies between 0 m and 21 m, and it is thinner at the northern part of the study area. The VSP well (HMU 13-2) is located where the Morrow A is thin. The VSP data, with higher resolution than the surface seismic data, provides the opportunity of delineating the Morrow A sandstone extension.

compressional dataset, and multioffset (or 3D) shear dataset. The main goal of the proposed processing flow (Table 3) is to preserve the original amplitude and its variations due to lithology changes.

One of the early steps in shear wave data processing is to reorient the horizontal components of all geophones toward the constant geographical coordinate system, commonly called "Receiver Reorientation." Hodogram analysis on direct compressional wave arrivals is one of the common approaches for finding the azimuth of each different geophone components in the borehole. Modeling showed that the hodogram analysis for a source with 488 m offset

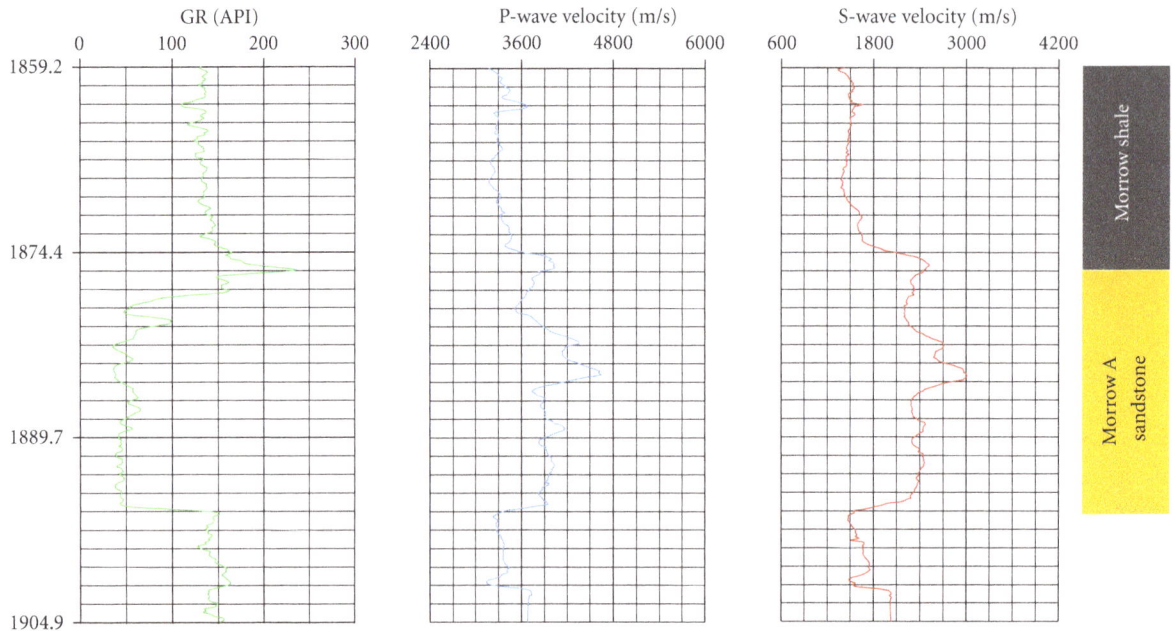

FIGURE 3: Gamma ray log and compressional (V_P) and shear (V_S) wave velocities at well HMU 24-4.

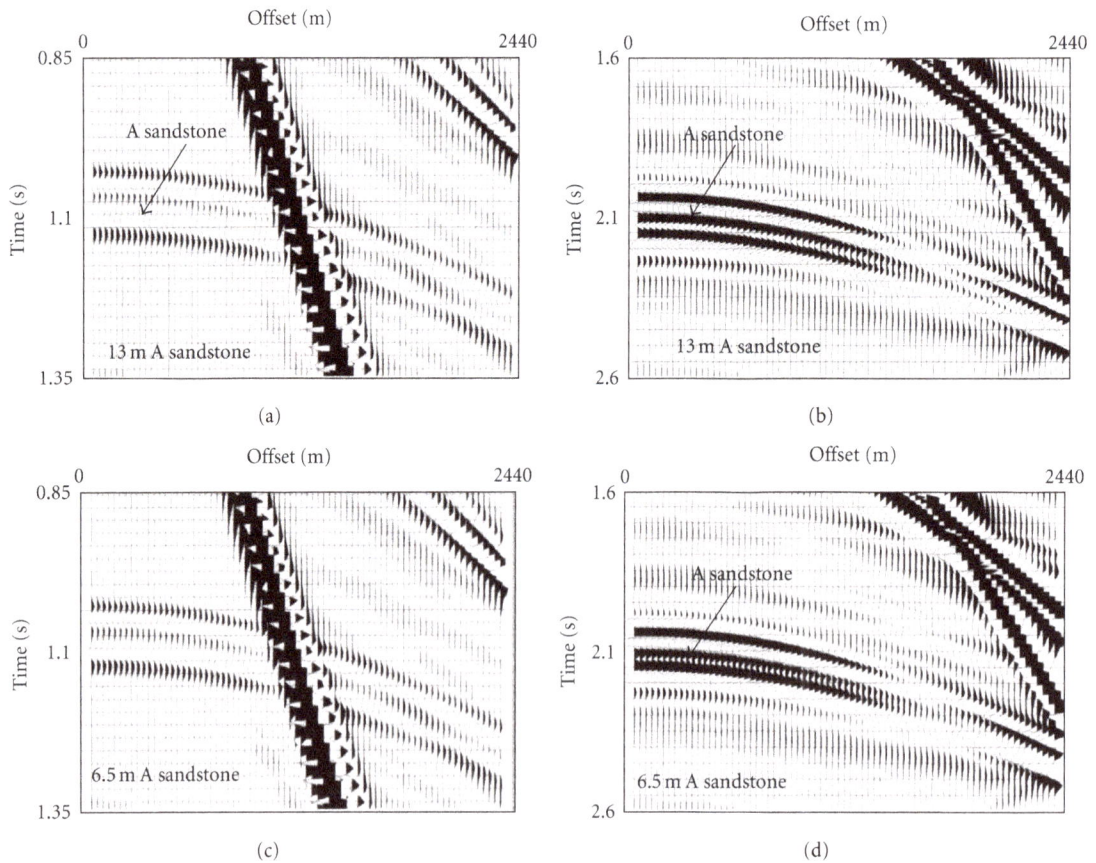

FIGURE 4: Synthetic gathers made for different thicknesses of the Morrow A sandstone in Postle Field. Compressional wave gathers (a) and (c), nonconverted shear wave gathers (b) and (d) for 13 m (a) and (b) and 6.5 m (c) and (d) thicknesses of the Morrow A sandstone. The synthetic gathers show that it is difficult to detect less than 13 m sandstone thickness using P-waves, because the peak doublet is absent. The nonconverted shear gather shows a distinct peak response even for 6.5 m of the Morrow A sandstone thickness [10].

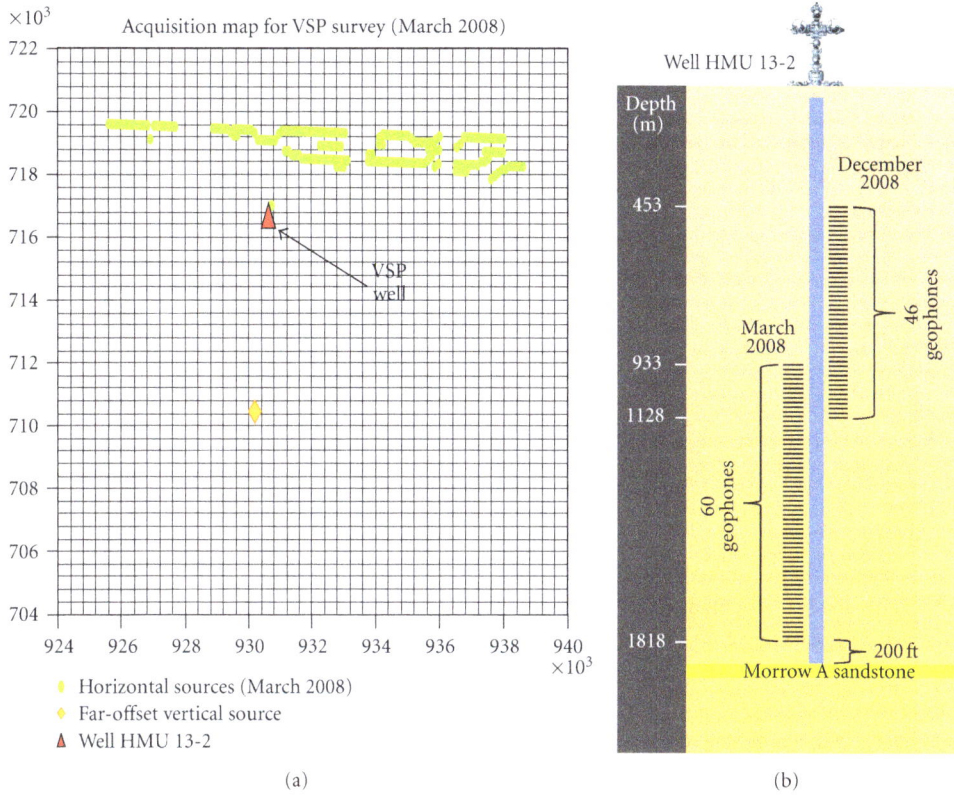

(a)

(b)

FIGURE 5: (a) The March 2008 survey data were used to image the Morrow A sandstone around the borehole. The zero-offset December 2008 survey data were also used to find the natural coordinate system in the field. (b) 3C geophone array in well HMU 13-2 used for the VSP surveys acquired in March and December 2008.

introduces up to 5 deg error in estimation of receiver azimuth. The error is expected to increase for more complex geological structures and noisy data. To reduce the uncertainty of the hodogram analysis, usually a number of shot gathers are analyzed. Since we had only one far-offset vertical source in our dataset, the error in the hodogram analysis and the receiver reorientation could have been considerable. Amplitude and time variations of the first shear wave arrivals in reoriented data (Figure 8(a)), using hodogram analysis of only the vertical source, confirmed the uncertainty in estimation of receivers azimuth. Uncertainty in receiver reorientation can diminish weak reflections from the reservoir or make invalid amplitude variations in the final VSP image. Because of this reason, we modified the receiver reorientation process by utilizing two more analyses: (1) hodogram analysis of compressional wave arrivals from a few horizontal vibrators and (2) 3C amplitude analysis. Hodogram analysis of the compressional wave generated by horizontal sources improved receiver reorientation results for some shallow geophones. Besides the mentioned methods, 3C amplitude analysis was done on the data which improved the results. In this approach, we rotated all three components of each geophone in a 3D half space and the RMS amplitude maps of the first arrivals of compressional and shear waves were projected onto two separate stereonets (Figure 7). The amplitude maps from 3C amplitude analysis are supposed to show the direction of maximum direct

compressional and shear wave energy in 3D space around the receivers. Therefore, accurate receiver reorientation should lead to the similar amplitude pattern on all geophones. Using this approach, some residual reorientation angles were calculated. Interpretation of the projected amplitude maps, in addition to previous hodogram analyses, provided more accurate rotation angle for aligning receivers (Figure 8(b)).

After the receiver reorientation, based on the polarity of the shear wave direct arrivals, the data polarity for different shots was fixed. Different receivers were rotated to the local coordinate system (radial-transverse), and then they were tilted in the direction of direct shear wave to maximize shear wave energy on the horizontal components. Many of the compressional wave arrivals and reflections disappeared on the horizontal components after this step.

4. Shear Wave Splitting

When a shear wave enters an anisotropic medium, it splits into two polarized shear waves that travel with different velocities [14, 15]. Shear wave splitting is an important phenomenon that requires attention during shear wave data processing and interpretation. Typically, when the medium is azimuthally anisotropic, shear wave data will be processed and presented in two modes: fast shear wave data (S_1) and slow shear wave data (S_2). Finding the fast and slow

TABLE 3: The proposed flow for processing the zero-offset and multioffset VSP datasets. A few processing techniques were introduced to preserve the original amplitude content of the data. (A) The zero-offset shear wave dataset, (B) the zero-offset compressional dataset, and (C) the multioffset shear wave dataset.

Processing steps	(A)	(B)	(C)
(1) Input VSP data and loading geometry on data headers	X	X	X
(2) Vibroseis sweep correlation and removing high frequency noises.	X	X	X
(3) P-wave first break arrivals		X	X
(4) Hodogram analysis on P-wave arrivals from a far vertical source			
(5) Hodogram analysis on P-wave arrivals from a few horizontal sources			X
(6) 3C amplitude analysis	X		X
(7) Receiver reorientation	X		X
(8) Fixing vibrator polarity	X		X
(9) Rotating data to the local coordinate system (radial-transverse)			X
(10) Calculating inclination of transmitted shear wave for each receiver			X
(11) Tilting each receiver based on calculated inclination			X
(12) 2C rotation analysis	X		
(13) Alford rotation analysis			X
(14) Rotating data to the natural coordinate system			X
(15) Creating velocity model and true amplitude recovery	X	X	X
(16) Applying median filter to isolate upgoing shear wave			X
(17) Applying dual median filter* to isolate desirable upgoing wave	X	X	
(18) Applying deconvolution filter on upgoing waves using downgoing waves	X	X	
(19) Aligning upgoing events and making corridor stack	X	X	
(20) Normalizing amplitude for different shots			X
(21) Applying VSP-CDP transformation on 2D VSP lines			X

* Dual median filter is explained in Section 3.

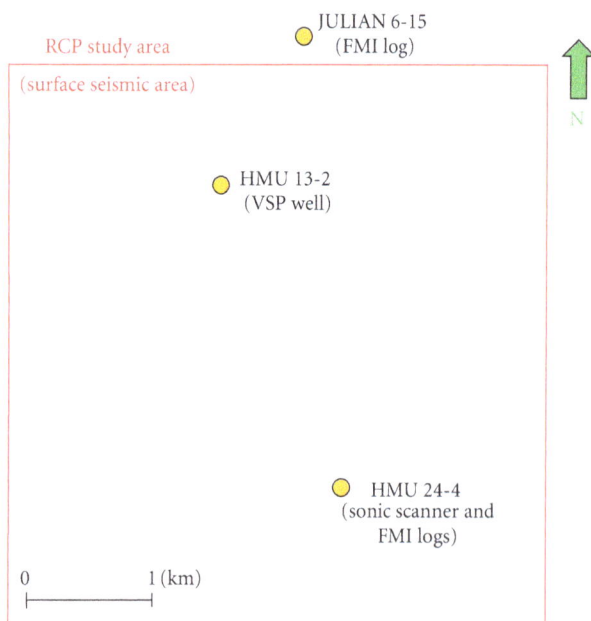

FIGURE 6: Map shows locations of the studied wells and the 3D 9C surface seismic area at Postle Field.

shear wave polarization directions (or the natural coordinate system) and their spatial and temporal variations has a significant impact on the quality of the final shear wave image. Therefore, information from different tools gives us

better insight about the reason for shear wave splitting, the polarization directions of fast and slow shear waves, and their temporal and spatial variations. Here, to show the importance of contributing different sources of information, we analyzed four different tools: full waveform sonic log (sonic scanner), image logs (FMI), multicomponent zero-offset VSP data, and multi-offset VSP data.

Slowness dispersion analysis (Figure 9(a)) on sonic scanner data from well HMU 24-4 (Figure 6) shows that the anisotropy in the study area, especially at the reservoir level, is mostly stress-induced anisotropy [16, 17]. In addition, the sonic scanner data suggest that the fast shear wave polarization direction is 126 deg (all the angles in this analysis are respect to north) at the reservoir level and 108 deg for the entire logged section (Figure 9(b)).

Since the origin of anisotropy in the field is presumed to be stress-induced anisotropy, fast and slow shear wave polarization directions should match maximum and minimum horizontal stress directions at the field. Drilling-induced fractures usually form in the direction of maximum horizontal stress [18], and they are detectable on micro-imaging logs (e.g., FMI logs). Figure 10 depicts the direction of some drilling-induced fractures in well HMU 24-4 and JULIAN 6-15 (Figure 6). Most of the drilling-induced fractures formed around 125 deg which is very close to the sonic scanner results. Although both image and sonic scanner logs are useful tools in determining the anisotropy

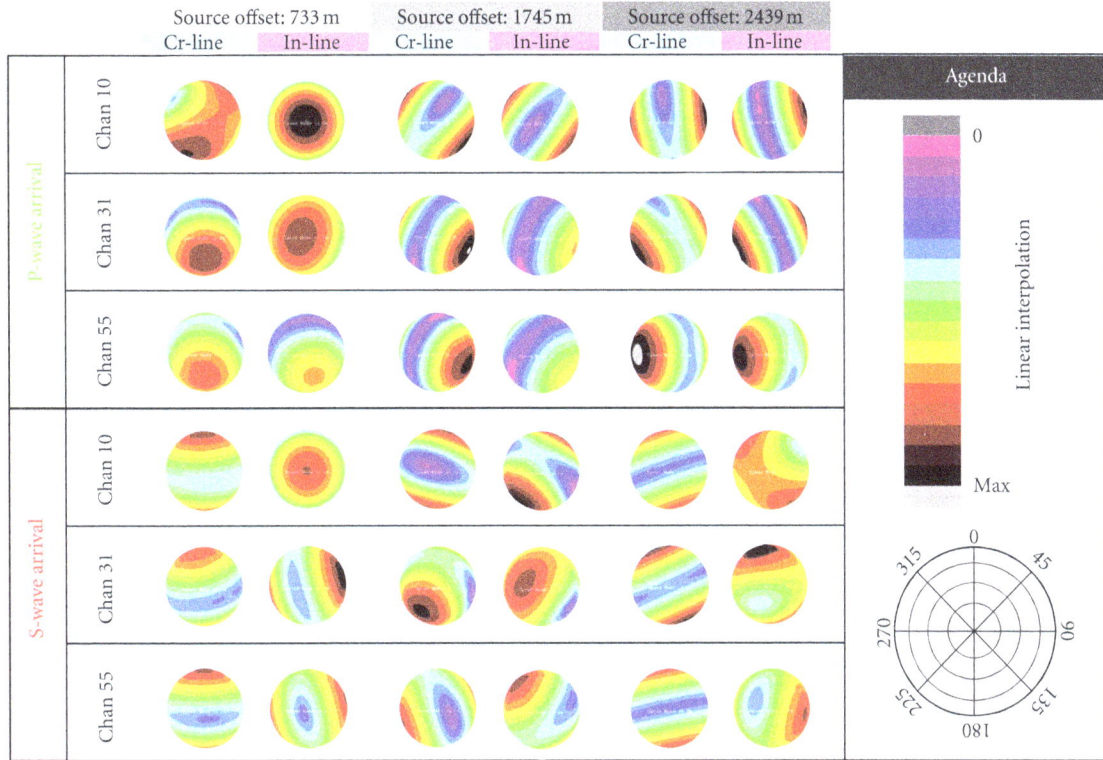

FIGURE 7: Projection of first-arrival compressional and shear wave amplitudes for H1 component of three different geophones (channel 10, 31, and 55) and three different horizontal-vibrator source offsets (733 m, 1745 m, and 2439 m). The pattern of these maps helped us to estimate the azimuth of each geophone component in the borehole. The results of the 3C amplitude analysis, in addition to the vertical and horizontal sources hodogram analyses, improved the receiver reorientation process.

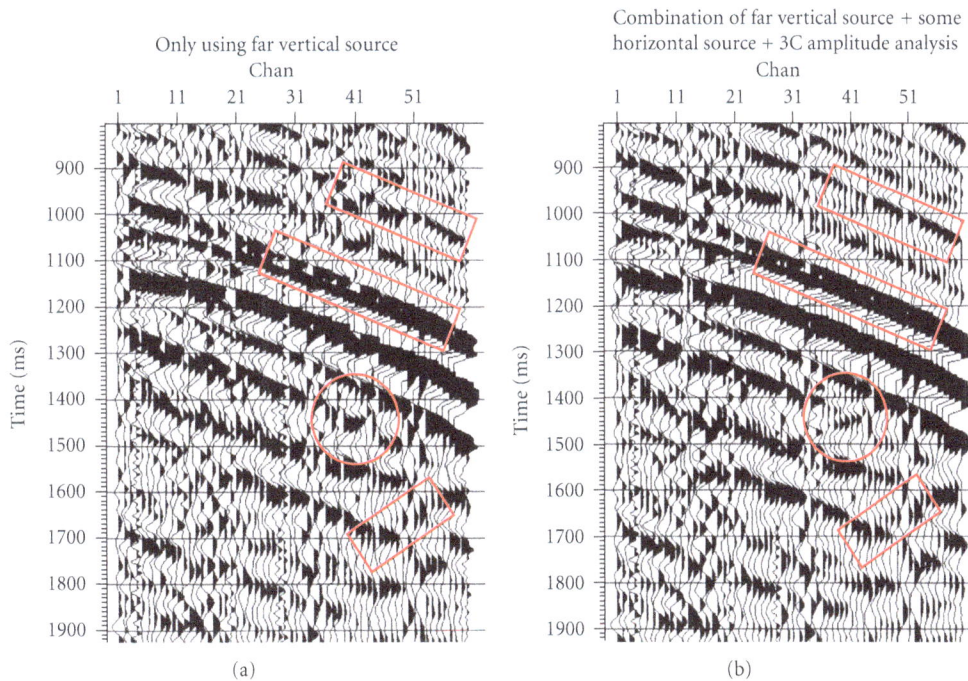

FIGURE 8: (a) shows a shear wave VSP shot gather after reorienting receivers based on the compressional wave hodogram analysis of only the far-offset vertical source. (b) depicts the same shot gather after utilizing three methods (i.e., compressional wave hodogram analysis of the far-offset vertical source, compressional wave hodogram analysis of a few horizontal sources, and the 3C amplitude analysis). Utilizing three methods in the receiver reorientation process caused remarkable improvements (note the red boxes and circles).

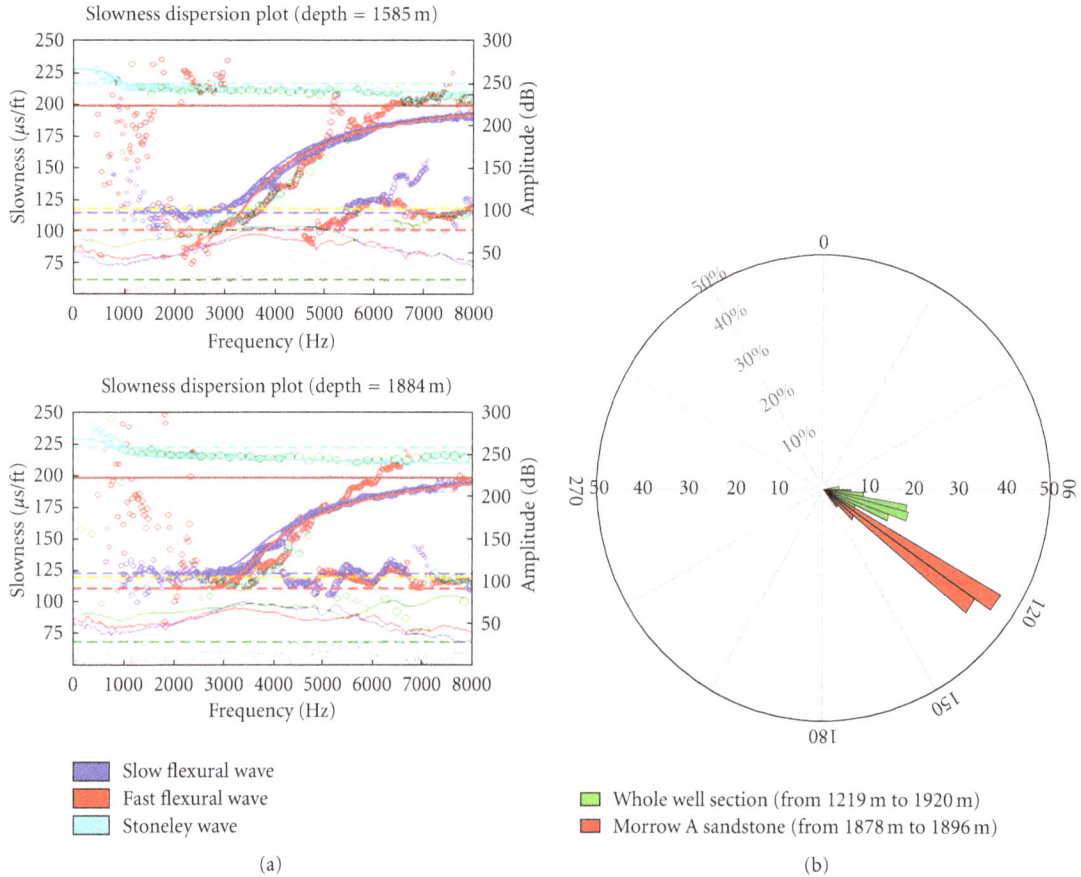

FIGURE 9: (a) Two flexural wave dispersion curves measured in well HMU 24-4. The red and blue circles represent fast and slow flexural waves, respectively. The cyan circles showes the Stoneley wave. The crossover pattern between the fast and slow dispersion curves show that the type of the anisotropy is stress-induced anisotropy. (b) The sonic scanner data suggest that the fast shear wave polarization azimuth is 126 deg at the reservoir level (from 1878 m to 1896 m) and 108 deg for the entire logged section (from 1219 m to 1896 m).

direction, both datasets are limited to the deep part of the well. In addition, the radius of investigation of these sources of information is limited to the vicinity of the VSP well. To find the azimuthal anisotropy direction at shallower layers and also away from the borehole, we did a comprehensive study on the shear wave VSP data.

The other source of information for determining the azimuthal anisotropy direction is zero-offset shear VSP data. There are two available zero-offset shear VSP datasets in well HMU 13-2: the March 2008 and December 2008 data. The geophones were located in the borehole from 933 m to 1818 m MD in March 2008 survey and from 453 m to 1128 m MD in the December 2008 survey (Figure 5). The significant depth spread helps to follow the azimuthal anisotropy direction from shallow layers to deep layers. The zero-offset VSP data (March 2008 data) were acquired using a horizontal vibrator that made shear waves polarized in only the E-W direction. Because of the single source polarization, 2C rotation analysis [19] was used to estimate the polarization directions of fast and slow shear waves (Figure 11(c)). The second zero-offset VSP data, December 2008, were acquired using one vibrator that made mutual shear wave motion in

the E-W and N-S directions. These crossed-dipole sources enabled us to apply Alford rotation analysis [20] to this data. The direct application of Alford rotation analysis is to find the natural polarization direction through a mathematical operation, assuming that the two split shear waves are orthogonally polarized and the anisotropy direction in the subsurface remains constant with depth [21]. The data were rotated from -90 deg to $+90$ deg (10 deg angle increment), and the RMS amplitude was calculated using a 200 ms time window around the first arrivals. Using (1), the final map (Figure 11(b)) was constructed to point out where the diagonal components (i.e., S11 and S22) had more energy than off-diagonal components (i.e., S12 and S21):

$$\text{Relative RMS Amplitude Map} = \frac{\text{RMS(S11)} + \text{RMS(S22)}}{\text{RMS(S12)} + \text{RMS(S21)}}. \tag{1}$$

On average, the March 2008 and December 2008 surveys suggested 134 deg as the fast shear wave polarization direction.

To understand spatial variation of the S11 polarization direction in the field, multioffset VSP data, using Alford

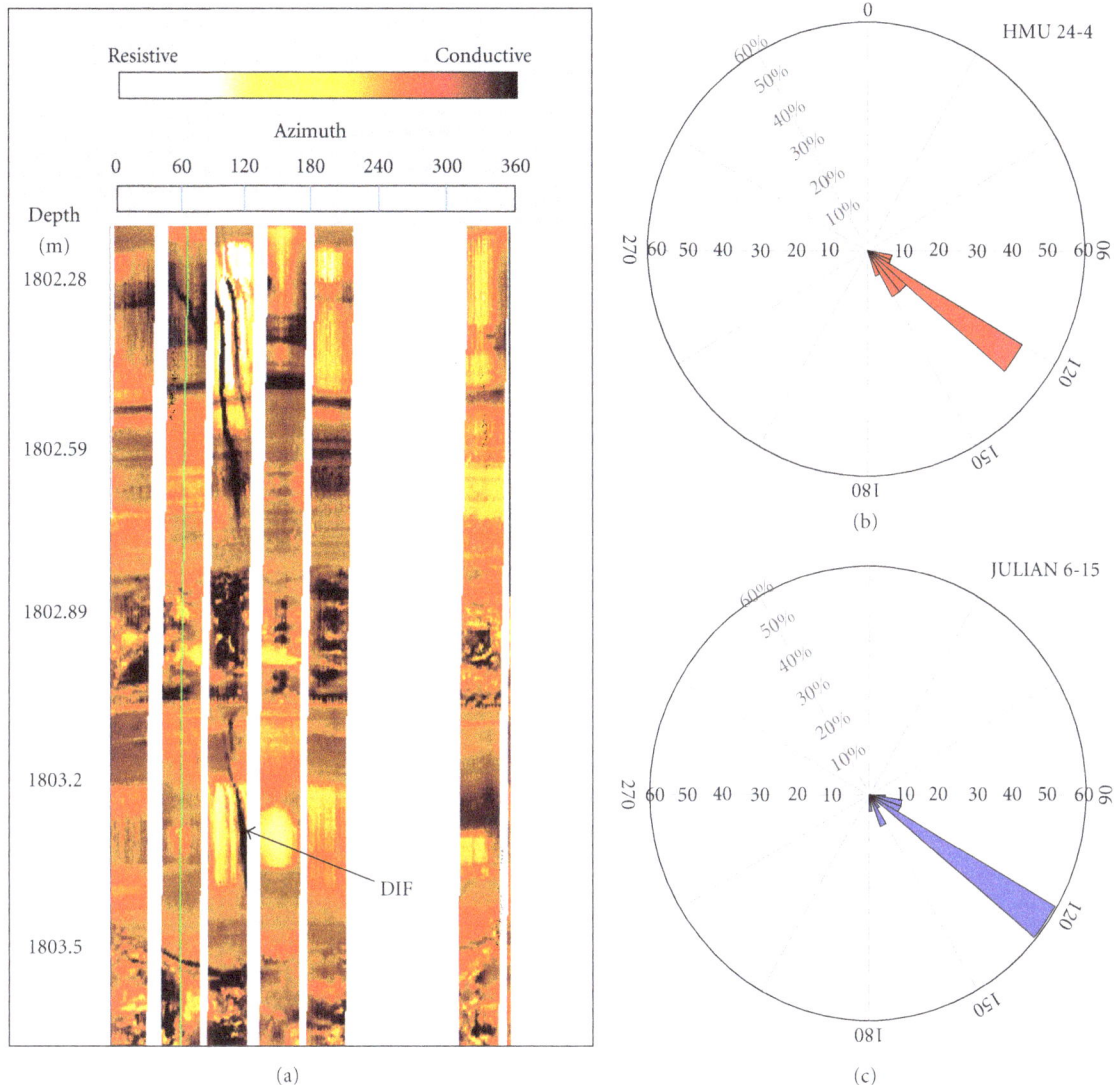

FIGURE 10: Drilling-induced fractures form in a drilled borehole in direction of the maximum horizontal stress. FMI logs could depict drilling-induced fractures on borehole wall clearly. (a) represents a part of the FMI log from the well HMU 24-4. The azimuth of the induced-drilling fractures (the high conductive vertical features) on the borehole wall was picked for two wells (HMU 24-4 and JULIAN 6-15). (b) and (c) show the azimuth of the drilling-induced fractures in the wells.

rotation analysis, were analyzed. Alford rotation analysis on multioffset VSP data (i.e., near-offset, mid-offset, and far-offset data) depicts that the fast shear wave polarization direction is approximately 138 deg (Figure 12). Similar to the zero-offset VSP results, multioffset VSP data show that the fast shear wave polarization direction does not change significantly with depth.

Analyzing the anisotropy direction using different tools (Figure 13) suggests a 30 deg change (from 108 deg to 138 deg) that may happen because of factors such as different radius of investigation, difference in measurement frequency, complex heterogeneity of the reservoir and overlying layers, among others. A discussion of these factors is outside the scope of this paper, but the data exhibited in Figure 13 indicated that analysis of different sources of information is required to understand azimuthal anisotropy.

5. Wavefield Separation and Imaging

Wavefield separation was the next step in the VSP processing flow. Generally, different wave types can be recognized in VSP data better and more accurately than in surface seismic data. Because of this ease of recognition, we are usually able to effectively filter undesirable waves from VSP data. Because VSP data in Postle Field will be used to image details inside the sand channels (e.g., braid bars and channel boundaries) around the borehole, we need to preserve discontinuities in the data. Median filtering offers the unique ability to preserve discontinuities in data without smearing discontinuities [22]. Although the median filter preserves discontinuities in the data, it diminishes some weak upgoing waves and introduces high frequency noise where upgoing waves interfere with strong downgoing waves. Because some

(a)

(b)

(c)

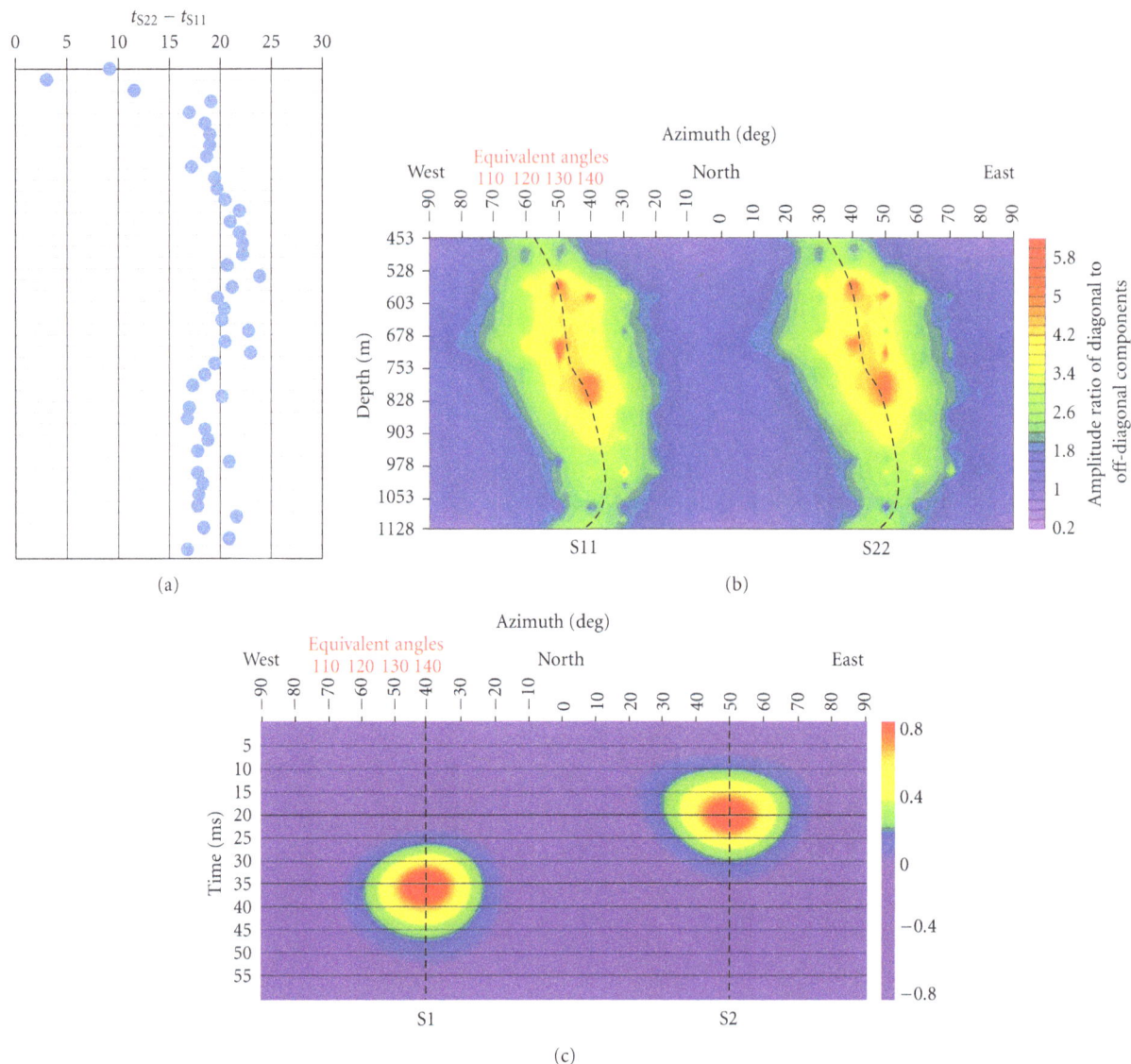

FIGURE 11: (a) depicts time difference between the fast and slow shear wave arrivals from the December 2008 zero-offset dataset. Alford rotation analysis (using (1) as criterion) on the December 2008 zero-offset VSP data (b) and 2C rotation analysis on the March 2008 zero offset VSP data (c) suggest that the fast shear wave polarization direction is approximately 134 deg.

geophones are close to the Morrow A sandstone, reflections (upgoing waves) from this horizon are covered by strong direct arrivals and their multiples (downgoing waves). As expected, the quality of the separated upgoing waves in the interference area is affected by the median filter and the wavefield separation process. To overcome this issue, synthetic zero-offset VSP data were built, and different parameters and approaches for wavefield separation were examined using different median filters.

In one approach, upgoing waves were aligned and they were separated using a median filter. The quality of these separated upgoing waves was not good, but after stacking this result with the separated upgoing waves using the common approach (i.e., removing downgoing waves from data), we obtained a higher signal-to-noise ratio and preserved weak reflections better (Figure 14). We called this method "Dual

Median Filtering." Because the method is adapted for zero-offset VSP data, we applied it to only the zero-offset VSP datasets. The multioffset VSP dataset was filtered by using a common median filtering. F-K filtering is another popular method for separating different wavefields but, due to the high level of noise in the data, discriminating different wavefields in the F-K domain is challenging. Also, F-K filtering cannot preserve sharp discontinuities in the data. After wavefield separation, the downgoing waves were used for making a deconvolution filter. Corridor stacks were made of the zero-offset compressional and shear wave datasets after applying the deconvolution filter (Figure 15).

VSP migration and VSP-CDP transformation are two common approaches for final VSP imaging. Lack of data fold in our VSP survey makes VSP migration method ineffective. An alternative for VSP migration in this case is the VSP-CDP

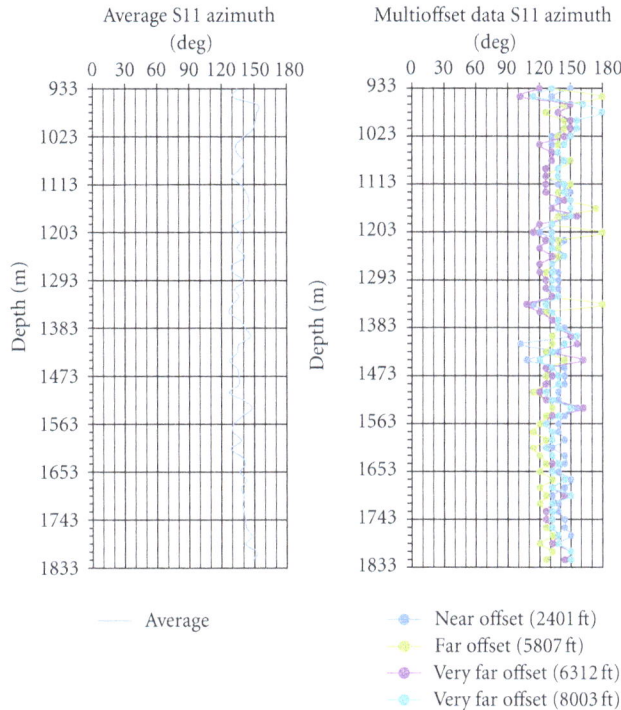

FIGURE 12: Shear VSP data with different source offsets (near-offset, mid-offset, and far-offset), using Alford rotation analysis, were analyzed.

transformation. A VSP-CDP transformation transforms VSP wave data from the depth-time domain to the offset-time domain [23]. This technique uses 2D ray tracing to map reflected signals into the offset-time domain [24]. After imaging the data using VSP-CDP transformation method, we calculated the fast shear wave RMS amplitude for a 30 ms window centered around the Morrow A sandstone (Figure 16).

6. Results and Discussion

Figures 15(c) and 15(d) compare the prestack zero-offset compressional and shear wave images as well as their corridor stacks. Although the Morrow A sandstone is visible on both datasets, it has higher amplitude when shear waves are used. It seems that the deconvolution filter has a considerable effect on compressional wave data and the Morrow A sandstone is not visible before applying this filter. Deconvolution filtering also assists us in resolving the Morrow A sandstone from the underlying layer in the shear wave data. Also, qualitatively, the zero-offset compressional and shear wave data have approximately the same temporal resolution.

Figure 16 displays the shear wave RMS amplitude map for a 30 ms window centered around the Morrow A sandstone from some shots on the northern part of the VSP well. Based on the Postle Field geology and the size of the high-amplitude areas, one hypothesis is that these features represent braid bars which are good quality and thick reservoir rocks. Also, the importance of the VSP data at this area is to map the edge of the channel (mainly NW-SE at this part), and some linear features could show the evolution of the channel edge during time too. At the far

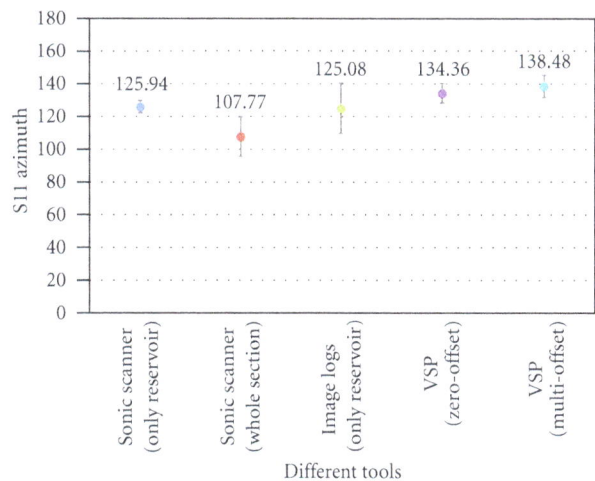

FIGURE 13: Fast shear wave polarization directions from different sources of information. The error bars indicate standard deviation of the estimated azimuths.

end of the map (data were recorded by geophones from 1 to 16, located from 933 m to 1158 m MD in the well), very high amplitude events could result from braid bars as well as strong noise (pay attention to arc shape of the high amplitude area). One reason for the high level of noise at this interval is a weak bond between the casing and the borehole wall. In addition, the presence of shale layers in the Morrow A sandstone can affect the RMS amplitude of the Morrow A. Some low-amplitude areas correspond to the presence of the mentioned shale layers and indicate low quality (or shaly sandstone) reservoir rocks. Unfortunately, as stated before,

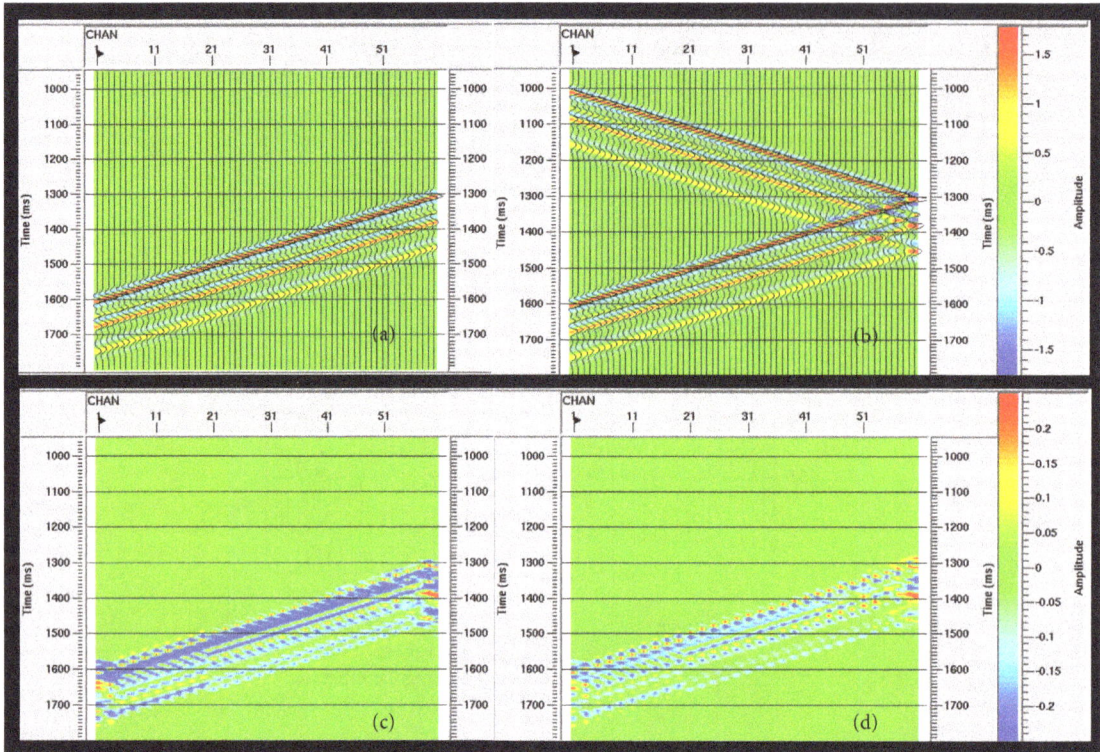

FIGURE 14: A synthetic zero-offset VSP shot gather was generated to compare the results after wavefield separation using median filter and dual median filter. (a) shows the synthetic upgoing wavefield, and (b) represents the synthetic zero-offset VSP shot gather after stacking with the downgoing wavefield. (c) and (d) depict the calculated residual amplitude (original upgoing wavefield-separated upgoing wavefield) after wavefield separation using median filter and dual median filter, respectively.

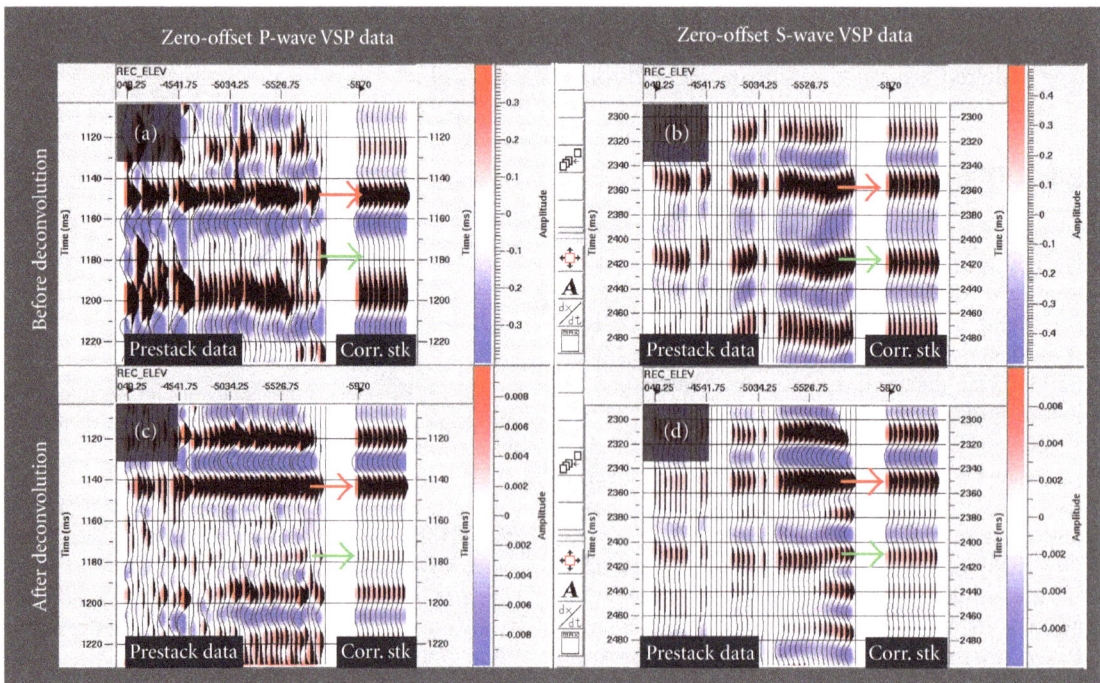

→ Atoka limestone
→ Morrow A sandstone

FIGURE 15: Zero-offset compressional (a) and (c) and shear (b) and (d) wave data before and after deconvolution. The green arrows show the Morrow A sandstone level, and the red arrows point to the Atoka limestone. Qualitatively, both datasets have similar temporal resolution, but the Morrow A event is more distinctive in the shear wave data than in the compressional wave data.

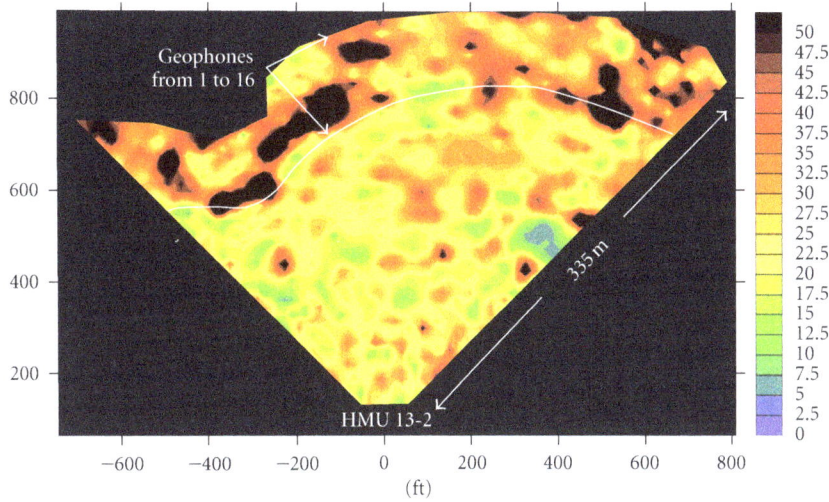

FIGURE 16: A 30 ms window centered around the Morrow A sandstone maps S11 RMS amplitude of the Morrow A sandstone on the northern part of the VSP well.

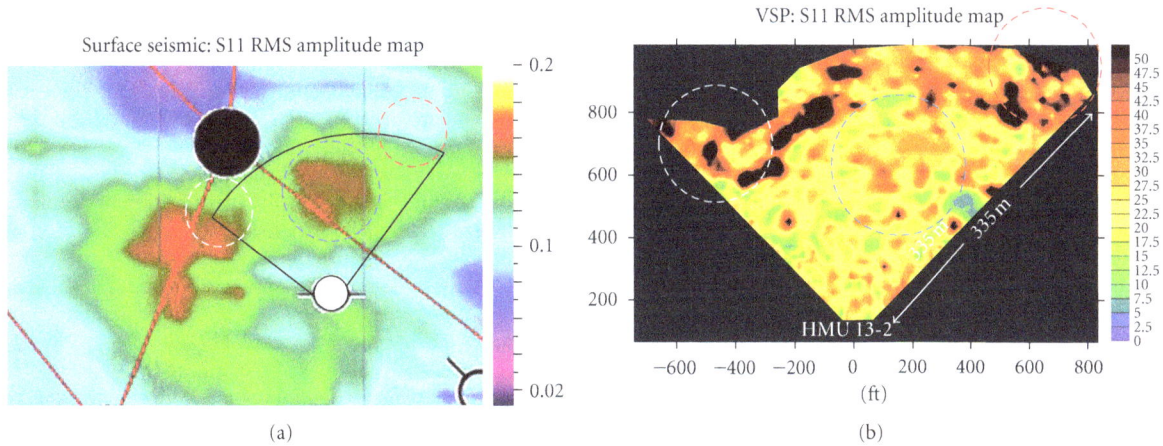

FIGURE 17: (a) S11 RMS amplitude map from the surface seismic data around well HMU 13-2. (b) S11 RMS amplitude map from the VSP data around well HMU 13-2. On both maps, there are two high-amplitude anomalies (white and blue circles). The white circle is on the edge of the VSP map where data are noisy and not reliable. But the blue circle highlights a coherent high amplitude area on both maps. The red circle shows high amplitude on the VSP data, but it is recognizable on the surface seismic data. Most likely, this high amplitude area (red circle) in the VSP data does not correspond to a geologic feature.

the data fold at this area is low and is the main limitation for the VSP image in this area. More 3D VSP coverage could image the Morrow A better and more reliably away from the VSP well.

Comparison of surface seismic and VSP RMS amplitude maps reveals advantages of the VSP data. Figures 17(a) and 17(b) represent the Morrow A sandstone S11 RMS amplitude map of the surface seismic (acquired on March 2008) and VSP data. Three amplitude anomalies are highlighted by circles showing similarities and differences between these two datasets. The most interesting anomaly is highlighted by blue circles on both images. The VSP data appear to map details better. The ability of the shear VSP data to delineate the extension of this anomaly reconfirms the advantages of the S-wave VSP technique. As mentioned before, due to high heterogeneity in the reservoir, preserving original amplitudes and amplitudes variations is crucial. The ability of VSP data,

after applying the proposed processing techniques in this paper, to delineate high-amplitude anomalies can reduce drilling risk.

7. Conclusions

Shear wave VSP data were processed using a modified processing flow. The main purposes of the proposed processing flow and some introduced techniques were to enhance the weak seismic response from the Morrow A sandstone in the presence of noise as well as to preserve amplitude variations caused by lithology changes (especially between the Morrow A sandstone and the Morrow shale). 3C amplitude analysis made VSP receiver reorientation more robust and avoided some processing artifacts. Also, estimating fast shear wave polarization direction using different tools addressed the scaling issue. The results of this analysis demonstrate that reliance on one source of information, either well logs or seismic data, can introduce significant uncertainty in final results. Dual median filtering was another method which was introduced to enhance VSP weak reflections where upgoing waves interfere with the strong downgoing waves.

Zero-offset VSP results reconfirmed the fact that the Morrow A sandstone signature is weak in compressional wave data. Also, it confirmed the ability of shear wave data to image the thin Morrow A sandstone. The Morrow A sandstone was mapped using the VSP-CDP transformation. High-amplitude anomalies in the shear wave RMS amplitude map could correspond to geologic features like braid bars, which fits the sediment deposition history of Postle Field. Comparison between surface seismic and VSP data showed the potential of VSP data in imaging detailed changes in subsurface rock properties and delineating anomalies with higher spatial resolution. Therefore, shear wave VSP data are able to map detailed geologic features where the heterogeneity in the Morrow A sandstone plays a major role.

Acknowledgments

The authors would like to acknowledge Tom Bratton from Schlumberger, Bruce Mattocks from CGGVeritas, and Richard Van Dok from HiPoint Reservoir Imaging for their advice. Also, they would like to thank their colleagues Dr. Robert Benson, Paritosh Singh, and Mohsen Minaei from RCP for their technical support. Special thanks to Whiting Petroleum for providing the multicomponent data. The authors would like to sincerely thank the Reservoir Characterization Project sponsors and students too.

References

[1] J. W. Benton, "Subsurface stratigraphic analysis, Morrow (Pennsylvanian), North Central Texas County, Oklahoma," *The Shale Shaker Digest*, vol. 21–23, pp. 21–23, 1973.

[2] D. C. Swanson, "Deltaic deposits in the Pennsylvanian upper Morrow formation of the Anadarko Basin, in Pennsylvanian sandstones of the mid-continent," *Tulsa Geological Society Special Publication*, no. 1, pp. 115–168, 1979.

[3] T. D. Jobe, *Optimizing geo-cellular reservoir modeling in a braided river incised valley fill: postle field, Texas County, Oklahoma [M.S. thesis]*, Colorado School of Mines, 2010.

[4] J. R. Halverson, "Seismic expression of the Upper Morrow Sands in the Western Anadarko basin," *Oil and Gas Journal*, vol. 86, no. 44, pp. 290–303, 1988.

[5] J. T. Noah, R. D. Teague, and G. Hoand, "Twin Morrow field: a case study," *The Leading Edge*, vol. 13, p. 2530, 1994.

[6] M. L. Willey, *Structural and stratigraphic controls on Morrow Sandstone reservoir distribution from 3-D seismic data, postle field, Texas County, Oklahoma [M.S. thesis]*, Colorado School of Mines, 2009.

[7] A. Robinson, *Acoustic impedance inversion for static and dynamic characterization of a CO2 EOR Project, Postle Field, Oklahoma [M.S. thesis]*, Colorado School of Mines, 2010.

[8] M. Minaei and T. L. Davis, "Increasing seismic resolution by post-stack processing procedures in Postle Field, Oklahoma," in *Proceedings of the 81st SEG Meeting*, pp. 1036–1040, San Antonio, Tex, USA, 2011.

[9] T. L. Davis, R. D. Benson, S. Wehner, M. Raines, and R. Freidline, "Seismic reservoir characterization of the Morrow A Sandstone, Postle Field, Oklahoma," in *Proceedings of the 80th SEG Meeting*, pp. 2256–2260, Denver, Colo, USA, 2010.

[10] P. Singh and T. L. Davis, "Advantages of shear wave seismic in Morrow Sandstone detection," *International Journal of Geophysics*, vol. 2011, Article ID 958483, 16 pages, 2011.

[11] A. V. Wandler, T. L. Davis, and P. K. Singh, "An experimental and modeling study on the response to varying pore pressure and reservoir fluids in the Morrow A Sandstone," *International Journal of Geophysics*, vol. 2012, Article ID 726408, 17 pages, 2012.

[12] B. A. Hardage, M. DeAngelo, and P. Murray, "Defining P-wave and S-wave stratal surfaces with nine-component VSPs," *The Leading Edge*, vol. 22, no. 8, pp. 720–729, 2003.

[13] P. Mazumdar and T. L. Davis, "Shear-wave sourced 3-D VSP depth imaging of tight gas sandstones in rulison field, Colorado," *CSEG Recorder*, pp. 20–26, 2010.

[14] H. H. Hess, "Seismic anisotropy of the uppermost mantle under oceans," *Nature*, vol. 203, no. 4945, pp. 629–631, 1964.

[15] S. Crampin, "Evaluation of anisotropy by shear-wave splitting," *Geophysics*, vol. 50, no. 1, pp. 142–152, 1985.

[16] B. K. Sinha and S. Kostek, "Stress-induced azimuthal anisotropy in borehole flexural waves," *Geophysics*, vol. 61, no. 6, pp. 1899–1907, 1996.

[17] N. Tamimi, A. V. Wandler, and T. Bratton, "Processing and preliminary interpretation of full sonic waveform data, HMU 24-4, postle Oilfield," *Reservoir Characterization Project Spring Meeting Report*, pp. 46–53, 2009.

[18] B. S. Aadnøy and J. S. Bell, "Classification of drilling-induced fractures and their relationship to in-situ stress directions," *Log Analyst*, vol. 39, no. 6, pp. 27–40, 1998.

[19] M. P. Harrison, *Processing of P-SV surface seismic data: anisotropy analysis, dip moveout, and migration [Ph.D. dissertation]*, University of Calgary, 1992.

[20] R. M. Alford, "Shear wave data in the presence of azimuthal anisotropy," in *Proceedings of the 56th SEG Meeting*, pp. 476–479, Dilley, Tex, USA, 1986.

[21] Z. Sun and M. J. Jones, "VSP multi-algorithm shear-wave anisotropy study," *CREWES Research Report*, vol. 5, pp. 601–622, 1993.

[22] J. H. Kommedal and B. A. Tjostheim, "A study of different methods of wavefield separation for application to VSP data," *Geophysical Prospecting*, vol. 37, no. 2, pp. 117–142, 1989.

[23] K. D. Wyatt and S. B. Wyatt, "Determining subsurface structure using the vertical seismic profile," in *Vertical Seismic Profiling, Part B—Advanced Concepts*, N. M. Toksoz and R.R. Stewart, Eds., Geophysical Press, 1984.

[24] S. J. Gulati, R. R. Stewart, J. Peron, and J. M. Parkin, "3C-3D VSP: normal moveout correction and VSPCDP transformation," *CREWES Research Report*, vol. 9, pp. 901–930, 1997.

16

Geoelectrical Surveys for Characterization of the Coastal Saltwater Intrusion in Metapontum Forest Reserve (Southern Italy)

Antonio Satriani, Antonio Loperte, Vito Imbrenda, and Vincenzo Lapenna

Institute of Methodologies for Environmental Analysis, CNR, 85050 Tito, Italy

Correspondence should be addressed to Antonio Satriani, satriani@imaa.cnr.it

Academic Editor: Sabatino Piscitelli

A geoelectrical survey was carried out in the Metapontum Forest Reserve located along the Ionian coast of the Basilicata region (Southern Italy). In this work we used the method of two-dimensional electrical resistivity tomography for obtaining high-resolution electrical images in the investigated site. In particular, three electrical resistivity tomography, all orthogonal to the coastline, in the investigated area were carried out. To complete and integrate the geophysical data, soil and groundwater samplings, seventeen and five, respectively, were analyzed using chemical physical techniques. Geoelectrical survey, supported by laboratory analysis of soil and water samples have revealed the presence of a process of saltwater in coastal Forest Reserve of Metapontum, which have caused the decline of the existing pine forest with the consequent erosion and desertification problems. The results have disclosed the way to identify and discriminate large areas affected by intensive soil salinization and high resolution electrical images of the subsurface electrical resistivity plays a key role in delineating the saltwater intrusion front in coastal areas. Furthermore, our integrated study represents a contribution to the future programs for the protection, planning, and management of the terrestrial and marine resources in this coastal area.

1. Introduction

Coastal areas are of great environmental, economic, social, and cultural relevance. Therefore, the implementation of suitable monitoring and protection actions is fundamental for their preservation and for assuring future use of this resource. Such actions have to be based on an ecosystem perspective for preserving coastal environment integrity and functioning and for planning sustainable resource management of both marine and terrestrial components (EU Recommendation on Integrated Coastal Zone Management—ICZM—European Commission EC/413/2002).

Post [1] has defined coastal aquifers as the subsurface equivalents of coastal areas where continental fresh groundwater and seawater meet. Coastal plains are often contaminated by salt waters and the process associated to the marine water entering an aquifer is generally called seawater intrusion. Different approaches have been adopted to estimate seawater intrusion. For example, many authors [2–5] have employed geochemical methods based on measures of electrical conductivity, of chloride concentration, and other cation and anion concentrations, parameters which generally highlight seawater contamination.

Nowadays great attention is focused on innovative geophysical methods for solving hydrogeological and environmental problems [6–10]. The geophysical methods for the subsurface exploration can contribute to the different phases of coastal environment characterization. Balia et al. [11] have applied a combination of geophysical techniques at the Muravera plain in southern Sardinia, Italy, which has been significantly affected by water salinization; Batayneh [12] used electrical resistivity methods for detecting subsurface fresh and saline water in the eastern Dead Sea coastal aquifers in Jordan; Bauer et al. [13] applied electrical resistivity tomography (ERT) to map the salinity distribution in the subsurface in the Okavango Delta, a large inland delta in

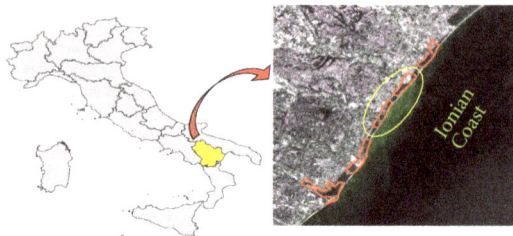

FIGURE 1: Location of the study area. The yellow ellipse highlights the natural reserve of Metapontum.

Botswana; while Koukadaki et al. [14] characterized the hydrogeology of a coastal aquifer using geophysical measurements and, in particular, electrical tomography. ERT, moreover, was applied to monitor seawater intrusion by Kruse et al. [15], Nowroozi et al. [16], Abdul Nassir et al. [17], and Sherif et al. [18]. Cimino et al. [19] carried out vertical electrical sounding in the coastal plain of Acquedolci (Northern Sicily- Italy) to assess seawater intrusion, while Khalil [20] applied a direct current resistivity geoelectric technique to delineate salt water intrusion from the Gulf of Suez. Moreover, Urish and Frohlich [21], Ebraheem et al. [22], and Choudhury and Saha [23] have performed geoelectrical surveys in different coastal area around the world.

In this paper we present the results of geoelectrical surveys, integrated with geochemical measurements, performed for the characterization of a coastal area along the Ionian side of the Basilicata region (Figure 1). This study area, located at the mouth of Bradano river (Metapontum, 40°22′ N, 16°51′ E), is mainly interested by a narrow shore (10–30 m) of fine sandy formations and has been affected by saltwater intrusion phenomena. This has caused the decline of the existing pine forest entailing significant ecological and environmental problems to the Metapontum Forest Reserve. Primary and secondary salinity of soil are linked to natural factors such as the climate, natural drainage patterns, topographic features, geological structure, distance to the sea, and by the indiscriminate exploitation of the soil and groundwater resources [1, 24, 25].

In order to manage rightly this kind of problem, it is necessary to have a thorough knowledge of the phenomenon, and the first objective of this study is to identify, through the use of different techniques and methodology, salt intake as the factor concurring or even determining the serious phenomenon of total decline and the deterioration of vegetation in the pine forest.

The geoelectrical investigations and chemical-physical analysis were carried out, to evaluate the seawater intrusion effect. In particular, geoelectrical surveys allowed to assess the extent and depth of the saltwater intrusion process, while the chemical-physical characterization was performed by considering some soil properties such as pH values, salt contents measured as the electric conductivity (ECe), exchangeable sodium, and cation exchange capacity. Moreover, this study aims to better delineate the saltwater intrusion front and to support the environmental protection and regeneration programs carried out by local and national

authorities for restoring the continuity of the woodland belt along the coast and consequently to obtain a reclassification of the landscape and protection of areas behind.

2. Study Area

The study area is located along the Ionian coastal plain, known as Metapontum plain. The area is made up of the regressive filling with marine sediments of a tectonic graben (Bradanic graben) during Pliocene to Pleistocene age. The area is bordered in the NE by the limestones of the Murgian foreland and in the SW by the clastic sediments of the Apennine chain. In the middle of the graben the full sedimentary sequence starts with sands, blue clays, covered by sands and gravels. The alluvial deposits in the estuarine environment are constituted of stratified sands to silty clays sediments. In this case, they are grey and very similar for aspect, but not for consistency, to the Pleistocene blue clays [4, 26, 27]. In particular, the Forest Reserve with an extent of about 240 hectares, situated in the north of the Ionian coastline between the mouth of rivers Bradano and Basento, is characterized by a superficial sandy substrate, and by clays and silts of river origin in the back dune zone (Figure 2).

The hydrogeology of the study area (Figure 3) is related to lithological and structural features of the same area, and they determine two types of aquifers.

An aquifer is constituted by marine terraces deposits and alluvial river valleys deposits, where the marine terraces aquifers show hydraulic conductivity from medium to high, equal or higher than the second type of aquifer constituted by coastal plain deposits.

The bottom of the this aquifer is a silty-clayey bed which lies under the sea level permitting the seawater intrusion, this phenomenon involves the coastal plain for a width of 1–1.5 km and it is less evident moving inwards where the altitude of the clayey bottom of the aquifer becomes progressively higher than the sea level [4].

The Metapontum coast is characterized by a strong erosive process. Marine erosion has really removed the dune behind the beach moving in the inland and causing the decline of the historical pinewood. Dunes and their vegetation are a natural defence against erosion by wind and sea and the variation of the line of the seashore is in close connection with the development of coastal dunes and their preservation [28]. Moreover, reduction of the coastal dune system in the study area facilitates the entry of salt water; dunes in fact, for their elevation and good infiltration capacity, give a sufficient freshwater recharge and a pressure above sea level, allowing hydrostatic control of saline intrusion [29].

Metapontum wooded strip is an artificial formation, planted in the first decade of the 50s in order to preserve the coast and the inland cultivated areas from salt sea wind. The forest is composed of Aleppo pine trees (*Pinus halepensis*) and domestic pine trees (*Pinus pinea*) prevalently. Secondary species are acacia saligna (*Acacia cianophylla*) and eucalyptus (*Eucalyptus globulus, E. camaldulensis*), besides other species that are typical of Mediterranean scrub as *Pistacia lentiscus, Juniperus macrocarpa* and *Juncos acutus*,

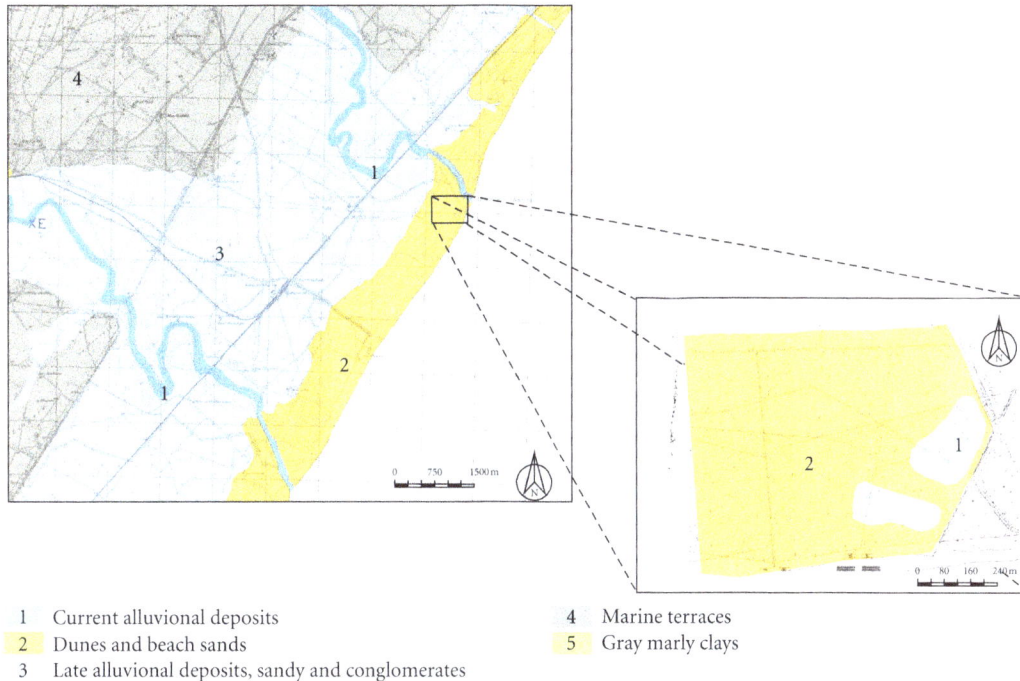

1	Current alluvial deposits		4	Marine terraces
2	Dunes and beach sands		5	Gray marly clays
3	Late alluvional deposits, sandy and conglomerates			

FIGURE 2: Geological map of the study area. (1) Current alluvional deposits; (2) dunes and beach sands; (3) late alluvional deposits sandy and conglomerates; (4) marine terraces; (5) gray marly clays.

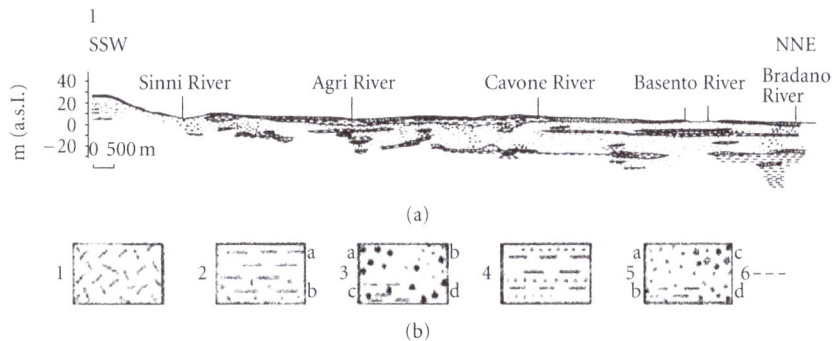

FIGURE 3: Schematic lithological section [4]. Legend: (1) soil; (2) clays or silty clays (yellow, brown, grey; (a)) locally sandy clays (b); (3) pebbles (a) in a sandy (b) or clayey matrix (c); pebbles locally cemented (d); (4) grey sands with clayey strata; (5) grey or yellow sands and silty sands (a), locally clayey sands (b) or with gravels (c), locally sandstone strata (d); (6) piezometric level.

often mixed with the type of vegetation which prefers sandy soil (called "psammofila"). The rearrangement of soil was done with the realization of "*baulature*," that is, elevation of soil in order to preserve tree roots from seawater intrusion.

On the basis of meteorological data, the climate of the area is semiarid with hot and dry summers; however, the evolution of the climate of the last period shows a trend to a reduction of annual precipitation and an increase in average temperatures [30].

3. Field Investigations and Sampling

In Figure 4 location of geoelectrical surveys and sampling points for chemical-physical analysis was indicated. The surveys have been performed during a geophysical campaigns

carried out in July, and in this time of year the investigated area, classified as climatic type semiarid, is characterized by drought.

3.1. Geoelectrical Surveys. Resistivity measurements were conducted as tomographies to determine variations with depth in soil resistivity.

The surface electrical resistivity tomographies are a useful tool in determining seawater intrusion in coastal areas for its capability to discriminate the large resistivity contrast between the presence of seawater that strongly reduces the resistivity values and saturated freshwater layers. Moreover, in order to evaluate if this contrast is significant and does not depend on geological layer, a validation of the geoelectrical

Figure 4: Location map of geophysical surveys and sampling points.

measurements with some chemical-physical analysis of soil samples has been performed.

Three electrical resistivity tomographies (ERT) were obtained using the georesistivity meter Syscal R2 (Iris Instruments) connected to a multielectrode system. We used a Wenner-Schlumberger configuration array with electrodes spacing on the surface of 10.00 meters. The electrical tomographies were all orthogonal to the coastline.

The Wenner-Schlumberger array provided horizontal and vertical resolution, and to obtain 2D resistivity models the data obtained were interpreted through the inversion algorithm, RES2DINV, proposed by Loke [31]. The measured values of apparent resistivity provide, in fact, a first preliminary image of the electrical subsurface structure denominated as the "pseudosection". In a second step, the apparent resistivity measurements are transformed into true resistivity values using the rapid inversion algorithm of Loke and Barker [31].

The inversion routine is based on a smoothness-constrained least-squares method inversion algorithm that divides the subsurface into rectangular blocks, and the resistivity of the blocks is adjusted to minimize iteratively the difference between the computed and the measured apparent resistivity values; the root mean square (RMS) error gives a measure of this difference [31, 32].

3.2. Soil and Water Sampling. In addition, chemical and physical analyses on soil and water samples were performed in order to validate and integrate geophysical data. Undis-

turbed soil samples, collected from the surface in the first 40 cm of soil using Soil Sample Ring Kits (Eijkelkamp Equipment Netherlands), were taken for chemical analyses, using standard analytical procedures (Official Gazette of Italy No 248 of 21 October 1999). Soil samples were analyzed for pH (in water and in $CaCl_2$), for cations (Na,K,Ca, and Mg), for sulphates, chlorurs and nitrates, for organic matter and organic carbon, for cation exchange capacity, for an electrical conductivity of a saturated paste extract (ECe) or its inverse, electrical resistivity, and for texture. Groundwater samples were collected from few shallow holes across the area, since the groundwater table in the study area was high.

4. Results

The 2D resistivity results obtained by the inversion algorithm performed with the software RES2DINV are shown as images in Figure 5.

Electrical resistivity values are related to geological parameters of the subsurface and, in particular, resistivity values are controlled by the types of rocks and fluid. Then, the high-resolution electrical images are a powerful tool to identify conductive zones for the saltwater intrusion phenomena. Figure 5 shows the three 2D electrical images obtained in the study area related with depth and true resistivity of the subsoil investigated.

ERT 1 with direction NNW towards SE obtained on a rather flat regular surface, showed a first zone in the shallow part with values of 5–15 Ωm which could be associated with

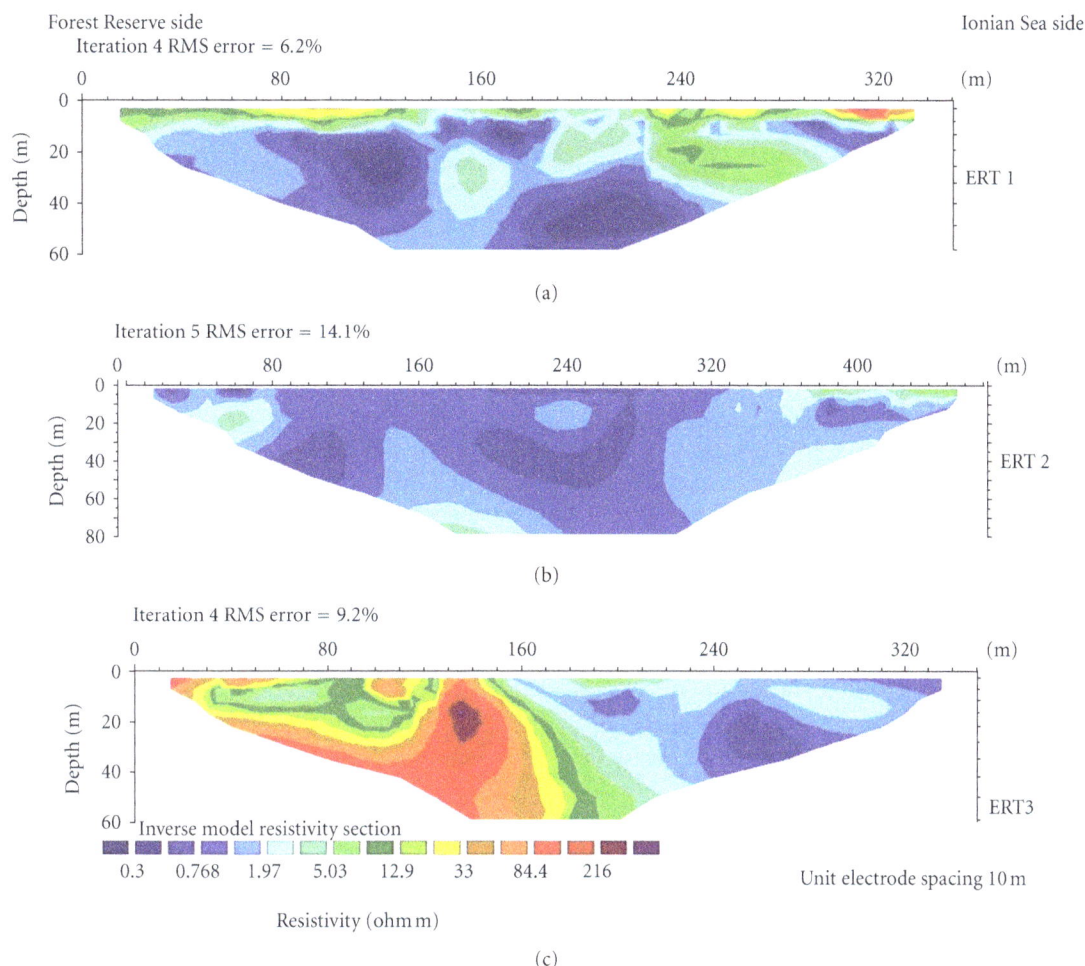

FIGURE 5: 2D electrical resistivity tomographies obtained in the investigated site. The position of each ERT is shown in Figure 3.

TABLE 1: Minimum and maximum values of chemical-physical data of soil samples collected in the study area.

Parameters	Minimum and Maximum Values
Na ($mg\,kg^{-1}$)	626–3425
K ($mg\,kg^{-1}$)	162–960
Ca ($mg\,kg^{-1}$)	917–2830
Mg ($mg\,kg^{-1}$)	203–905
Nitrogen ($g\,kg^{-1}$)	0.6–1.4
Organic carbon ($g\,kg^{-1}$)	4.9–21.5
Organic matter ($g\,kg^{-1}$)	8.4–37.1
pH (H_2O)	8.6–9.8
pH ($CaCl_2$)	7.9–8.9
Electrical conductivity (ECe) (Siemens m^{-1})	0.153–1.484
Electrical resistivity (Ωm)	6.536–0.674
Cation exchange capacity (meq/100 gr)	7.4–17.2
Chlorurs ($mg\,kg^{-1}$)	162–5894
Nitrate ($mg\,kg^{-1}$)	<0.1
Sulphate ($mg\,kg^{-1}$)	15.3–441
Sand %	33.3–85.2
Silt %	7.1–35.7
Clay %	7.5–38.2

a layer filled with inert material, and a second one with lower resistivity values ($\rho < 1.5\,\Omega m$) reflecting saturated strata.

ERT 2 is nearly parallel to the direction of ERT 1, it runs from NW to SE. As shown in Figure 5, the geoelectrical image presents a lateral variation in resistivity distribution, with higher electrical resistivity values (about $1–10\,\Omega m$) towards SE in the first part of the section (between 0 and 110 m). Lower electrical resistivity values were observed towards sea side, and these values reduced are maybe due to saturated strata for a subsurface seawater flow zone.

ERT 3 in direction of WNW-SE shows a significant trend of resistivity, in fact, the resistivity section is characterized by two zones with different resistivity values: the first zone, towards the inland of the Forest Reserve and throughout the thickness of soil explored, shows moderate to high resistivity values $10–220\,\Omega m$, while the second zone, towards the sea, shows very low resistivity values.

As expected, the 2D electrical resistivity tomographies show that resistivity values decrease in depth and from inland towards the Ionian Sea.

We have checked, one by one, the apparent resistivity values obtained during the field survey, and the topographic correction has been introduced. Only the apparent resistivity

TABLE 2: Electrical proprieties and pH values of groundwater samples.

Samples	pH	Electrical conductivity (ECe) Siemens m^{-1} to 25°C	Electrical resistivity Ωm
G1	7.7	0.68	1.470
G2	7.7	0.81	1.234
G3	7.8	1.92	0.520
G4	7.3	2.92	0.342
G5	7.5	7.27	0.137
Sea water	8.2	5.66	0.176

measurements affected by a standard deviation lower than 5% have been considered. In a second step we have increased the number of iterations of the inversion software routine. After this procedure the RMS of the electrical imaging is lower than 15%. If we take into account the strong variability of the resistivity values from ~1000 Ωm for sandy layers to ~1 Ωm of fully saltwater saturated zones, this error can be considered acceptable.

We used the RES2DINV, that is, a well-assessed algorithm for 2D resistivity data inversion, and there is a large scientific literature in which the RMS of the 2D electrical imaging is greater than 10%. Electrical imaging of complex geological environments are generally affected by RMS greater than 15% [10, 33] and in many cases is not possible to reduce the RMS without introducing artifacts.

In literature, salt water resistivity values below 1.0 Ωm were reported, in fact, seawater has an average resistivity of 0.2 Ωm [16, 34], while the resistivity of a layer saturated by saline water and dissolved solids is in the range of 8 to 50 Ωm [13, 16, 35, 36]. Therefore, based on these values of resistivity of layers saturated by saline water and some dissolved solids reported in the literature, resistivity data obtained in this work from geoelectric investigations, highlight the presence of strata saturated with seawater.

In Tables 1 and 2 the results of laboratory analysis for the physical and chemical parameters of soil and groundwater samples were summarised. Laboratory analyses permit to suppose that high pH values are determined by the accumulation of sodium and magnesium on the soil exchange complex. In fact, in the study area, pH values (in $CaCl_2$) have a range between 7.9 and 8.9 in soil samples, and between 7.3 and 7.8 in groundwater samples. A sample of sea water, instead, reported pH value of 8.2. Important differences were resulted in the conductivity (ECe) expressed in Siemens m^{-1} or its inverse, electrical resistivity, expressed in Ohm m. The mean ECe value on soil samples was equal to 0.613 Siemens m^{-1}, corresponding to 1.631 Ωm with a range from 0.153 to 1.484 Siemens m^{-1}, equal to a range of resistivity from 6.536 to 0.674 Ωm while in water samples the mean ECe value was equal to 2.720 Siemens m^{-1}, corresponding to 0.368 Ωm, and with a range from 0.68 to 7.27 Siemens m^{-1}, equal to a range of resistivity from 1.470 to 0.137 Ωm.

On the other hand, in seawater samples ECe value was equal to 5.660 Siemens m^{-1}, corresponding to 0.177 Ωm. Furthermore, the major ions were Na^+, Ca^{2+}, and Cl^-. Therefore, according to the classification NRCS [37], the area investigated presents from inland to coast, zones slightly saline, zones moderately saline, and zones close to highly

saline. The high concentration of soluble salts determines high osmotic pressure that makes very difficult to absorb water and nutrients from the roots of plants, creating the limiting factor to the diffusion and survival of forest species. The salinity distribution in the subsurface of study area is reflected by the vegetation cover, in fact, the zone with soil salinity is covered by salt-tolerant grasses [30]. On the basis of the geoelectrical measurements and of the chemical-physical analysis, the presence of zones with very low electrical resistivity values, can be associated to the intrusion of salt water into the subsoil. In the coastal environment, like the study area, hydraulic connection with seawater and inland gradient that move water to inland from seawater source, are such conditions as to cause seawater intrusion [2].

5. Conclusions

Geoelectrical surveys reveal the presence of two principal zones characterized by different resistivity values. The former is a strip back dune, closer to the sea, with a higher clay content and located at an altitude below sea level. It is submerged during the winter by water for a poor drainage, and it is characterized by lower resistivity values. The second zone is characterized by higher resistivity values and has a quite good vegetative cover for the best environmental conditions, greater distance from the sea, and elevation above sea level. In this area the spontaneous emergence of new plant pine is frequently observed for the fact that soil conditions are favourable as there is no seawater intrusion yet.

Results presented in this paper showed that the process of salt water is active on coastal Forest Reserve of Metapontum, and geoelectrical surveys have proved to be precious tools. Geoelectrical surveys have highlighted significant variations in subsurface resistivity associated with lithological characteristics of underground layers and variations in water saturation. In particular, these surveys have produced images of resistivity of the soil investigated, where the spatial distribution of brackish and saline water in the pine forest and resistivity variation with depth is very clear. These observations are also confirmed from laboratory analysis on soil and groundwater samples.

Acknowledgments

The authors would like to thank the *Corpo Forestale dello Stato* for the availability and logistic support during the measures in the Reserve. The authors are also grateful to Dr. Achille Palma, Metapontum Agrobios, for the soil chemical analyses. This paper was supported by MIUR in the framework of "Tecnologie per le Osservazioni della Terra e i Rischi Naturali -TeRN" project (PON 2004–2006; Obiettivo Realizzativo OR2).

References

[1] V. E. A. Post, "Fresh and saline groundwater interaction in coastal aquifers: is our technology ready for the problems ahead?" *Hydrogeology Journal*, vol. 13, no. 1, pp. 120–123, 2005.

[2] C. P. Petalas and I. B. Diamantis, "Origin and distribution of saline groundwaters in the upper Miocene aquifer system, coastal Rhodope area, northeastern Greece," *Hydrogeology Journal*, vol. 7, no. 3, pp. 305–316, 1999.

[3] C. Petalas and N. Lambrakis, "Simulation of intense salinization phenomena in coastal aquifers-the case of the coastal aquifers of Thrace," *Journal of Hydrology*, vol. 324, no. 1–4, pp. 51–64, 2006.

[4] M. Polemio, P. P. Limoni, D. Mitolo, and F. Santaloia, "Characterisation of the Ionian-Lucanian coastal aquifer and seawater intrusion hazard," in *Proceedings of the 17th Salt Water Intrusion Meeting*, pp. 422–434, Delft, The Netherlands, 2002.

[5] M. A. Somay and U. Gemici, "Assessment of the salinization process at the coastal area with hydrogeochemical tools and geographical information systems (GIS): Selçuk plain, Izmir, Turkey," *Water, Air, and Soil Pollution*, vol. 201, no. 1–4, pp. 55–74, 2009.

[6] D. W. Steeples, "Engineering and environmental geophysics at the millennium," *Geophysics*, vol. 66, pp. 31–35, 2001.

[7] V. Lapenna, P. Lorenzo, A. Perrone, S. Piscitelli, E. Rizzo, and F. Sdao, "2D electrical resistivity imaging of some complex landslides in the Lucanian Apennine chain, southern Italy," *Geophysics*, vol. 70, no. 3, pp. B11–B18, 2005.

[8] S. Pantelis, M. Kouli, F. Vallianatos, A. Vafidis, and G. Stavroulakis, "Estimation of aquifer hydraulic parameters from surficial geophysical methods: a case study of Keritis Basin in Chania (Crete—Greece)," *Journal of Hydrology*, vol. 338, no. 1-2, pp. 122–131, 2007.

[9] D. Chianese and V. Lapenna, "Magnetic probability tomography for environmental purposes: test measurements and field applications," *Journal of Geophysics and Engineering*, vol. 4, no. 1, article 08, pp. 63–74, 2007.

[10] V. Naudet, M. Lazzari, A. Perrone, A. Loperte, S. Piscitelli, and V. Lapenna, "Integrated geophysical and geomorphological approach to investigate the snowmelt-triggered landslide of Bosco Piccolo village (Basilicata, southern Italy)," *Engineering Geology*, vol. 98, no. 3-4, pp. 156–167, 2008.

[11] R. Balia, E. Gavaudò, F. Ardau, and G. Ghiglieri, "Geophysical approach to the environmental study of a coastal plain," *Geophysics*, vol. 68, no. 5, pp. 1446–1459, 2003.

[12] A. T. Batayneh, "Use of electrical resistivity methods for detecting subsurface fresh and saline water and delineating their interfacial configuration: a case study of the eastern Dead Sea coastal aquifers, Jordan," *Hydrogeology Journal*, vol. 14, no. 7, pp. 1277–1283, 2006.

[13] P. Bauer, R. Supper, S. Zimmermann, and W. Kinzelbach, "Geoelectrical imaging of groundwater salinization in the Okavango Delta, Botswana," *Journal of Applied Geophysics*, vol. 60, no. 2, pp. 126–141, 2006.

[14] M. A. Koukadaki, G. P. Karatzas, M. P. Papadopoulou, and A. Vafidis, "Identification of the saline zone in a coastal aquifer using electrical tomography data and simulation," *Water Resources Management*, vol. 21, no. 11, pp. 1881–1898, 2007.

[15] S. E. Kruse, M. R. Brudzinski, and T. L. Geib, "Use of electrical and electromagnetic techniques to map seawater intrusion near the Cross-Florida Barge Canal," *Environmental and Engineering Geoscience*, vol. 4, no. 3, pp. 331–340, 1998.

[16] A. A. Nowroozi, S. B. Horrocks, and P. Henderson, "Saltwater intrusion into the freshwater aquifer in the eastern shore of Virginia: a reconnaissance electrical resistivity survey," *Journal of Applied Geophysics*, vol. 42, no. 1, pp. 1–22, 1999.

[17] S. S. Abdul Nassir, M. H. Loke, C. Y. Lee, and M. N. M. Nawawi, "Salt-water intrusion mapping by geoelectrical imaging surveys," *Geophysical Prospecting*, vol. 48, no. 4, pp. 647–661, 2000.

[18] M. Sherif, A. El Mahmoudi, H. Garamoon, and A. Kacimov, "Geoelectrical and hydrogeochemical studies for delineating seawater intrusion in the outlet of Wadi Ham, UAE," *Environmental Geology*, vol. 49, no. 4, pp. 536–551, 2006.

[19] A. Cimino, C. Cosentino, A. Oieni, and L. Tranchina, "A geophysical and geochemical approach for seawater intrusion assessment in the Acquedolci coastal aquifer (Northern Sicily)," *Environmental Geology*, vol. 55, no. 7, pp. 1473–1482, 2008.

[20] M. H. Khalil, "Geoelectric resistivity sounding for delineating salt water intrusion in the Abu Zenima area, west Sinai," *Egypt Journal of Geophysics and Engineering*, vol. 3, pp. 243–251, 2006.

[21] D. W. Urish and R. K. Frohlich, "Surface electrical resistivity in coastal groundwater exploration," *Geoexploration*, vol. 26, no. 4, pp. 267–289, 1990.

[22] A. A. M. Ebraheem, M. M. Senosy, and K. A. Dahab, "Geoelectrical and hydrogeochemical studies for delineating groundwater contamination due to salt-water intrusion in the northern part of the Nile Delta, Egypt," *Ground Water*, vol. 35, no. 2, pp. 216–222, 1997.

[23] K. Choudhury and D. K. Saha, "Integrated geophysical and chemical study of saline water intrusion," *Ground Water*, vol. 42, no. 5, pp. 671–677, 2004.

[24] G. Spilotro, F. Canora, F. Caporale, G. Leandro, and N. Vignola, "Hydrogeology and groundwater salinization in the ionian coastal plane of the Basilicata Region," in *Proceedings of the 17th Salt Water Intrusion Meeting*, pp. 422–434, Delft, The Netherlands, 2002.

[25] E. Amezketa, "An integrated methodology for assessing soil salinization, a pre-condition for land desertification," *Journal of Arid Environments*, vol. 67, no. 4, pp. 594–606, 2006.

[26] N. Ciaranfi, M. Maggiore, P. Pieri, L. Rapisardi, G. Ricchetti, and N. Walsh, "Considerazioni sulla neotettonica della Fossa bradanica," in *Nuovi Contributi Alla Realizzazione Della Carta Neotettonica d'Italia*, vol. 251, pp. 73–96, C.N.R. Progetto Finalizzato Geodinamica, Napoli, Italy, 1979.

[27] M. Tropeano, L. Sabato, and P. Pieri, "The Quaternary "Post-turbidite" sedimentation in the South-Apennines Foredeep (Bradanic Trough-Southern Italy)," *Italian Journal of Geosciences Bull*, vol. 1, pp. 449–454, 2001.

[28] E. Valpreda and U. Simeoni, "Assessment of coastal erosion susceptibility at the national scale: the Italian case," *Journal of Coastal Conservation*, vol. 9, no. 1, pp. 43–48, 2003.

[29] B. M. S. Giambastiani, M. Antonellini, G. H. P. Oude Essink, and R. J. Stuurman, "Saltwater intrusion in the unconfined coastal aquifer of Ravenna (Italy): a numerical model," *Journal of Hydrology*, vol. 340, no. 1-2, pp. 91–104, 2007.

[30] A. Satriani, A. Loperte, T. Simoniello, M. D'Emilio, C. Belviso, and V. Lapenna, "A multidisciplinary approach for studying

the forest reserve of Metapontum (southern Italy) affected by saltwater intrusion phenomena," *EGU Geophysical Research Abstracts*, vol. 9, 2007.

[31] M. H. Loke and R. D. Barker, "Rapid least-squares inversion of apparent resistivity pseudosections by a quasi-Newton method," *Geophysical Prospecting*, vol. 44, no. 1, pp. 131–152, 1996.

[32] Y. Sasaki, "Resolution of resistivity tomography inferred from numerical simulation," *Geophysical Prospecting*, vol. 40, no. 4, pp. 453–463, 1992.

[33] A. Perrone, G. Zeni, S. Piscitelli et al., "Joint analysis of SAR interferometry and electrical resistivity tomography surveys for investigating ground deformation: the case-study of Satriano di Lucania (Potenza, Italy)," *Engineering Geology*, vol. 88, no. 3-4, pp. 260–273, 2006.

[34] D. S. Parasnis, *Principle of Applied Geophysics*, Chapman & Hall, London, UK, 1986.

[35] W. De Breuk and G. De Moor, "The water table acquifer in the eastern coastal area of Belgium," *Bulletin of the International Association of Scientific Hydrology*, vol. 14, pp. 137–155, 1969.

[36] A. A. R. Zohdy, P. Martin, and R. J. Bisdorf, "A study of seawater intrusion using direct-current soundings in the southeastern part of the Oxnard Plain, California," Open-File Report 93-524, Geological Survey, 1993.

[37] Soil Survey Division Staff, *Soil Survey Manual USDA Agric Handbook 18*, U.S. Government Printing Office, Washington, DC, USA, 1993.

17

Electric Signals on and under the Ground Surface Induced by Seismic Waves

Akihiro Takeuchi,[1] Kan Okubo,[2] and Nobunao Takeuchi[3]

[1] Earthquake Prediction Research Center, Institute of Oceanic Research and Development, Tokai University, 3-20-1 Orido, Shimizu-ku, Shizuoka 424-8610, Japan
[2] Division of Information and Communications Systems Engineering, Tokyo Metropolitan University, 6-6 Asahigaoka, Hino 191-0065, Japan
[3] Research Center for Prediction of Earthquakes and Volcanic Eruptions, Graduate School of Science, Tohoku University, 6-6 Aza-aoba, Aramaki, Aoba-ku, Sendai 980-8578, Japan

Correspondence should be addressed to Akihiro Takeuchi, atakeuchi@sems-tokaiuniv.jp

Academic Editor: Laurence Jouniaux

We constructed three observation sites in northeastern Japan (Honjo, Kyowa, and Sennan) with condenser-type large plate electrodes ($4 \times 4\,m^2$) as sensors supported 4 m above the ground and with pairs of reference electrodes buried vertically at 0.5 m and 2.5 m depth (with a ground velocity sensor at Sennan only). Electrical signals of an earthquake (M6.3) in northeastern Japan were detected simultaneously with seismic waves. Their waveforms were damped oscillations, with greatly differing signal amplitudes among sites. Good positive correlation was found between the amplitudes of signals detected by all electrodes. We propose a signal generation model: seismic acceleration vertically shook pore water in the topsoil, generating the vertical streaming potential between the upper unsaturated water zone and the lower saturated water zone. Maximum electric earth potential difference was observed when one electrode was in the saturated water zone, and the other was within the unsaturated water zone, but not when the electrodes were in the saturated water zone. The streaming potential formed a charge on the ground surface, generating a vertical atmospheric electric field. The large plate electrode detected electric signals related to electric potential differences between the electrode and the ground surface.

1. Introduction

Many researchers throughout the world have explored abnormal electromagnetic phenomena preceding earthquakes, such as electromagnetic emissions [1] and ionospheric disturbances [2]. However, most such reports describe methods that are retrospective. Scientific proof of the precursors is apparently still elusive. Earthquake prediction by detection of such precursors cannot be realized easily at this stage. To make steady progress in the scientific study of seismo-electromagnetic precursors, we believe that it is important to prove, first of all, the existence of phenomena that occur at the occurrence of earthquakes and at the arrival of seismic waves and to evaluate these phenomena quantitatively.

First, the piezoelectric effects of quartz and electrokinetic effects of pore water will be regarded as the matter that couples seismic waves and electromagnetic phenomena. However, Ogawa and Utada previously simulated an electric signal induced by seismic waves in a piezoelectric body lying at shallow depth and concluded that the signals would be weak unless the ground motion was unusually strong [3]. On the other hand, Pride proposed, theoretically, coupling of electromagnetics and acoustics of porous media saturated with electrolyte [4] although the purpose of that study was not earthquake prediction. Bordes and coworkers confirmed such coupled phenomena using an experimental apparatus built in an ultrashielded chamber in a deep underground laboratory [5]. Garambois and Dietrich detected electric signals induced by a dynamite shot in the field and analyzed

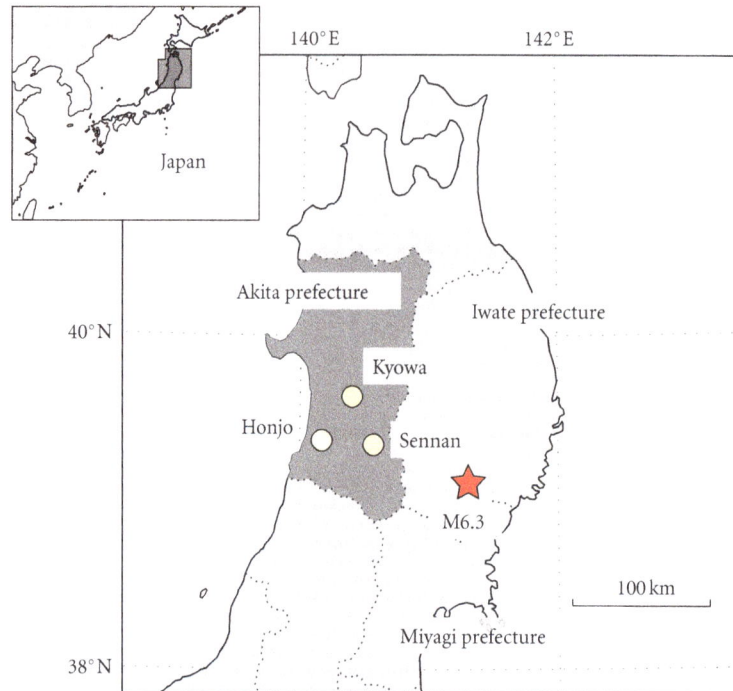

FIGURE 1: Location of three observation sites (Honjo, Kyowa, and Sennan) and the epicenter of an earthquake (M6.3) in northeastern Japan.

them in light of Pride's theory [6]. Consequently, it is expected that electromagnetic phenomena appear under the ground surface at the arrival of "natural" seismic waves, which are driven by the electrokinetic effect of pore water.

We have conducted observations of seismo-electromagnetic phenomena under the ground surface using electrodes buried in the ground. Additionally, we used condenser-type plate electrodes supported above the ground surface. This paper describes the simultaneous detection of electrical signals by electrodes of both types at the arrival of natural seismic waves and discusses their generation models based on their streaming potential under several underground environmental circumstances, although results remain qualitative at this stage.

2. Observation Sites and System

Our first observation sites were in Miyagi Prefecture in northeastern Japan [7, 8]. These sites had sensors of two types: (i) condenser-type electrodes supported above the ground surface by insulator pillars and (ii) reference electrodes buried horizontally and vertically under the ground surface. They detected electric signals at the arrival of seismic waves. The next site was located in Akita Prefecture in northeastern Japan [9]. This site had a similar set of sensors that also detected electric signals. We expected a relation between the signals and pore water in the ground and proposed a prototype model to explain the signals induced by the arrival of seismic waves. However, we had never detected a signal simultaneously at more than two sites until 2001. Therefore, we were unable to discuss a detailed model that was probably strongly related to the observation conditions.

In 2003, we prepared three new observation sites in Akita Prefecture in northeastern Japan. Figure 1 depicts the locations of the three observation sites.

(1) The Honjo observation site (N39°23′, E140°04′) was located in a green belt on the Honjo Campus of Akita Prefectural University. To construct the campus, a hill surrounded by rice fields was leveled. The topsoil in this area was brown forest soil or gley soil with silt stones and conglomerates. No tall building existed near this site.

(2) The Kyowa observation site (N39°40′, E140°23′) was located in a garden yard of a small recreation house of Akita Prefectural University on a flat land in the middle of mountains. The topsoil in this area was brown forest soil with acid tuff and sandstone. The area around this site was dotted with low-rise houses.

(3) The Sennan observation site (N39°23′, E140°30′) was located on the sports ground of a former primary school surrounded by rice fields. The topsoil in this area was gley soil or sand-rich sediments. No tall building existed around this site.

Figure 2 presents a schematic of the observation system. They basically had sensors of two types: (i) a pair of reference electrodes (RE-5; M. C. Miller Co. Inc.) to detect the vertical component of the earth potential difference (EPD), and (ii) a large plate electrode to detect the signal related to the vertical component of the atmospheric electricity (AE). No tall building existed around the large plate electrodes. Therefore, we can reduce the atmospheric electricity disturbing because of such buildings. Only the Sennan observation site also

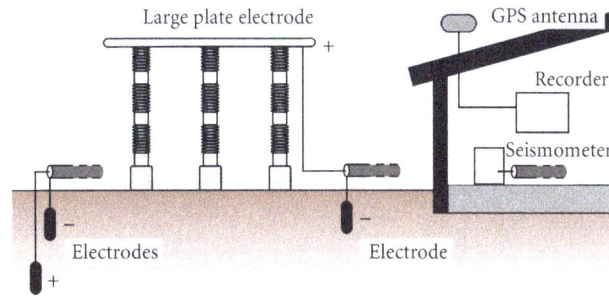

FIGURE 2: Schematic of the observation system. The pair of reference electrodes measured the vertical component of the earth potential difference (EPD). The positive electrode was buried 2.5 m deep; the negative one was buried 0.5 m deep. The aluminum large plate electrode measured signals related to the vertical component of the atmospheric electricity (AE). The plate area was $4 \times 4\,\text{m}^2$. It was supported by five insulators at a height of about 4 m. The velocity sensor (only at Sennan), which was fixed on the concrete base of a barn near the sensors, measured the up-down component of the ground surface velocity.

had a velocity sensor (L-22D; Mark Products LLC) to detect the up-down ground motion. All sensors were connected to a data recorder (DR-1021; DKK-TOA Corp.) with coaxial cables. A PC controlled the recorder and stored the data on its HD at 4 Hz sampling at the Honjo site and at 10 Hz sampling at the Kyowa and Sennan sites. The data clock was synchronized to within 1 ms of the time accuracy using a GPS unit.

Results confirmed that no disturbance appeared in EPD data by minute displacement of the electrodes buried in the ground or by movement of the coaxial cables. Based on our observation experiences [7, 8], we selected the electrode pairs buried vertically rather than horizontally, which can minimize the effects of artificial noise superimposed on the natural telluric currents flowing horizontally. The large plate electrodes were tested under various weather conditions such as heavy snow from August 2000. Thereafter, ordinary observations in the three sites started in May 2001. Although the system worked well during 2001–2006, it became too old to use in 2007: it was difficult to maintain and repair. Therefore, we dismantled the system in 2009.

3. Observation Results

Small earthquakes occurred in and around Akita Prefecture, sometimes causing weak electric signals at the Sennan observation site at the arrival of seismic waves [9]. When a major earthquake occurred in Iwate Prefecture (Figure 1), it caused electric signals at all three sites. We specifically examine this earthquake, which was the only major earthquake occurring around this area during 2001–2006. It occurred at 22:02 JST (= UT + 9 hr) on December 2, 2001. Its magnitude was 6.3 on the Japan Meteorological Agency (JMA) scale. The epicenter was (N39°23′, E141°16′), with focal depth of 130 km. The focal depth of this earthquake was so great that the hypocentral distances of the three sites were almost identical. In general, local seismic intensity depends strongly on the site environments and the propagation paths of seismic waves. However, JMA seismometers around our three observation sites indicated the local seismic intensity of 3-4

on the JMA scale that is classified into 10 ranks. Therefore, we can assume that the local seismic intensities at the sites were similar.

Figure 3 portrays plots of raw data before and after the arrival of seismic waves at each site. The upper rows show raw EPD, the middle ones show raw AE, and the lower one shows the raw ground velocity at the Sennan site. The time range is 22:00–22:05 JST. Dotted lines show the time of origin of the earthquake. Vibrating waveforms are confirmed, although some are small against background variations. These background variations were probably the result of local variations of atmospheric electricity, power line noise, and wind. To clarify the waveforms that are of interest for this study, we adopted the moving average method. In this study, the time period of 1 s was used for calculation of the moving average. The 1 s moving averages were subtracted from the original data, yielding the remnant waveforms as shown in Figure 4. The result clarifies that the damped oscillations are induced simultaneously in EPD and AE at the arrival of seismic waves, although some of them still include background noise, probably from power lines. Spectrum analyses show that both signals have a peak at about 1 Hz. The outline waveforms of the induced EPD and AE signals are similar to that of the ground velocity at the Sennan site. Those at Honjo and Kyowa will be also similar to those of the ground velocity at each site.

Figure 5 shows the relation between the average amplitudes of the induced EPD and AE signals ($\overline{\text{EPD}}$ and $\overline{\text{AE}}$, resp.). They are defined as shown below

$$
\text{average}
= \sqrt{\frac{1}{30(s)} \sum_{\substack{\text{peak time }(t=0(s))}}^{t=30(s)} (\text{induced signal} + \text{background}(t))^2}
$$
$$
- \sqrt{\frac{1}{30(s)} \sum_{\substack{\text{no propagation }(t'=0(s))}}^{t'=30(s)} (\text{background}(t'))^2}.
$$

$$\tag{1}$$

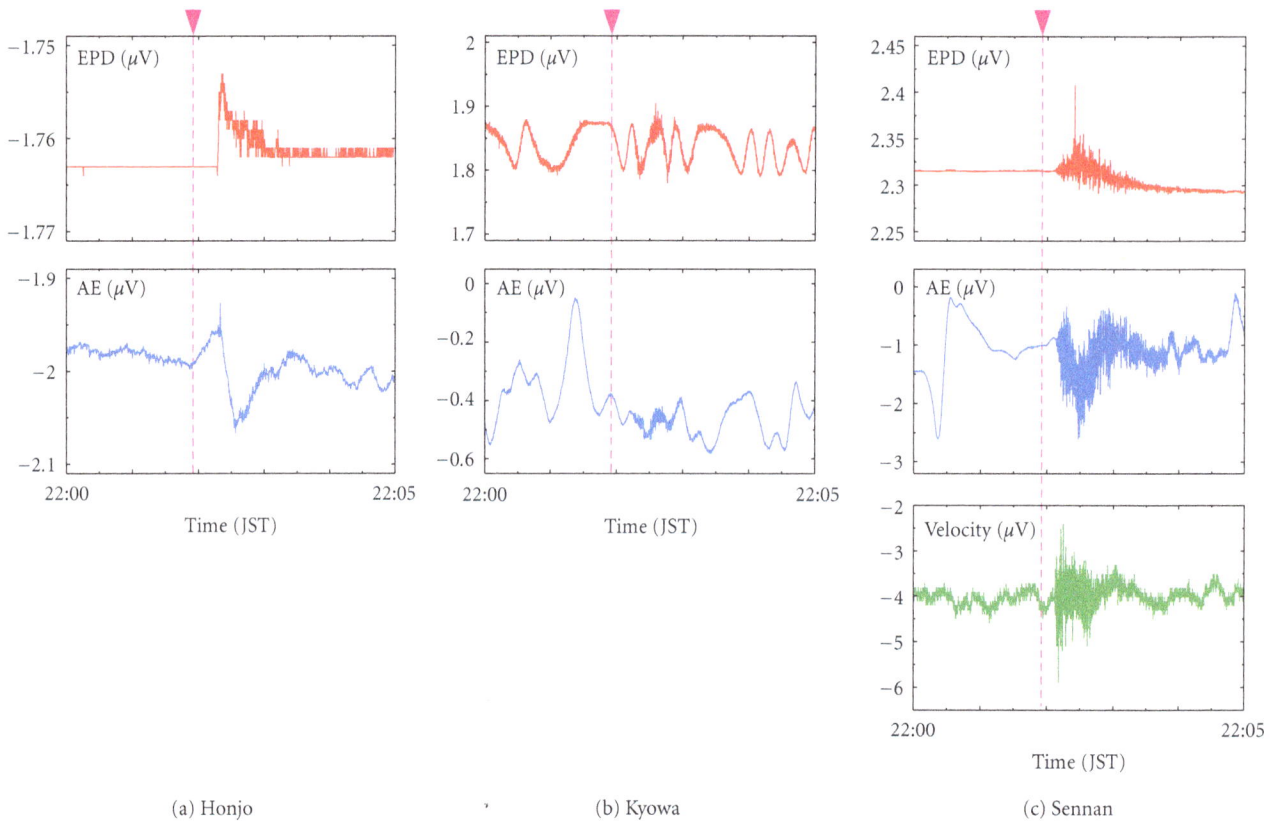

(a) Honjo   ,   (b) Kyowa   (c) Sennan

FIGURE 3: Raw data recorded before and after the arrival of seismic waves from a M6.3 earthquake: (a) at Honjo, (b) at Kyowa, and (c) at Sennan. Upper row graphs show raw data of the earth potential difference (EPD). Middle ones show raw data of the signals related to the atmospheric electricity (AE). Lower one shows raw data of the ground velocity (only at Sennan). Broken lines show the origin time. The time is in JST (= UT + 9 hr).

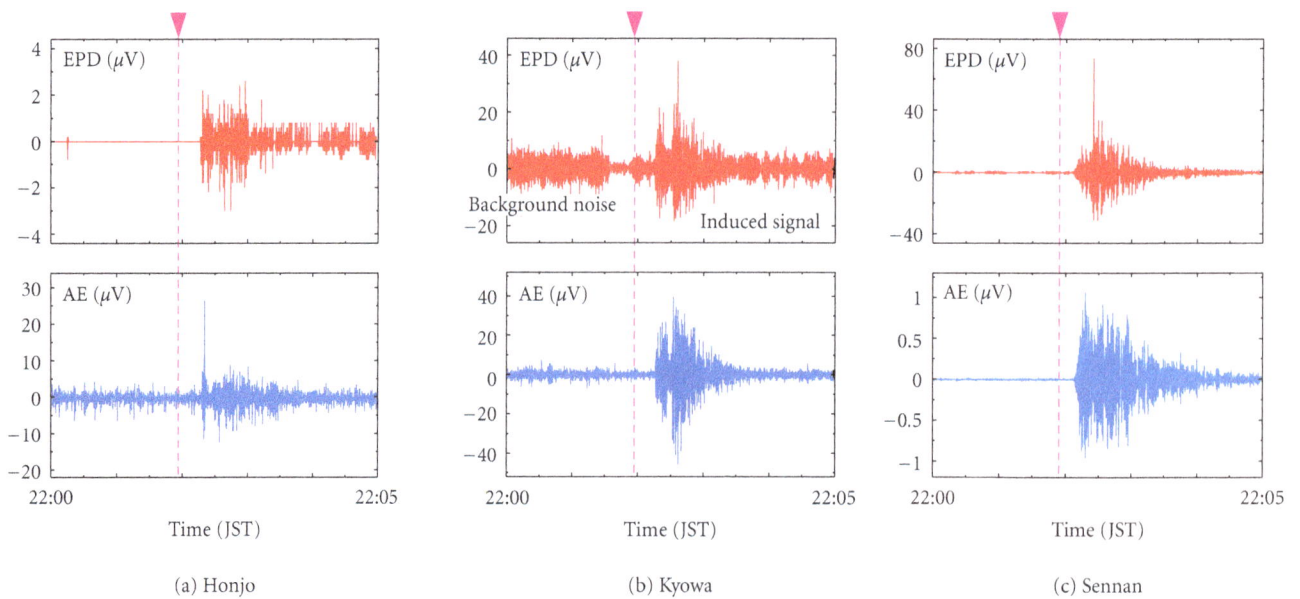

(a) Honjo   (b) Kyowa   (c) Sennan

FIGURE 4: Waveforms extracted from raw data presented in Figure 3 using the 1 s moving average method: (a) at Honjo, (b) at Kyowa, and (c) at Sennan. Broken lines mean the origin time. The time is in JST (= UT + 9 hr).

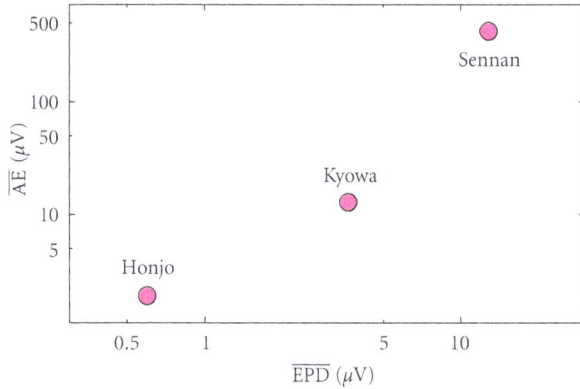

FIGURE 5: Positive correlation between the average of the induced earth potential difference ($\overline{EPD}$) and the average of the induced atmospheric electricity ($\overline{AE}$).

The summation in the first root is over a period of 30 s after the peak time of the induced EPD and AE signals; another in the second root is over a period of 30 s when no seismic wave propagated. Figure 5 depicts a positive relation between $\overline{EPD}$ and $\overline{AE}$. Although the local seismic intensity is probably similar among the three sites as described above, the $\overline{EPD}$ and $\overline{AE}$ at the Honjo site are smaller; those at the Sennan site are larger.

4. Discussion

4.1. Generation Mechanism of the Induced EPD Signals. We can list two effects as possible mechanisms generating the induced EPD signals. The first is a piezoelectric effect of quartz grains involved in the topsoil under the large plate electrodes. However, all observation sites were located on the topsoil, which was wet sometimes. Electric polarizations at each small quartz grain will be quite small. Moreover, polarization vectors will be random. Therefore, the vector summation is almost negligible. Furthermore, as described in the Introduction, an early simulation by Ogawa and Utada indicated that an electric signal induced by piezoelectric effect is expected to be weak [3]. In conclusion, the possibility of piezoelectricity is discounted. The second is an electrokinetic effect of pore water in the topsoil under the large plate electrodes. Our earlier observation data detected by similar sensors at a different site in Miyagi Prefecture and at Sennan site revealed a linear relation in log-log plots between the local seismic intensity and $\overline{EPD}$ [7, 9]. Moreover, in those observations, the EPD waveform resembles that of the ground acceleration. Therefore, the force applied to pore water varies similarly to the EPD waveform. These results show that $\overline{EPD}$ increases proportionally with the pressure difference in pore water. In this paper, we specifically examine the streaming potential caused by electrokinetic effect of pore water under seismic pressure.

From the viewpoint of geohydraulics, the water content generally increases with increased depth in the topsoil to be with pores. The upper is called the unsaturated water zone. The lower is called the saturated water zone [10, 11]. Next,

we consider only the vertical component. As portrayed in Figure 6(a), we assume the network of pores in the topsoil as a bundle of tubes connecting the air and the saturated water zone, as is often assumed also for the study of electric fields induced by the vertical fluid flow [12, 13]. The upper part is a fine tube; the lower is a capillary tube. The lower part of the capillary tubes holds the pore water, called the capillary saturated water zone. The lower part is the saturated water zone, where the network of pore water is mutually and completely connected. It is noteworthy that such a bundle of tubes is not generated by seismic waves, but is instead a representation of the network of pores. Here, we assume that positive ions are predominant in water in the capillary tubes, although negative ions adhere to the tube inner walls. Maldistribution of positive and negative ions engenders the formation of electric potential difference between the center and wall of the tubes, called the zeta potential. Because this maldistribution depends on the soil type and the pore water pH [12, 14], positive ions will adhere to the walls under different conditions.

Based on field observations and theoretical studies, it is well known that the seismic acceleration is coupled with the pore pressure [15, 16]. This coupled dynamics must be valid in our geohydraulic system depicted in Figure 6(a). The wavelength of the seismic waves is much longer than the tube dimensions. Therefore, we can assume that the acceleration is uniform in our geohydraulic system. As indicated in Figure 6(b), when acceleration is inflicted upward in this system, the water in the capillary tubes flows upward along the tubes. In the real soil layer, soil grains function also as obstacles to the vertical water flow, so that a portion of the flow will bend in horizontal directions. Such horizontal flows deviated from each vertical flow will cancel each other. As a result, the vertical component of the flow remains. The tubes in our model represent the flow routes of the remnant vertical flow of pore water. Pride used a similar assumption to derive the coupled electromagnetics and acoustics of porous media from first principles [4].

The number of cations exceeds the number of anions in the diffuse layer of the electrical double layer in this system. Therefore, upward flow of the water engenders a positive electrification in the upper range of the tubes. However, the lower range of the tubes is negatively charged. Consequently, the vertical electric dipoles are formed and an electric potential difference (called streaming potential) appears between the upper level of the water flowing up and the saturated water zone. Because no more charge occurs above the upper level of the water, the electric potential in this range is almost constant or decreases because of a small amount of pore water in this range. When the positive and negative terminals of electrodes are buried, respectively, in the saturated water zone and the unsaturated water zone, they will detect a negative induced EPD signal. As Figure 6(c) shows, when acceleration is inflicted downward in this system, the opposite phenomena occur: the electrode will detect a positive induced EPD signal. Consequently, when seismic waves vertically oscillate this geohydraulic system, the electrodes detect an induced EPD signal that also oscillates.

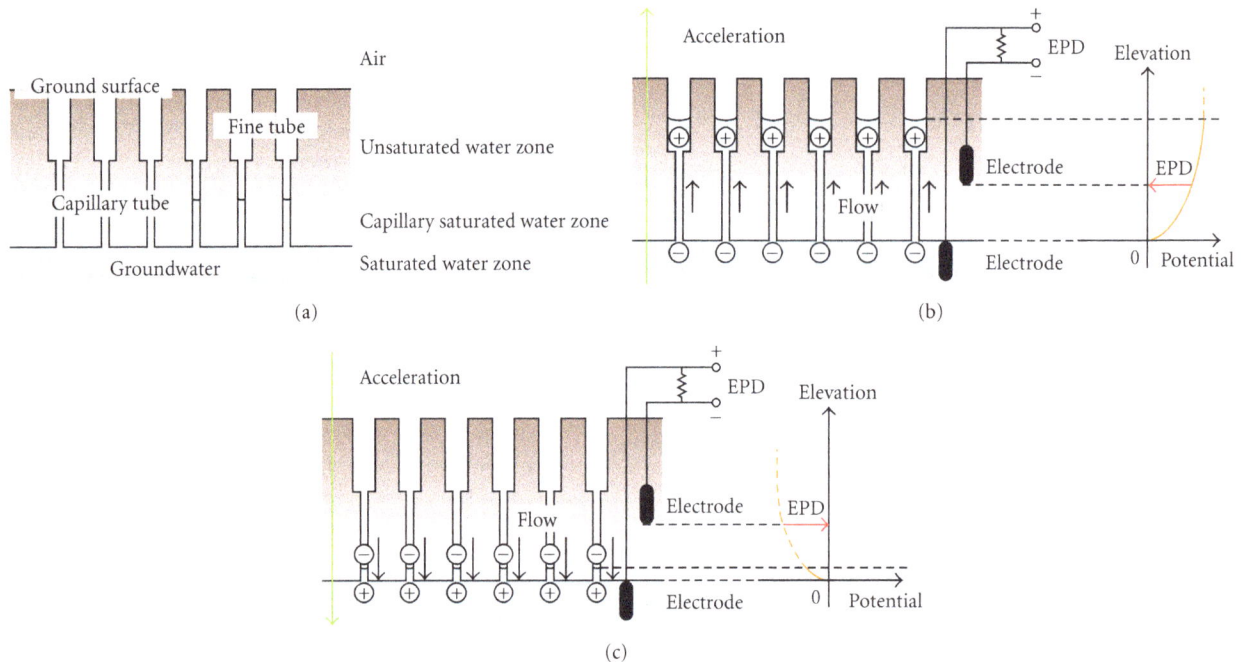

(a)

(b)

(c)

FIGURE 6: Generation model of the induced earth potential difference (EPD). (a) Model of the topsoil from the viewpoint of water saturation/unsaturation. (b) Generation of a negative EPD signal induced by streaming potential because of the upward flow of water along the tubes. (c) Generation of a positive EPD signal induced by streaming potential because of the downward flow of water along the tubes.

It is known that there are the underground electric fields produced by electric dipoles existing at the boundary between the capillary saturated water zone and the saturated water zone (the so-called water table) [17]. The dipoles are perpendicular to the boundary, and their polarity and amplitude depend, respectively, on the zeta potential and the conductivity of the saturated water zone. However, they do not reflect a transient effect, but a DC effect. Therefore, the transient movement of a seismic wave cannot move this boundary at this time scale. Consequently, the underground electric field produced by these dipoles will not contribute to the induced EPD signals in our observation system.

The amplitude of EPD signals will depend on the electrode positions in this geohydraulic system. Next we assume the six cases presented in Figure 7.

(a) When both electrodes are located mutually parallel in the unsaturated water zone (Figure 7(a)), the induced EPD signals will be slight. Even if the streaming potential appears at the electrodes, the potential difference between them will be small. For example, in our previous similar observations at a different site [7], the vertical induced EPD signal was 10 times larger than the horizontal one, in which each electrode was separated by 2 m.

(b) When one electrode is located in the unsaturated water zone and another is in the saturated water zone (Figure 7(b)), the induced EPD signals will be large. The electrode pair is located at both ends of electric polarization. Therefore, the electrodes detect the

potential difference most effectively. Consequently, the amplitude becomes the maximum.

(c) When both electrodes are located in the saturated water zone (Figure 7(c)), the induced EPD signals will be slight. Because the pore water network is completely connected, this zone is sufficiently conductive. Most of the streaming potential will be canceled. For example, in our previous similar observation at a different site [7], two electrodes were buried at depths of 10 m and 12 m, probably in the saturated water zone. They detected only very small induced EPD signals.

(d) When one electrode is located deep in the unsaturated water zone and another is deep in the saturated water zone (Figure 7(d)), the induced EPD signals will be large. This case resembles case (b). For example, in our previous similar observations at a different site [8], a pair of electrodes was buried at depths of 7 m and 10 m. A borehole survey confirmed that the upper level of the saturated water zone was at 9 m depth. This site detected sufficiently large induced EPD signals against background variations.

(e) When both electrodes are located in the unsaturated water zone far from the deep saturated water zone (Figure 7(e)), the induced EPD signals will be very small. No pore water flows up in the upper range of the thick unsaturated water zone. Therefore, little potential difference appears between the electrodes.

(f) When one electrode is in the unsaturated water zone and another is near or above the saturated water zone

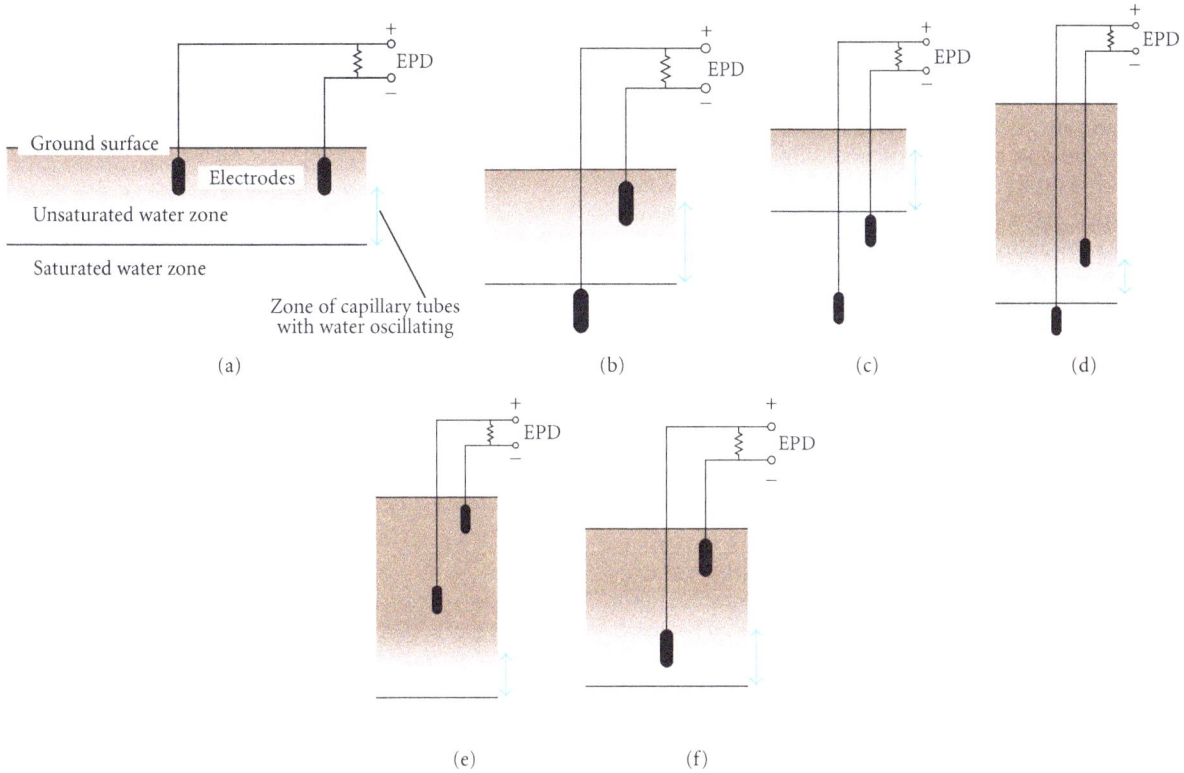

FIGURE 7: Various cases of locations with the electrode pair buried and with saturated/unsaturated water zone levels. (a) Both electrodes are mutually parallel in the unsaturated water zone. (b) One electrode is buried in the unsaturated water zone; another is buried in the saturated water zone. (c) Both electrodes are buried in the saturated water zone. (d) One electrode is deeply buried in the unsaturated water zone; another is buried deeply in the saturated water zone. (e) Both electrodes are buried in the unsaturated water zone far from the deep saturated water zone. (f) One electrode is buried in the unsaturated water zone; another is buried near but above the saturated water zone.

(Figure 7(f)), the induced EPD signals will be not so large. Until the lower electrode is buried in the water flowing up, no sufficient potential difference appears between the electrodes.

Because this model is based on streaming potential, $\overline{EPD}$ depends on the zeta potential, that is, the soil type. However, as described above, the topsoil is fortunately similar among site areas: brown forest soil or gley soil. Therefore, we need not be so sensitively concerned about the soil type in this study. Nevertheless, according to reports of earlier studies, the apparent electrokinetic coefficient of a (un-)saturated medium reaches the maximum under the 70% saturation state, not under a perfectly saturated state [18]. This fact is concordant with our six models presented in Figure 7 because our model for the induced EPD signal is based on the electrokinetic effect, which is strongly related to the electrokinetic coefficient. For example, the saturated state of case (c) is 100%. Therefore, the apparent electrokinetic coefficient is less than that in case (b). This engenders a decrease of $\overline{EPD}$ in case (c). Here, we can expect that the significant difference of the $\overline{EPD}$ among the three sites mainly results from differences in the position of the buried electrodes and saturated water zone.

(1) The Honjo observation site was located 6 m above the surrounding rice fields. The saturated water zone is expected to be a few meters deeper than the fields. Therefore, this site corresponds to case (e): the $\overline{EPD}$ was small.

(2) The Kyowa observation site was located on a flat land in the middle of mountains. Therefore, this site will correspond to case (f): the $\overline{EPD}$ was not so large.

(3) The Sennan observation site was located on the same level as the surrounding rice fields. The saturated water zone will be a few meters deeper than the fields. Therefore, this site will correspond to case (b): the $\overline{EPD}$ was large.

In case (b), the $\overline{EPD}$ will be roughly equal to the streaming potential, as defined simply by

$$\frac{\overline{EPD}}{L} = \left| C\frac{P_{bottom} - P_{top}}{L} \right|, \tag{2}$$

where L (ca. 1-2 m) is the capillary water height, C signifies the apparent electrokinetic coefficient, P_{bottom} denotes the seismic pressure applied at the bottom of the capillary water, and P_{top} stands for the pressure at the top. The value of C is obtained as

$$C = \frac{\varepsilon_r \varepsilon_0 \zeta}{\eta \sigma}, \tag{3}$$

where ε_r denotes the relative permittivity of capillary water, ε_0 ($= 8.8 \times 10^{-12}$ F/m) represents the permittivity of vacuum, ζ is the zeta potential, η stands for the viscosity of capillary water, and σ signifies the electric conductivity of the capillary water. We have no concrete values for these parameters for our observation sites except σ of approximately 0.01 S/m. Therefore, we simply apply standard values that are often used for order estimation: ε_r of about 80, $|\zeta|$ of about 10^{-2} V, and η of approximately 10^{-4} Pa s [19]. Using these values, we obtain $|C|$ of approximately 10^{-5} V/Pa. Moreover, Allègre and coworkers made a compilation that $|C|$ is of approximately 10^{-6} V/Pa in the case of sands and sand stones at pH 7-8 [18]. Therefore, we adopt $|C|$ of approximately 10^{-6}–10^{-5} V/Pa for this study. On the other hand, the P_{bottom} is obtained from

$$P_{\text{bottom}} = \frac{ma}{S}, \tag{4}$$

where m ($= m^*SL$, with $m^* = 10^3$ kg/m^3) represents the mass of the capillary water, a is the seismic acceleration, and S denotes the cross section of the capillary tube. The local seismic intensity 3 in the JMA scale has a of about 10^{-1} m/s^2 at ca. 1 Hz. Consequently, we obtain P_{bottom} of ca. 10^2 Pa. By contrast, P_{top} is approximately 0 Pa because nothing is on the top surface of the capillary water. Finally, we obtain $\overline{\text{EPD}}$ of about 10^{-4}–10^{-3} V, which is not so different from the value obtained from the observations at Sennan site.

4.2. Generation Mechanism of the Induced AE Signals. As shown in Figure 4, the induced AE signals appear in combination with the induced EPD signals. Moreover, as shown in Figure 5, a good positive correlation exists between $\overline{\text{AE}}$ and $\overline{\text{EPD}}$. Therefore, the generation mechanism of the induced AE signals must be coupled with that of EPD: both mechanisms are strongly related to streaming potential. Figure 8 presents a schematic diagram of a possible generation model of the induced AE signals in case (b) shown in Figure 7. As described in Figure 6, oscillation of the streaming potential appears in the unsaturated water zone. This is equivalent to oscillation of the electric dipoles vertical to the ground immediately under the ground surface. Their electric charges on the ground surface generate an atmospheric electric field vertical to the ground. Results show the appearance of oscillation of the electric potential difference between the large plate electrode and the ground surface. Electric charges move between the electrode and the ground through the recorder to cancel the potential oscillation. This movement is detected as oscillation of the induced AE signal.

Here, it is noteworthy that the recorded signal is not equal to the potential difference between the large plate electrode and the ground surface; the electric potential superimposed on the background variation is caused by a transient current passing through the internal resistance of the recorder. To obtain the original waveform of the induced AE signals, we need the integral of the recorded waveforms using different coefficients for each frequency. However, obtaining the coefficients is difficult. We have no concrete values to calculate the original AE signals at the present stage.

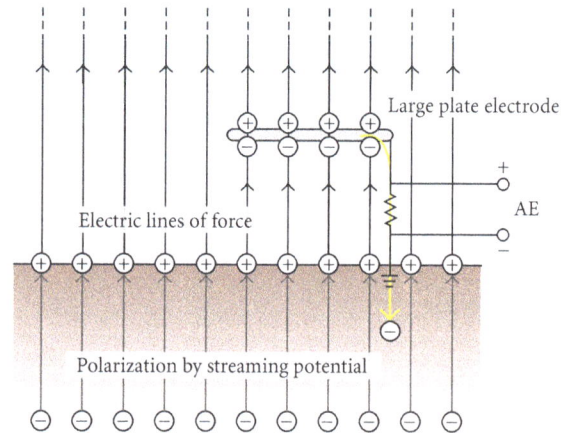

FIGURE 8: Generation model of induced atmospheric electricity (AE) coupled with generation of streaming potential in the topsoil.

Therefore, we must compare only the outwards of the signal amplitudes, as described in Section 3.

According to our model presented in Figure 8, a greater acceleration and a larger electrokinetic coefficient engender the higher charge density on the ground surface under certain good underground conditions such as those in case (b) of Figure 7. This engenders the larger atmospheric electricity. The Sennan observation site will correspond to this case. If this surface charge is sufficiently large, say ca. 5×10^{-5} C/m^2 [20], to cause dielectric breakdown of the air on the ground surface under a certain special condition, a dim light of corona discharge might appear in the atmosphere. This might be a possible mechanism of luminous phenomena accompanied with earthquakes, so-called earthquake lightning. If this field can be sufficiently large, say ca. 1 kV/m [21, 22], in an extensive area under a certain special condition, then it might disturb plasma in the ionosphere. However, the charge density on the ground surface cannot be large in case (e) of Figure 7 because a portion of the electric field vertical to the ground will be absorbed in the thick upper unsaturated (but not dry) water zone. This engenders a decrease of the charge density on the ground surface. Consequently, only the small induced AE signal appears. The Honjo observation site corresponds to this case.

5. Conclusions

The Honjo, Kyowa, and Sennan observation sites in northeastern Japan detected the induced earth potential difference (EPD) signal and the induced atmospheric electricity (AE) signals at the arrival of seismic waves from a M6.3 earthquake near the sites. The signal waveforms were damped oscillations in combination with the seismic waves. Although the signal amplitudes were different at each site, good positive correlation was found between the averaged amplitudes of EPD ($\overline{\text{EPD}}$) and AE ($\overline{\text{AE}}$) signals. According to our coupled generation model of the electric signals based on streaming potential, the amplitude of the vertical atmospheric electric field is influenced strongly by the depth of the saturated water zone and the location of electric polarization because of electrokinetic effects. This model can explain the site dependence of $\overline{EPD}$ and $\overline{AE}$. According to this model, no guarantee exists that the amplitude of the vertical atmospheric electric field is strongest at the epicenter.

Although only one major earthquake simultaneously shook the observation sites during operations, this observation advanced our prototype generation model. The model proposed in this paper presents consideration of the actual underground environmental circumstances. Although it remains merely qualitative, we believe that it is a great step up to steady progress in the study of seismoelectromagnetic precursors. The next step is elucidation of the dynamic electric fields in the ground and accompanying magnetic fields, which are both induced by seismic waves. To do so, observation sites in the ground [23–26] that can detect three-dimensional dynamic electric/magnetic signals are also needed.

Acknowledgments

The authors thank Mr. Hiromi Ishikawa of Yurtec Corp., Akita, Japan, for his support in designing and constructing the large plate electrodes. They also thank two anonymous reviewers for their valuable comments on the paper.

References

[1] M. Hayakawa, Y. Kasahara, T. Nakamura et al., "A statistical study on the correlation between lower ionospheric perturbations as seen by subionospheric VLF/LF propagation and earthquakes," *Journal of Geophysical Research A*, vol. 115, no. 9, Article ID A09305, 2010.

[2] K. I. Oyama, Y. Kakinami, J. Y. Liu, M. Kamogawa, and T. Kodama, "Reduction of electron temperature in low-latitude ionosphere at 600 km before and after large earthquakes," *Journal of Geophysical Research A*, vol. 113, no. 11, Article ID A11317, 2008.

[3] T. Ogawa and H. Utada, "Electromagnetic signals related to incidence of a teleseismic body wave into a subsurface piezoelectric body," *Earth, Planets and Space*, vol. 52, no. 4, pp. 253–260, 2000.

[4] S. Pride, "Governing equations for the coupled electromagnetics and acoustics of porous media," *Physical Review B*, vol. 50, no. 21, pp. 15678–15696, 1994.

[5] C. Bordes, L. Jouniaux, S. Garambois, M. Dietrich, J. P. Pozzi, and S. Gaffet, "Evidence of the theoretically predicted seismomagnetic conversion," *Geophysical Journal International*, vol. 174, no. 2, pp. 489–504, 2008.

[6] S. Garambois and M. Dietrich, "Seismoelectric wave conversions in porous media: field measurements and transfer function analysis," *Geophysics*, vol. 66, no. 5, pp. 1417–1430, 2001.

[7] N. Takeuchi, N. Chubachi, and K. Narita, "Observations of earthquake waves by the vertical earth potential difference method," *Physics of the Earth and Planetary Interiors*, vol. 101, no. 1-2, pp. 157–161, 1997.

[8] N. Takeuchi, N. Chubachi, K. Narita, N. Honma, and T. Takahashi, "Characteristics of vertical earth potential difference signals," *Transactions of IEE Japan*, vol. 117-C, no. 5, pp. 554–560, 1997 (Japanese).

[9] K. Okubo, M. Takayama, and N. Takeuchi, "Electrostatic field variation in the atmosphere induced by earth potential difference variation during seismic wave propagation," *IEEE Transactions on Electromagnetic Compatibility*, vol. 49, no. 1, pp. 163–169, 2007.

[10] G. N. Smith, *Elements of Soil Mechanics for Civil and Mining Engineers*, Collins, London, UK, 5th edition, 1982.

[11] D. F. McCarthy, *Essentials of Soil Mechanics and Foundations: Basic Geotechnics*, Prentice Hall, Upper Saddle River, NJ, USA, 7th edition, 2006.

[12] T. Ishido and J. Muzutani, "Experimental and theoretical basis of electrokinetic phenomena in rock-water systems and its applications to geophysics," *Journal of Geophysical Research*, vol. 86, no. B3, pp. 1763–1775, 1981.

[13] M. D. Jackson, "Multiphase electrokinetic coupling: insights into the impact of fluid and charge distribution at the pore scale from a bundle of capillary tubes model," *Journal of Geophysical Research*, vol. 115, Article ID B07206, 17 pages, 2010.

[14] X. Guichet, L. Jouniaux, and N. Catel, "Modification of streaming potential by precipitation of calcite in a sand-water system: laboratory measurements in the pH range from 4 to 12," *Geophysical Journal International*, vol. 166, no. 1, pp. 445–460, 2006.

[15] R. Muir-Wood and G. C. P. King, "Hydrological signatures of earthquake strain," *Journal of Geophysical Research*, vol. 98, no. 12, pp. 22,035–22,068, 1993.

[16] R. Yan, Y. Chen, F. W. Gao, and F. Q. Huang, "Calculating Skempton constant of aquifer from volume strain and water level response to seismic waves at Changping seismic station," *Acta Seismologica Sinica*, vol. 21, no. 2, pp. 148–155, 2008.

[17] A. Revil, V. Naudet, J. Nouzaret, and M. Pessel, "Principles of electrography applied to self-potential electrokinetic sources and hydrogeological applications," *Water Resources Research*, vol. 39, no. 5, pp. SBH31–SBH315, 2003.

[18] V. Allègre, L. Jouniaux, F. Lehmann, and P. Sailhac, "Streaming potential dependence on water-content in Fontainebleau sand," *Geophysical Journal International*, vol. 182, no. 3, pp. 1248–1266, 2010.

[19] H. Mizutani, T. Ishido, T. Yokokura, and S. Ohnishi, "Electrokinetic phenomena associated with earthquakes," *Geophysical Research Letters*, vol. 3, no. 7, pp. 365–368, 1976.

[20] D. A. Lockner, M. J. S. Johnston, and J. D. Byerlee, "A mechanism to explain the generation of earthquake lights," *Nature*, vol. 302, no. 5903, pp. 28–33, 1983.

[21] S. A. Pulinets, K. A. Boyarchuk, V. V. Hegai, V. P. Kim, and A. M. Lomonosov, "Quasielectrostatic model of atmosphere-thermosphere-ionosphere coupling," *Advances in Space Research*, vol. 26, no. 8, pp. 1209–1218, 2000.

[22] Y. Rapoport, V. Grimalsky, M. Hayakawa et al., "Change of ionospheric plasma parameters under the influence of electric field which has lithospheric origin and due to radon emanation," *Physics and Chemistry of the Earth*, vol. 29, no. 4–9, pp. 579–587, 2004.

[23] K. Okubo, S. Sato, T. Ishii, and N. Takeuchi, "Observation of atmospheric electricity variation signals during underground seismic wave propagation," *Transactions on Electrical and Electronic Engineering*, vol. 1, no. 2, pp. 182–187, 2006.

[24] K. Okubo, N. Takeuchi, M. Utsugi, K. Yumoto, and Y. Sasai, "Direct magnetic signals from earthquake rupturing: Iwate-Miyagi earthquake of M 7.2, Japan," *Earth and Planetary Science Letters*, vol. 305, no. 1-2, pp. 65–72, 2011.

[25] A. Takeuchi, K. Okubo, S. Watanabe, Y. Nakamura, and N. Takeuchi, "Electric and ionic environmental circumstances interacting at Hosokura underground mine in northeast Japan," *IEEJ Transactions on Fundamentals and Materials*, vol. 129, no. 12, pp. 870–874, 2009.

[26] A. Takeuchi, Y. Futada, K. Okubo, and N. Takeuchi, "Positive electrification on the floor of an underground mine gallery at the arrival of seismic waves and similar electrification on the surface of partially stressed rocks in laboratory," *Terra Nova*, vol. 22, no. 3, pp. 203–207, 2010.

18

Changes in Electrokinetic Coupling Coefficients of Granite under Triaxial Deformation

Osamu Kuwano and Shingo Yoshida

Earthquake Research Institute, University of Tokyo, 1-1-1 Yayoi, Bunkyo-ku, Tokyo 113-0032, Japan

Correspondence should be addressed to Osamu Kuwano, kuwano@eri.u-tokyo.ac.jp

Academic Editor: Tsuneo Ishido

Electrokinetic phenomena are believed to be the most likely origin of electromagnetic signals preceding or accompanying earthquakes. The intensity of the source current due to the electrokinetic phenomena is determined by the fluid flux and the electrokinetic coupling coefficient called streaming current coefficient; therefore, how the coefficient changes before rupture is essential. Here, we show how the electrokinetic coefficients change during the rock deformation experiment up to failure. The streaming current coefficient did not increase before failure, but continued to decrease up to failure, which is explained in terms of the elastic closure of capillary. On the other hand, the streaming potential coefficient, which is the product of the streaming current coefficient and bulk resistivity of the rock, increased at the onset of dilatancy. It may be due to change in bulk resistivity. Our result indicates that the zeta potential of the newly created surface does not change so much from that of the preexisting fluid rock interface.

1. Introduction

Electrokinetic phenomena occur when an electrolyte flows along charged solid surfaces. For several decades, these phenomena have been of interest to geophysicists in many subfields. Observed self-potential has been associated with geothermal fields (e.g., [1, 2]), volcanic activity and topography (e.g., [3–6]), and shallow ground water flow (e.g., [7, 8]). In numerical modelings, quantitative interpretation of self-potential observed in geothermal and volcanic areas and modelings in hydrogeophysics have been studied (e.g., [9–13]). Electrokinetic phenomena are also believed to be the most likely origin of the observed electromagnetic signals preceding or accompanying earthquakes. Mizutani et al. [14] first proposed a model: during dilatancy stage, which is assumed to precede earthquakes [15, 16], pore pressure in the dilatant region decreases and water flows into this region from the surrounding area, generating electromagnetic precursors to earthquakes due to electrokinetic phenomena.

To provide an appropriate interpretation of field observations, a better understanding of the physics of electrokinetic effect at the level of the rock-fluid interface and at the level of the rock sample is required. In laboratory experiments, zeta potential and streaming potential coefficients, fundamental quantities that characterize the electrokinetic effect, were measured for crushed rocks (e.g., [5, 17–21]) and for natural intact rocks (e.g., [22–29]) to determine the electrokinetic parameters as a function of pH, resistivity, permeability, or temperature. Jouniaux and Pozzi [23] measured the streaming potential coefficients of Fontainebleau sandstones under triaxial stress up to failure. They reported a large increase of the streaming potential coefficient beginning at about 75% of the yield stress. Yoshida [27] measured electric current and electric potential during rock deformation and found that the streaming current flowed before main failure, showing good correlation with dilatancy rate and water flow rate. In his study, however, changes of coupling coefficient (streaming potential coefficient or streaming current coefficient) during deformation were not measured.

Jouniaux and Pozzi [23] suggested that the increase of the streaming potential coefficient is due to an increase of zeta potential. An increase of the streaming potential coefficient, however, is also caused by an increase of bulk resistivity. In this study, by measuring not only the streaming potential coefficient but also the streaming current coefficient which has different dependence on bulk resistivity, we investigate what causes changes in the coupling coefficient during rock deformation.

2. Electrokinetic Phenomena

It is well known that when in contact with an electrolyte, the surface of rock-forming minerals are charged and surrounded by an equivalent amount of ionic charge of opposite sign from the electrolyte. The overall arrangement of the electric charge on the solid surface together with the balancing charge in the bulk liquid phase is often referred to as an electric double layer. Electrokinetic phenomena are induced by the relative motion between the fluid and the rock which develops an electric double layer. When the fluid in such a system moves due to a pressure gradient, the charges in the fluid are transported in the direction of fluid motion, resulting in an electric current. In a porous medium the electric current density $\mathbf{i}$ (in A/m^2) and fluid flux $\mathbf{j}$ (i.e., flow velocity, in m/s) are described by the following relations [30, 31]:

$$\mathbf{i} = -\frac{\sigma_f + \sigma_s}{F}\operatorname{grad}\phi + \frac{\epsilon\zeta}{\mu F}\operatorname{grad}P_p,$$
$$\mathbf{j} = \frac{\epsilon\zeta}{\mu F}\operatorname{grad}\phi - \frac{k}{\mu}\operatorname{grad}P_p, \tag{1}$$

where σ_f and σ_s are the electrical bulk and surface conductivities, ϵ is the dielectric constant of the fluid, ζ is the zeta potential (the potential at the slipping plane near the boundary), k is the permeability, μ is the viscosity of the fluid, P_p is the pressure of the fluid, and ϕ is the streaming potential. The first term of (1) represents Ohm's law, and the second term represents streaming current which can be derived by considering the product of the charge density (proportional to $\epsilon\zeta$) with the flow velocity of the viscous fluid (proportional to $\operatorname{grad}P_p/\mu$). The first term of (1) shows the macroscopic conductivity of rock (reciprocal of bulk-resistivity), which is expressed as

$$\sigma_R = \frac{\sigma_f + \sigma_s}{F}. \tag{2}$$

Considering the capillary model [17, 32], we define formation factor F as

$$F = \frac{\sigma_{\text{eff}}}{\sigma_R} = \frac{T^2}{\eta}, \tag{3}$$

where η is the porosity, T is the tortuosity, and σ_{eff} is the effective conductivity defined as

$$\sigma_{\text{eff}} = \sigma_f + \sigma_s. \tag{4}$$

The surface conductivity σ_s (in S/m) is related to the specific surface conductance Σ_s (in S) by $\sigma_s = 2\Sigma_s/m$, where m is

hydraulic radius for the capillary model. The permeability of the capillary model is represented as

$$k = \frac{\eta}{T^2}\frac{m^2}{b}, \tag{5}$$

where b is a constant related to the shape of pore; $b = 8$ for capillaries with a circular cross-section. We refer to $-\mathbf{i}/\operatorname{grad}P_p$ under $\operatorname{grad}\phi = 0$ in (1) as the streaming current coefficient C_c:

$$C_c = \frac{\epsilon\zeta}{\mu F} = \frac{\eta}{T^2}\frac{\epsilon\zeta}{\mu}. \tag{6}$$

Both C_c and k are functions of the fluid path network, and the dependence of C_c and k on η and T are the same. However, dependencies on m are different for C_c and k. The difference by m^2 can be understood if we note that the volume flow rate of a viscous fluid through a tube is proportional to the square of the cross-sectional area of the tube, while the amount of the transport electric charges distributed along the boundary are proportional to circumference length (i.e., proportional to the radius) and the flow velocity around the boundary is also proportional to the radius.

If there are no external current sources and no leaking current, the streaming current (due to $\operatorname{grad}P_p$) would be balanced by the conduction current (due to $\operatorname{grad}\phi$), so

$$\Delta\phi = \frac{\epsilon\zeta}{\sigma_{\text{eff}}\mu}\Delta P_p, \tag{7}$$

which is the Helmholtz-Smoluchowski equation. The ratio $\Delta\phi/\Delta P_p$ is referred to as the streaming potential coefficient,

$$C_p = \frac{\Delta\phi}{\Delta P_p} = \frac{\epsilon\zeta}{\sigma_{\text{eff}}\mu} = \frac{C_c}{\sigma_{\text{eff}}/F} = \frac{C_c}{\sigma_R}. \tag{8}$$

In general geometry, divergence of the total current is zero, but the zero total current condition (7) is not always satisfied.

3. Experimental Methods

In this study, we measured the streaming current (or streaming potential), permeability, and dilatancy of the rock specimen simultaneously and continuously during rock deformation test. We used the triaxial apparatus which was specially designed to investigate the electrical behavior during rock deformation and failure [27]. In this apparatus, the rock specimen is electrically isolated from the surroundings by inserting alumina plates. The pore fluid tubes of stainless steel inside the vessel are also isolated from the outside fluid tubes by using insulating tubes through the vessel closure as shown in Figure 1. This apparatus has two options for force loading: hydraulic loading with servo valves and a screwed pump with a servo motor. In the present experiment, we used the screwed pump for deformation test at a strain rate of approximately 10^{-7}/s. This apparatus is equipped with up to 11 feedthroughs that Nishizawa [33] developed on the basis of Bridgeman's self-sealing mechanism. During the

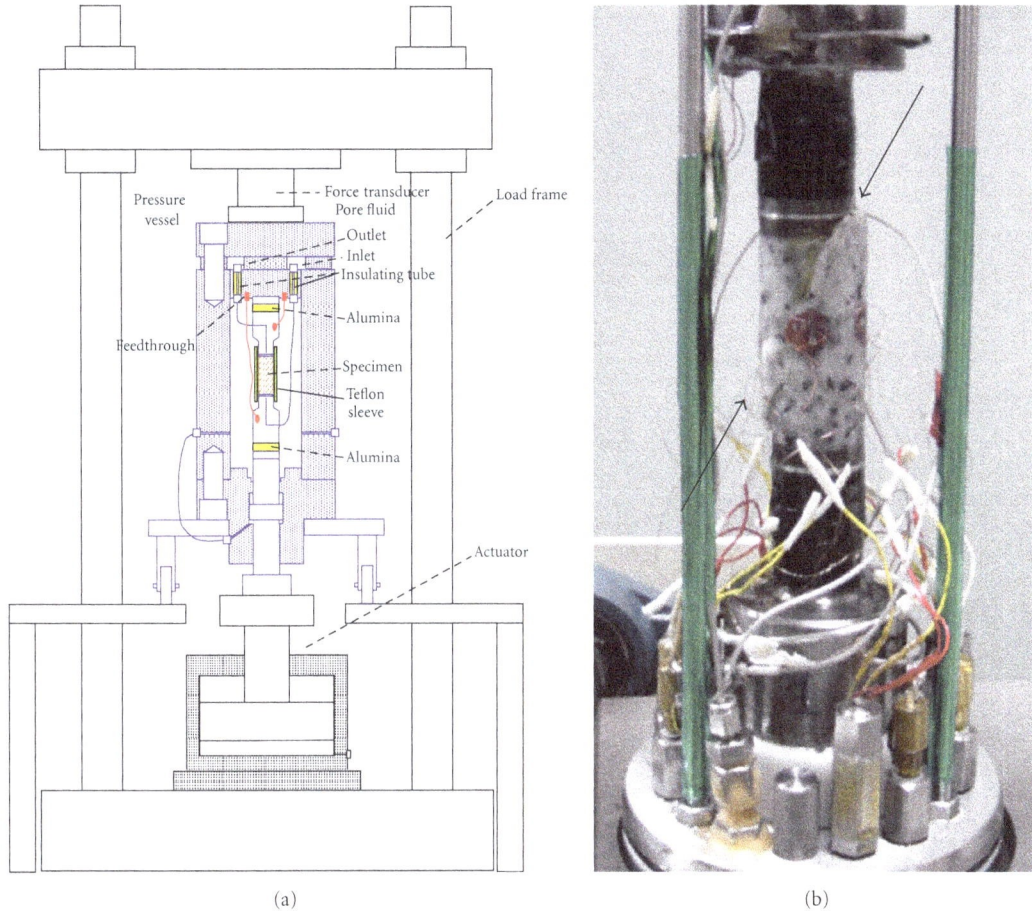

FIGURE 1: (a) Schematic diagram of apparatus. A rock specimen is electrically isolated by inserting alumina plates and by using insulating tubes of pore water through vessel closure. (b) Picture of the specimen after experiment. A failure plane is indicated by black arrows.

experiment, we measured differential axial stress, axial piston displacement, pore fluid volume, pore pressure change, electric signals (either current or potential), and local strains of the specimen, at a sampling rate of 10 Hz.

In the present study, we usedcoarse grained Inada granite (from a locality in Ibaraki, Japan), which has been often used as a standard specimen in rock mechanics (e.g., [27, 29, 34]). We used two specimens which were cored cylindrically 24 mm in diameter and 60 mm in length. One specimen was used in the electric potential measurement and denoted G01. The other was used in the electric current measurement and denoted G02. The porosity of both specimens was approximately 1%. The rock specimens were air-dried and degassed under a vacuum for 12 hours before being saturated with 10^{-3} M KCl solution for 2 days. The conductivity of the fluid was 14mS/m. Then, the specimens were placed between the stainless steel end plugs and jacketed in a Teflon sleeve.

To measure the axial strain e_z and the circumferential strain e_h, strain gauges were mounted at four positions on the cylindrical surface of the Teflon sleeve. A volume change of the specimen ΔVol is estimated as ΔVol $= $ Vol$^0(\tilde{e}_z + 2\tilde{e}_h)$, where $\tilde{e}_z$ and $\tilde{e}_h$ are averaged strains for four positions, and Vol^0 is the initial volume of the specimen. The cross-section of the specimen is assumed to remain circular. This

assumption is not satisfied when a fault plane is formed and large localized deformation occurs. A volume change due to dilatancy is obtained by subtracting elastic deformation from the volume change.

To measure the permeability continuously during the deformation experiments [35], we adopted the sinusoidal oscillation method [36–39]. The method is based on the measurement of an attenuation and a phase retardation of an oscillation of the pore-fluid pressure as it propagates through the specimen. In its application, a sinusoidal pressure oscillation is imposed at one end of the specimen and a pressure response is monitored at the other end as illustrated in Figure 2. The permeability is calculated from the measured attenuation factor R between downstream and upstream pore-fluid-pressure sinusoidal waves, and the phase lag δ, using the following relation (calculations detailed in [37]),

$$P_{p1} = RP_{p2} \sin(\omega t - \delta),$$

$$R \cdot \exp(i\delta) = \frac{1}{\cosh[\psi(1 + i)] + \gamma\psi(1 + i)\sinh[\psi(1 + i)]},$$

(9)

where P_{p1} and P_{p2} are upstream and downstream pore fluid pressure, respectively, R is the attenuation factor, δ is the

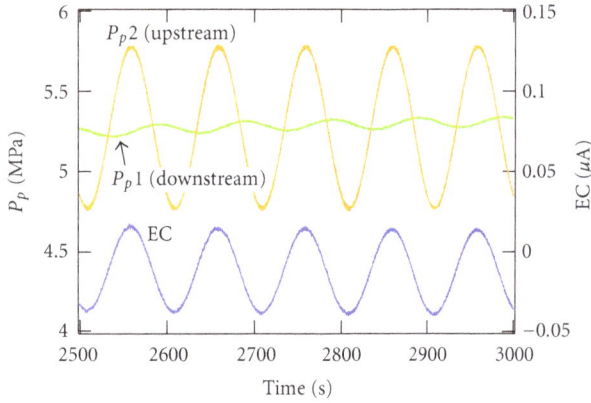

FIGURE 2: An example of measurement using the sinusoidal oscillation method. Upstream and downstream fluid pressure are shown as orange and green lines, respectively. In this example, electric current measured using stainless-steel end plugs equipped at each end of specimen as electrode is also shown (blue line). The streaming current coefficient is calculated from the amplitude of fluid-pressure difference ($P_{p1} - P_{p2}$) and the amplitude of the electric current.

phase lag, ω is an angular frequency, and i is the imaginary unit. From the measurements of R and δ, the value of dimensionless parameters ψ and γ can be evaluated, and then we obtain permeability k,

$$k = \frac{\pi \mu f B_d L}{A \psi^2 \gamma}, \tag{10}$$

where f is a frequency of oscillation, A is the cross-sectional area, L is the length of the specimen, and B_d is the storage of the downstream reservoir ($B_d = 0.0038 \, \text{cm}^3/\text{MPa}$ in our apparatus). During the experiments, the upstream pore-fluid pressure (P_{p2}) was sinusoidally oscillated at a frequency of 0.01 Hz by computer-controlled servo-mechanism. The amplitude of the sinusoidal pressure variation was 0.5 MPa. The upper specimen face is connected to the downstream reservoir of a volume of 9.52 cm^3. To calculate the permeability from the measured attenuation factor and phase lag, we used a custom-made code following Takahashi [35].

To reveal the physical mechanism causing the observed changes of electrokinetic properties of rock-water system, it is convenient to measure both of the two coupling coefficients, namely, the streaming potential coefficient and the streaming current coefficient, because the streaming potential coefficient involves bulk resistivity and the source current. We measured electric current or potential between the upper and bottom faces of the specimen with an electrometer (Keithley 6517) by using the stainless-steel end plugs placed at the both ends of the specimen as electrodes. Electric potential was measured in the experiment using specimen G01 and electric current was measured in another experiment using specimen G02, respectively, because an electric potential and an electric current could not be measured at the same time. The sinusoidal oscillation method [29] enables us to measure the streaming potential (or current) induced by pore-fluid movement continuously

during a rock deformation test. Figure 2 shows a typical data obtained by the oscillation test at a frequency of 0.01 Hz. The resultant electric current showed sinusoidal variations. By reading the amplitude of the sinusoidal variation of the electric current (ΔI_s) and the amplitude of the sinusoidal variation of the fluid pressure difference ($\Delta P_p = P_{p1} - P_{p2}$), we evaluated the generated current per unit change in the fluid pressure $\Delta I_s / \Delta P_p$. The streaming current coefficient C_c was obtained using the relation

$$C_c = \frac{\Delta I_s / S}{\Delta P / L} = \frac{\Delta I_s}{\Delta P} \frac{L}{S} \, [\text{A/mMPa}]. \tag{11}$$

Similarly, from measurements of the amplitude of electric potential variation ($\Delta \phi$), we estimated the streaming potential coefficient as

$$C_p = \frac{\Delta \phi}{\Delta P_p} \, [\text{V/MPa}]. \tag{12}$$

Frequency effect on the coupling coefficient [40–42] for the Inada granite with the present experiment system has been reported in [27]. Coupling coefficient does not depend on frequency in such a low frequency range (0.01–1 Hz) for the intact Inada granite. Although we cannot rule out the possibility of the changes of frequency dependence on the coupling coefficient during deformation, we focus on the continuous measurement during deformation to fix the frequency to 0.01 Hz and do not discuss the frequency dependence in the present study.

4. Results

Here, we show the results of the two experiments. Experimental conditions for G01 and G02 were the same. The confining pressure P_c was kept at 15 MPa. The pore-fluid pressure at each end of the specimen was set to 5.3 MPa at the beginning of the experiment. Then, pore fluid pressure of the bottom face of the specimen was sinusoidally oscillated at a frequency of 0.01 Hz with an amplitude of 0.5 MPa throughout the rest of the experiment. Experiments were conducted under room temperature ($25 \pm 1°C$).

Figure 3 shows the results of the run G01, in which electric potential was monitored. The axial loading rate was $5.1 \times 10^{-7}/\text{s}$. A shear failure plane was found in the postexperimental sample (Figure 1(b)). From Figure 3(a) showing the differential axial stress and the displacement, it can be seen that dynamic failure occurred at $t = 37,323 \, \text{s}$. Some small releases of axial stress occurred around $t = 6,000$ and $17,000 \, \text{s}$. These small stress releases may be due to poor initial setting of the apparatus. Thus, we do not analyze the data before these stress changes ($t \sim 20,000 \, \text{s}$). The pore-fluid pressure P_{p1} and P_{p2} are shown in Figure 3(b). The pore-fluid pressure of the bottom face of the specimen appears to be a thick line in this scale because it is sinusoidally oscillated at a frequency of 0.01 Hz and amplitude of 0.5 MPa. Figure 3(e) shows the volume change of the specimen obtained from strain measurements and the dilatancy calculated by subtracting the elastic deformation from the volume change. In Figure 3(b), the volume change

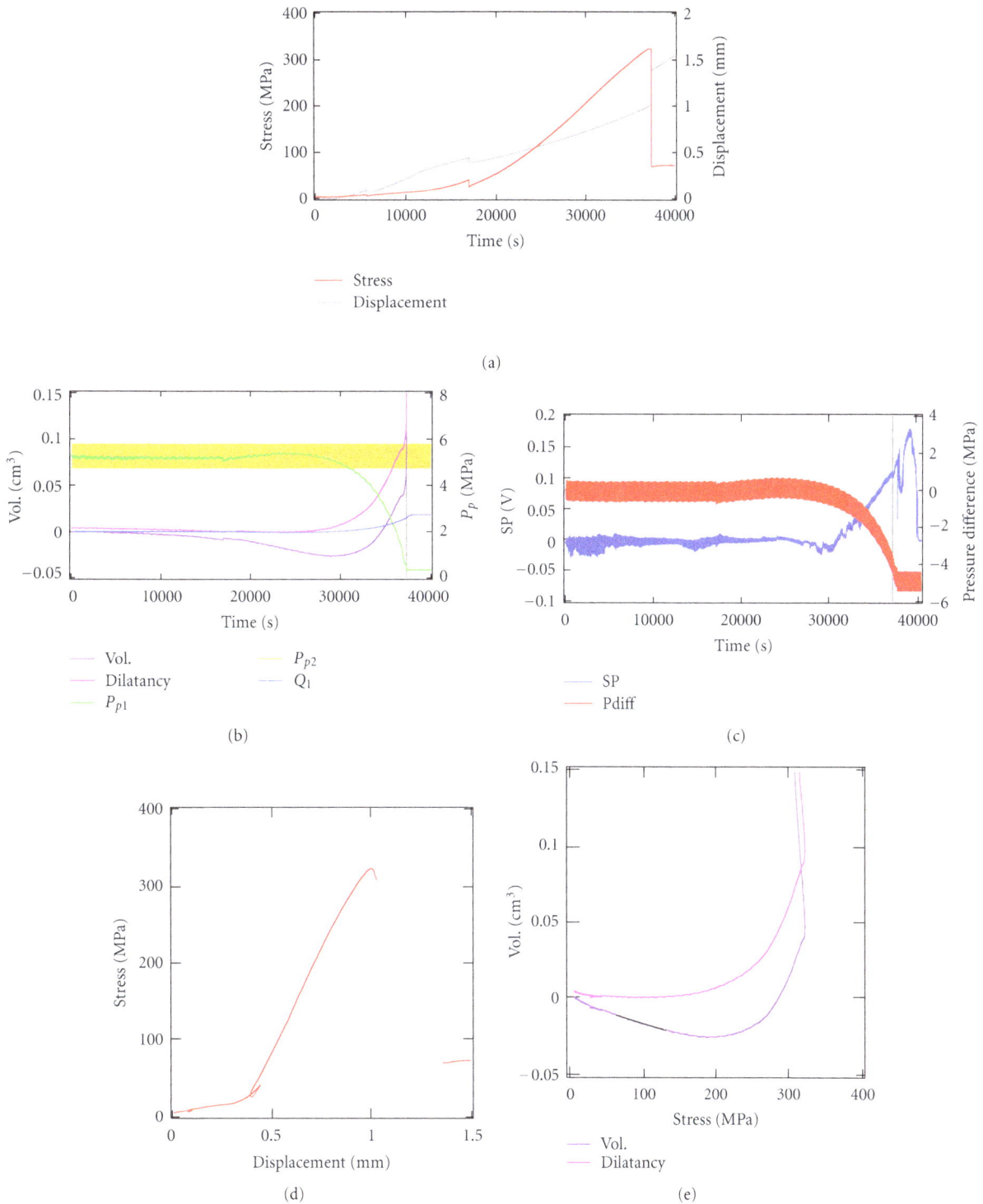

FIGURE 3: Result of deformation test for initially intact Inada granite (G01) at $P_c = 15\,\text{MPa}$, $P_p = 5\,\text{MPa}$. (a) Differential stress and displacement. (b) Volume change of the specimen obtained from the average of the strain measurements at four positions, dilatancy, pore fluid pressures, and water volume. (c) Streaming potential (SP) and pore-fluid pressure difference between bottom and top faces of the specimen. (d) Axial stress versus axial displacement. In this experiment, failure stress was 324 MPa and Young's modulus was 32 MPa. (e) Volume change of the specimen and dilatancy versus the axial stress. The elastic deformation is indicated by the thick black line in this figure.

(a)

(b)

(c)

(d)

(e)

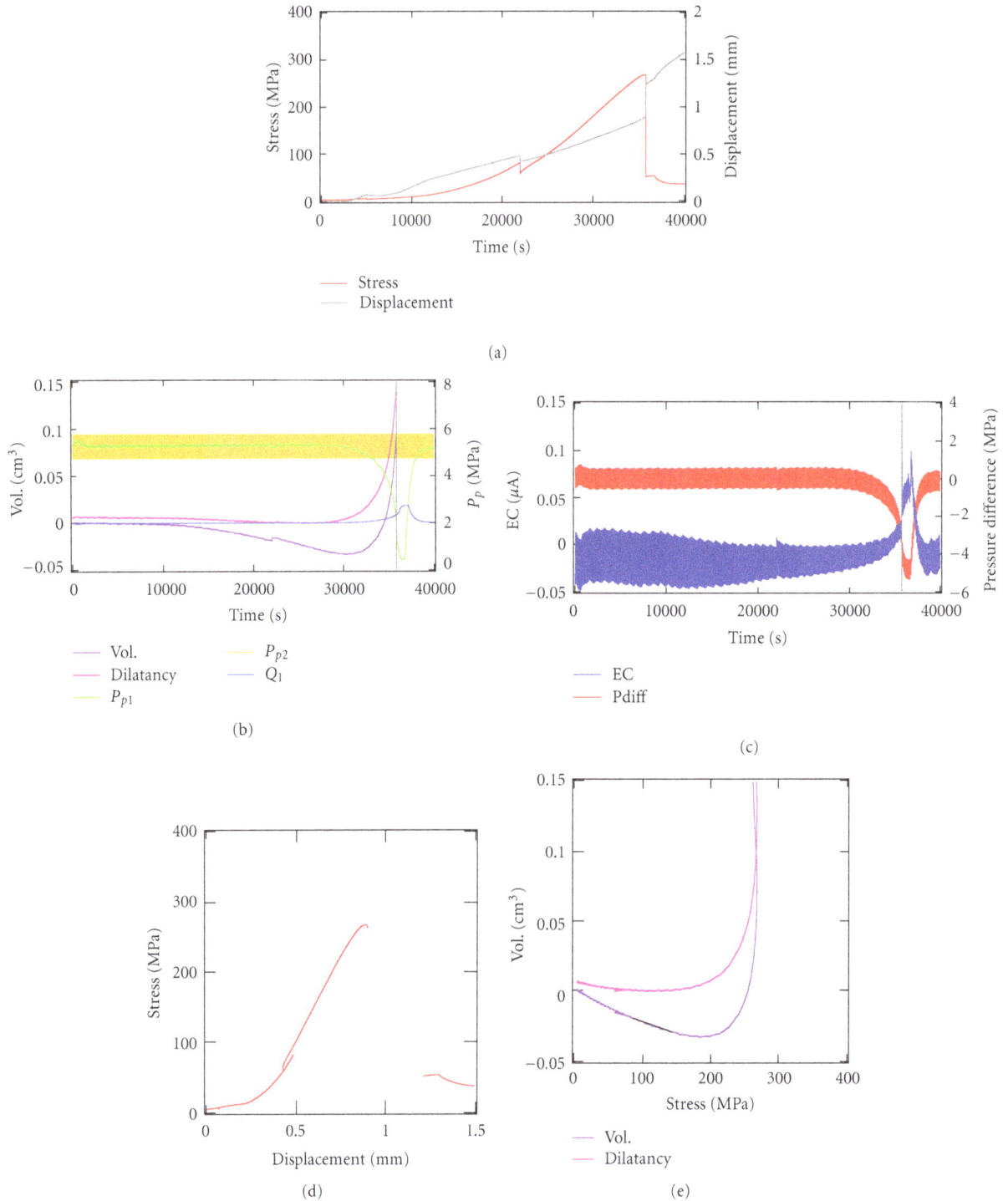

FIGURE 4: Result of deformation test for initially intact Inada granite (G02) at $P_c = 15\,\text{MPa}$, $P_p = 5\,\text{MPa}$. In this experiment, streaming current was measured instead of potential. This result also shows good correlation between the pressure difference and the streaming current. Failure stress was 268 MPa and Young's modulus was 29 MPa.

and the dilatancy are plotted. When dilatancy began (around $t = 27,000\,\text{s}$), the downstream pore pressure P_{p1} began to drop, indicating that the pore pressure in the specimen dropped and water flowed into the specimen. From the change of P_{p1}, we calculated the water volume Q_1, which

flows into the specimen from the downstream reservoir, as $Q_1 = -B_d(P_{p1} - P_{p0})$, where P_{p0} is an initial pore pressure. Although we attempted to estimate the water volume Q_2, which flows into the specimen from the upstream side, using the displacement of the piston of the pore water

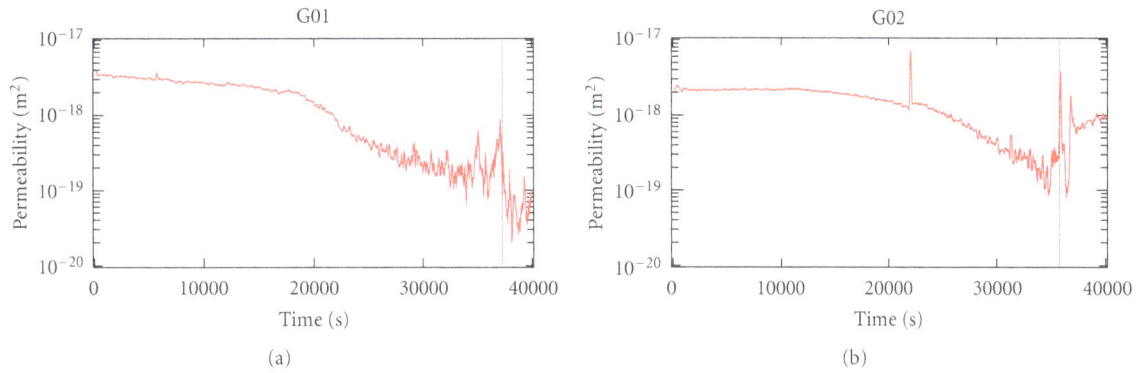

FIGURE 5: Permeability calculated from measured amplitude ratio and phase rag between upstream and downstream pore pressure (P_{p2}, P_{p1}). Vertical gray lines indicate the time of failure.

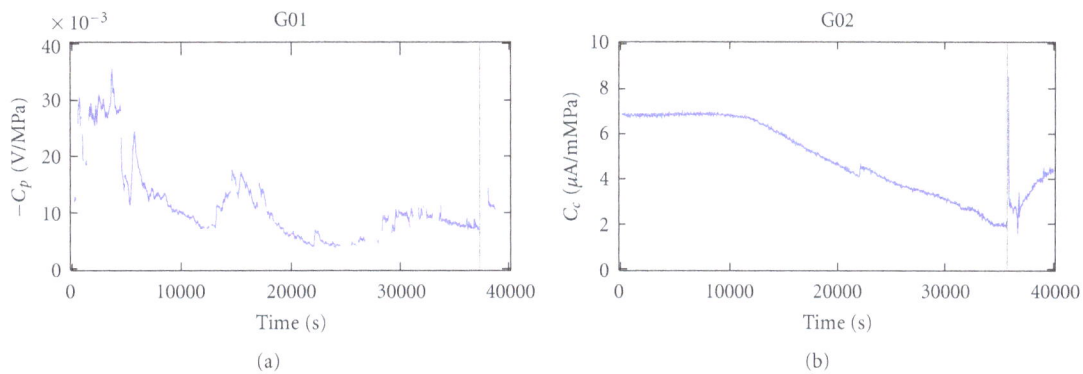

FIGURE 6: Coupling coefficients, C_p (G01) and C_c (G02), calculated from measured amplitude ratio of potential (or current) to pore pressure difference. Vertical gray lines indicate the time of failure.

intensifier, we could not estimate Q_2 due to the leakage of the water at the upstream side. Therefore, we show only Q_1 in Figure 3(b). In Figures 3(b) and 3(c), we can see good correlation among the trends of dilatancy, the pore pressure difference, and the streaming potential (SP). Details of this "DC" relation are discussed later. Here, we focus on the results of "AC" measurement based on the imposed sinusoidal oscillation of the pore pressure.

Figure 4 shows the other run (G02) in which the electric current was measured. A similar result with the run G01 was obtained, showing good correlation among dilatancy, pore pressure difference, and the streaming current (EC). The axial loading rate was 5.5×10^{-7}/s. As in G01, a discrete shear plane was found in the postexperimental sample, indicating that the main failure (around $t = 35,788$ s) involved the formation of such a failure plane. When dilatancy began (around $t = 28,650$ s), the pore pressure of the downstream P_{p1} began to decrease. Some small stress releases were also observed at about $t = 6,000$ and $22,000$ s due to the setting of the apparatus similarly to the former experiment.

Figure 5 shows the permeability of G01 and G02. The permeability was initially of the order of 10^{-18} m^2. With the increase of the axial loading, the permeability decreased to 10^{-19} m^2. Then, just before the failure, the permeability increased to $\sim 10^{-18}$ m^2 in the both experiments. There are a

lot of studies dealing with permeability-porosity relationship and permeability-stress relationship (e.g., [43–45]). In our experiment, permeability reduction is approximately an exponential function of effective mean stress [45] and mainly attributed to elastic crack closure [43]. The permeability increase indicates enhanced connection of cracks. Although the dilatancy should involve the creation of microcracks, the permeability continued to decrease with progressive loading, indicating that microcracks were not fully interconnected or not fully saturated with pore fluid.

5. Discussion

The coupling coefficients C_p (or C_c) were calculated from amplitude of pore pressure difference and potential (or current). The polarity of obtained coupling coefficients were negative in the present experiments, indicating negative zeta potential, as expected for granites. The values of the coupling coefficients are shown in absolute values hereafter. Figure 6 shows the coupling coefficients of G01 and G02. To remove the data which are not suitable for calculating streaming potential coefficient, we evaluated a signal quality by $P(0.01 \, \text{Hz})/\sum P(f)$, where $P(f)$ is the power spectrum of potential variation. The data with the signal quality lower than 0.98 were not used. We can see variations of C_p around

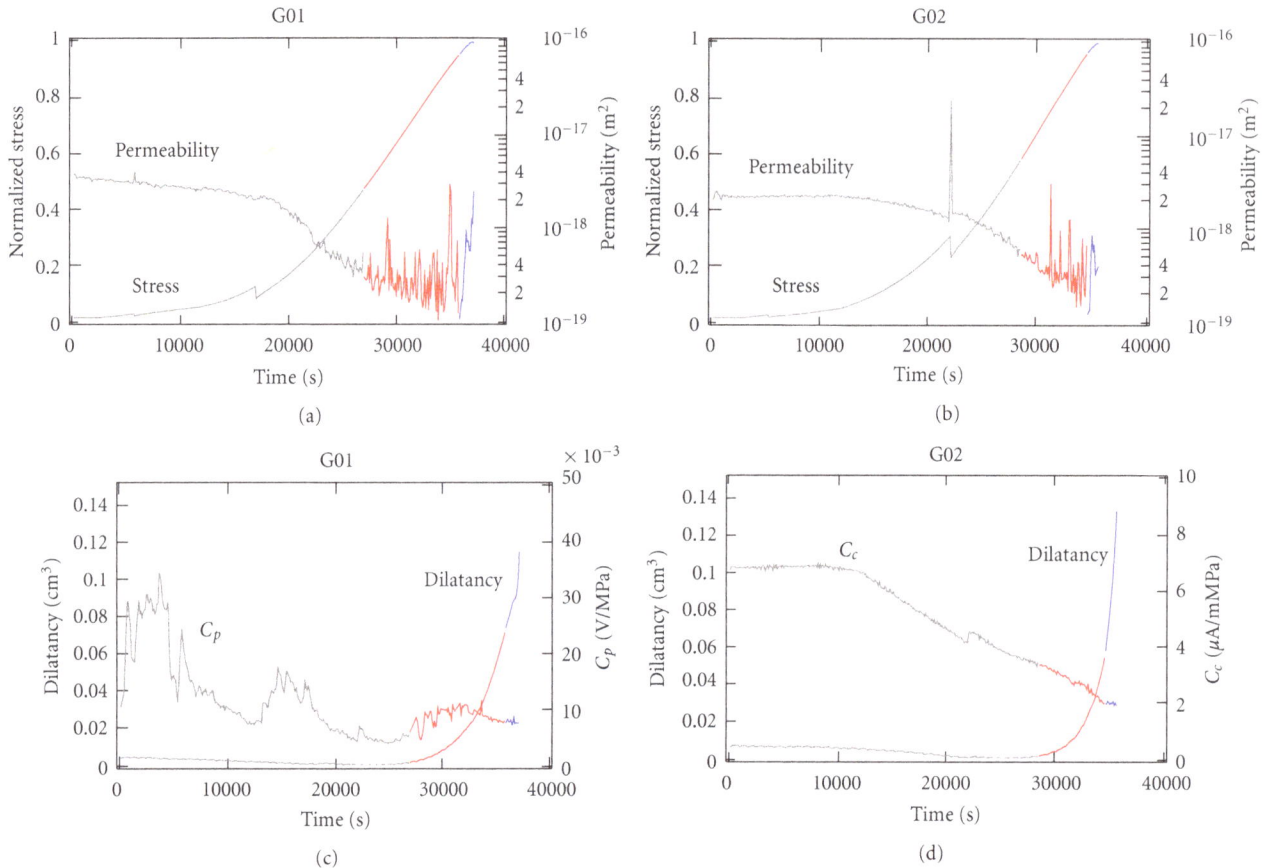

FIGURE 7: Normalized stress, permeability, dilatancy, and coupling coefficient versus time. The gray, red, and blue lines indicate the stages A, B, and C, respectively. Three stages which are divided by the beginning of dilatancy (G01 $t = 27,000$ s, G02 $t = 28,650$ s) and the beginning of permeability increase (G01 $t = 36,000$ s, G02 $t = 34,750$ s).

$t = 6,000$ and $17,000$ s in Figure 6(b). These variations were due to small stress releases resulting from the setting of the apparatus as mentioned before. A fluctuation of C_c around $t = 22,000$ s was due to the same reason. The streaming potential coefficient C_p decreased with loading (Figure 6) and then increased by a factor of two at the onset of dilatancy (around $t = 27,000$ s). Note that C_p did not continue to increase with dilatancy increase but C_p increased just at the onset of the dilatancy. In contrast, the streaming current coefficient C_c continued to decrease during the loading until the time of failure, not particularly affected by dilatancy. It is noted that the observed change in C_c indicates that the source current density did not increase during the deformation, and therefore observed increase in C_p is attributed to bulk resistivity (see (8)).

The estimation of the zeta potential is done from the measured streaming current coefficient. The streaming current coefficient C_c of Inada granite before loading is approximately 7μA/mMPa. The formation factor F of Inada granite under atmospheric pressure was estimated to be 1100 from the measurement of the resistance of the rock sample saturated with KCl solution with a high conductivity (0.2–1.1 S/m). Inserting these values into (6), we obtained the zeta potential as -11 mV, which is slightly smaller than previously reported value of granite [17, 18, 29].

To investigate the evolution of the coupling coefficients in detail, we divide experiments into three stages; stage A: from the start of experiment to the beginning of dilatancy, stage B: from the beginning of dilatancy to the beginning of permeability increase, stage C: from the beginning of permeability increase to the failure. Figure 7 indicates these stages in different colors. We defined the normalized stress as the stress normalized by the failure stress. Figure 8(a) shows the streaming potential coefficient C_p and the dilatancy of G01 as a function of the normalized stress. We can see that the dilatancy and increase of streaming potential coefficient began at 47% of yield stress. Figure 8(b) shows the relation between the streaming potential coefficient and the dilatancy. Increase of the streaming potential coefficient (5 V/MPa to 10V/MPa) occurred at the onset of the dilatancy. Figure 8(c) shows C_c and the dilatancy of G02 as a function of the normalized stress. We can see the dilatancy began at 58% of the failure stress. The streaming current coefficient C_c continued to decrease at an approximately constant rate unrelated to the dilatancy. Relation between the C_c and the dilatancy is shown in Figure 8(d). When the permeability increase (stage C) began, C_c stopped to decrease and remained roughly constant during the stage C up to failure.

Figure 9 shows the relation between the streaming current coefficients and the permeability. The streaming current

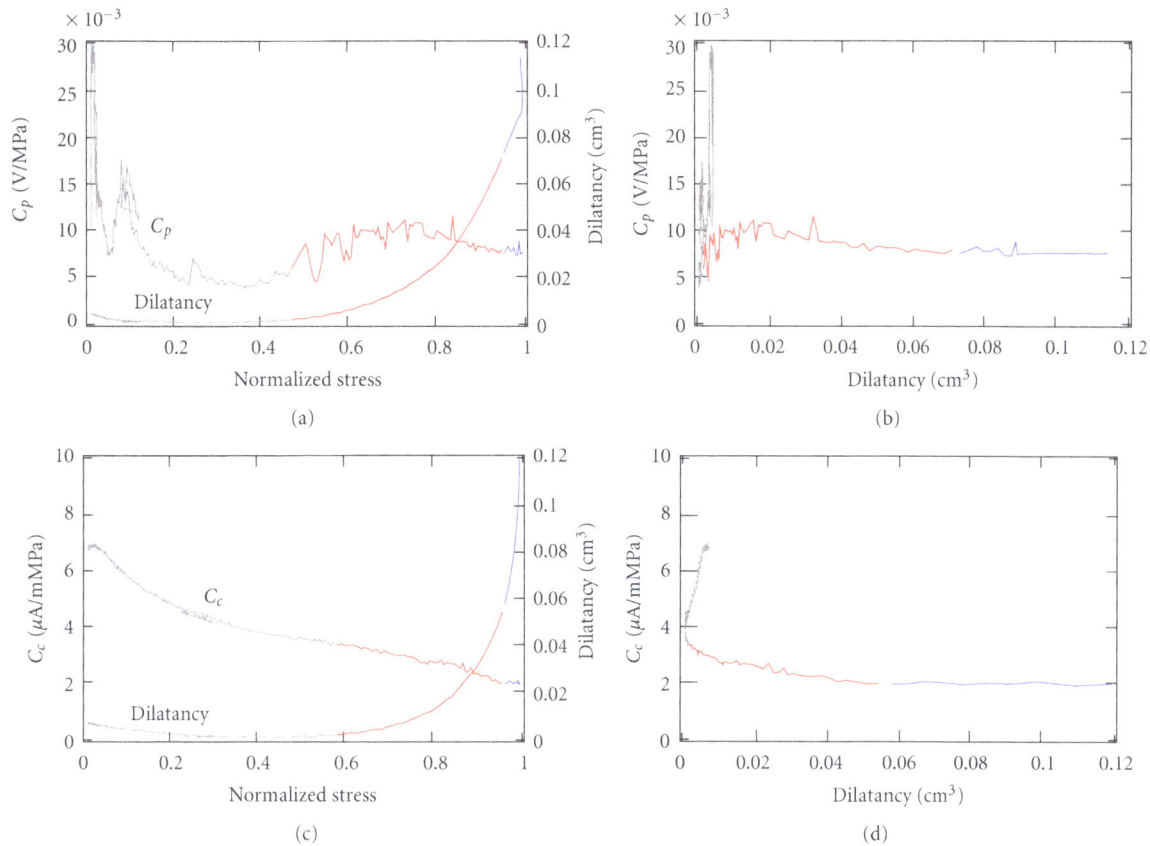

FIGURE 8: (a) C_p and dilatancy versus normalized stress. (b) C_p versus dilatancy. (c) C_c and dilatancy versus normalized stress. (d) C_c versus dilatancy.

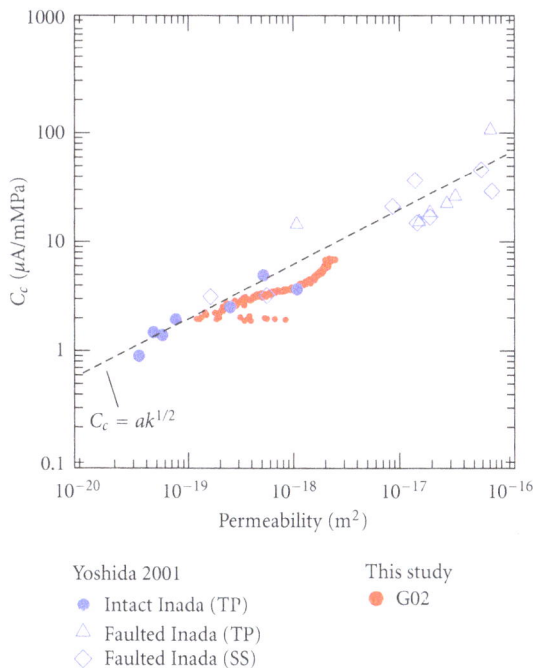

FIGURE 9: The streaming current coefficients C_c as a function of the permeability. Blue symbols are data of Inada granite obtained by Yoshida [27].

coefficient C_c was approximately proportional to the square root of the permeability. This dependence can be explained by assuming that m^2 is proportional to $1/F = \eta/T^2$ in (5) and (6) for the capillary model [17, 27, 32]. This assumption is supported by the experimental results that $\log k$ is linearly related to $\log F$ with slope of ~ -2 for granite reported by Walsh and Brace [32].

The fact that the streaming current coefficient did not increase indicates that the zeta potential did not increase throughout the deformation test. Furthermore, there is a possibility of decrease of the zeta potential, because C_c does not increase with the permeability increase in stage C. If bulk resistivity increases at the onset of the dilatancy, the streaming potential coefficient C_p, which is the product of C_c and bulk resistivity (see (8)), will increase. Figures 3(b) and 4(b) show the volume of water flow from the downstream Q_1, which is much smaller than dilatancy. The ratio of Q_1 to dilatancy is approximately 0.1 to 0.2, indicating the possibility of the undersaturation of the pore, although the water flow from the upstream is not included. We mention the possibility of bulk-resistivity change here. To understand the observed change of C_p, we would require measurements of the bulk-resistivity changes during the deformation. On the basis of recent studies (e.g., [46–48]), however, the magnitude of C_p decreases with decreasing water saturation S_w in most situations even though substantial increase of bulk-resistivity takes place at the same time.

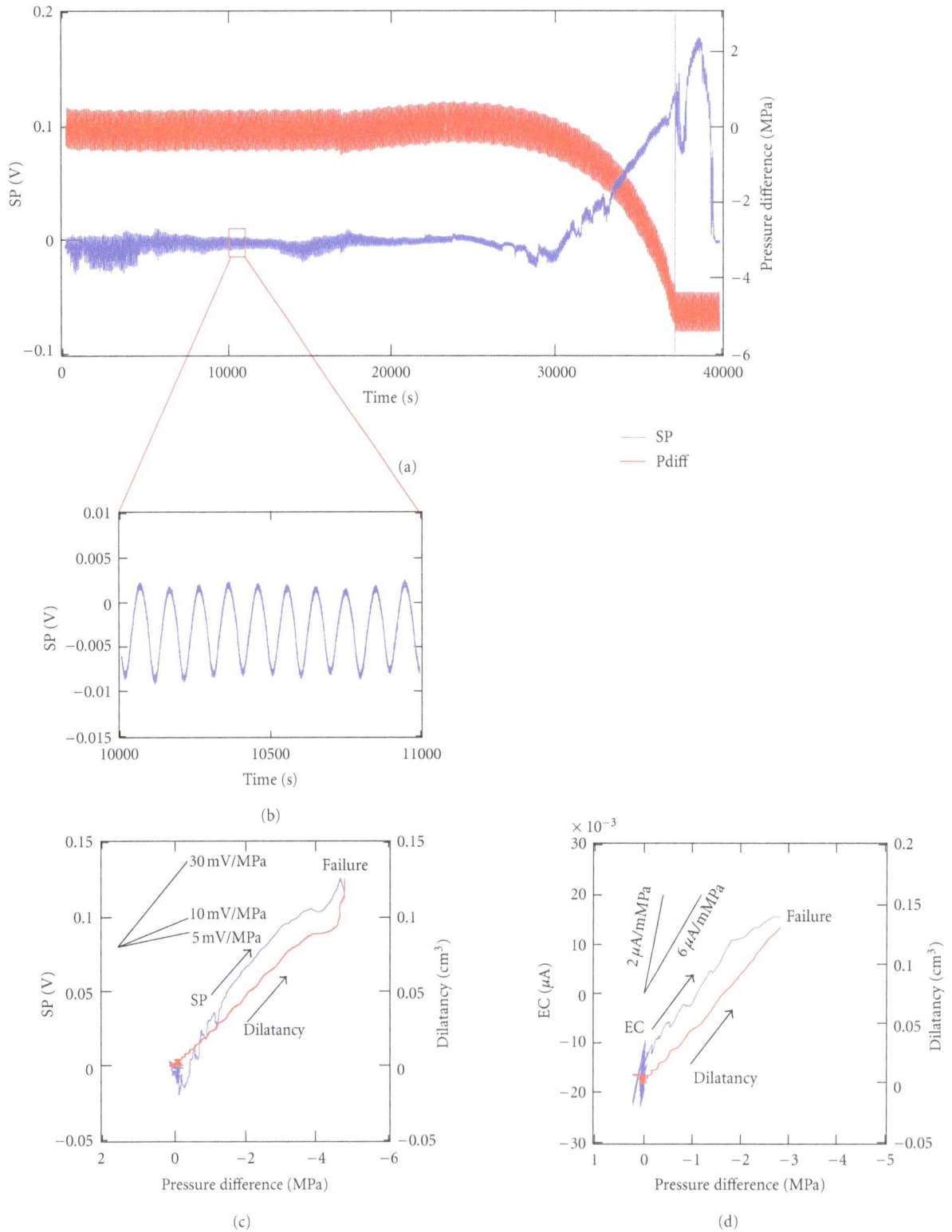

(a)

(b)

(c)   (d)

FIGURE 10: The streaming potential (SP) and the pore pressure difference (Pdiff). Long-term trend of these values are used in the "DC" measurements of electrokinetic phenomena (a). Amplitude and phase of the sinusoidal variation caused by imposed pore pressure oscillation are used for the "AC" measurements of electrokinetic phenomena (b). The DC coupling coefficients and the dilatancy as a function of the pressure difference for G01 (c) and G02 (d). Blue and red lines indicate coupling coefficients and dilatancy, respectively.

The S_w dependence of the coupling coefficients is important also for modeling of field self-potential data related to unsaturated flow in volcanic areas (e.g., [49, 50]) and shallow groundwater systems. We need further study to clarify the S_w change during dilatancy stage and its effect on the coupling coefficients of low-permeability rocks such as granite used in this study.

Here we compare the coupling coefficients obtained from the AC measurements and those obtained from the DC measurements. We can see the DC electrokinetic effect in Figure 10(a). Figure 10(b) shows the AC electrokinetic effect discussed earlier. The DC coupling coefficients and the dilatancy are shown as a function of the pressure difference in Figures 10(c), 10(d). The dilatancy showed a linear relation with the pressure difference in both experiments. The magnitude of the DC streaming current coefficient C_c was approximately 6 μA/mMPa before dilatancy began and then decreased to 2 μA/mMPa after dilatancy began. These values agree well with the C_c obtained from the AC measurement (see Figure 7). On the other hand, the magnitude of the DC streaming potential coefficient C_p during dilatancy was approximately 30 mV/MPa, which was three-times as large as that from the AC measurement (see Figure 7). At the present stage, we do not fully understand the discrepancy between AC and DC streaming potential coefficients. We need further study including a frequency dependence of the specific resistivity of the rock specimen.

6. Conclusions

Jouniaux and Pozzi [23] suggested that the onset of increase in the streaming potential coefficient corresponded to the onset of shear localization and that this increase was due to an increase of the zeta potential in the shear zone as new surfaces were created and connected. Although Jouniaux and Pozzi [23] suggested a possibility of the enhancement in the zeta potential on the newly created surface, there exists some uncertainties. They measured the streaming potential, which was the product of the streaming current density and the specific resistivity of the specimen. Therefore, one cannot deny a possibility that the observed increase of the streaming potential was due to an increased bulk resistivity rather than an enhanced zeta potential.

In our experiment, the C_p increased but the C_c did not increase, indicating that the source current density did not increase during the deformation. Such an increase in C_p due to increase in bulk resistivity cannot be the source of the electric signals unless the increase in bulk resistivity occurs broadly in the observation field. Variation of the zeta potential according to the deformation stage makes it difficult to interpret the self-potential data quantitatively. Results of the present experiments, however, indicate the zeta potential does not vary so much throughout all the deformation stage of the rock up to failure.

Acknowledgments

The authors thank the Editor T. Ishido and the anonymous reviewers for constructive comments which greatly improved the paper. This paper was partially supported by JSPS KAKENHI(05J11788).

References

[1] R. F. Corwin and D. B. Hoover, "Self-potential method in geothermal-exploration," *Geophysics*, vol. 44, no. 2, pp. 226–245, 1979.

[2] T. Ishido, H. Mizutani, and K. Baba, "Streaming potential observations, using geothermal wells and in situ electrokinetic coupling coefficients under high temperature," *Tectonophysics*, vol. 91, no. 1-2, pp. 89–104, 1983.

[3] T. Ishido, T. Kikuchi, N. Matsushima et al., "Repeated self-potential profiling of izu-oshima volcano, Japan," *Journal of Geomagnetism and Geoelectricity*, vol. 49, no. 11-12, pp. 1267–1278, 1997.

[4] J. Zlotnicki, Y. Sasai, P. Yvetot et al., "Resistivity and self-potential changes associated with volcanic activity: the July 8, 2000 Miyake-jima eruption (Japan)," *Earth and Planetary Science Letters*, vol. 205, no. 3-4, pp. 139–154, 2003.

[5] H. Hase, T. Ishido, S. Takakura, T. Hashimoto, K. Sato, and Y. Tanaka, "Zeta potential measurement of volcanic rocks from Aso caldera," *Geophysical Research Letters*, vol. 30, no. 23, pp. 2–4, 2003.

[6] K. Aizawa, "A large self-potential anomaly and its changes on the quiet Mt. Fuji, Japan," *Geophysical Research Letters*, vol. 31, no. 5, article L05612, p. 4, 2004.

[7] F. Perrier, M. Trique, B. Lorne, J. P. Avouac, S. Hautot, and P. Tarits, "Electric potential variations associated with yearly lake level variations," *Geophysical Research Letters*, vol. 25, no. 10, pp. 1955–1958, 1998.

[8] M. Trique, F. Perrier, T. Froidefond, J. P. Avouac, and S. Hautot, "Fluid flow near reservoir lakes inferred from the spatial and temporal analysis of the electric potential," *Journal of Geophysical Research B*, vol. 107, no. 10, pp. 1–28, 2002.

[9] W. R. Sill, "Self-potential modeling from primary flows," *Geophysics*, vol. 48, no. 1, pp. 76–86, 1983.

[10] T. Ishido and J. W. Pritchett, "Numerical simulation of electrokinetic potentials associated with subsurface fluid flow," *Journal of Geophysical Research B*, vol. 104, no. 7, pp. 15247–15259, 1999.

[11] A. Revil, H. Schwaeger, L. M. Cathles, and P. D. Manhardt, "Streaming potential in porous media 2. Theory and application to geothermal systems," *Journal of Geophysical Research B*, vol. 104, no. 9, pp. 20033–20048, 1999.

[12] S. R. Pride, F. Moreau, and P. Gavrilenko, "Mechanical and electrical response due to fluid-pressure equilibration following an earthquake," *Journal of Geophysical Research B*, vol. 109, no. 3, article B03302, p. 10, 2004.

[13] M. R. Sheffer and D. W. Oldenburg, "Three-dimensional modelling of streaming potential," *Geophysical Journal International*, vol. 169, no. 3, pp. 839–848, 2007.

[14] H. Mizutani, T. Ishido, T. Yokokura, and S. Ohnishi, "Electrokinetic phenomena associated with earthquakes," *Geophysical Research Letters*, vol. 3, no. 7, pp. 365–368, 1976.

[15] A. Nur, "Dilatancy, pore fluids, and premonitory variations of T_s-T_p travel times," *Bulletin of the Seismological Society of America*, vol. 62, no. 5, pp. 1217–1222, 1972.

[16] C. H. Scholz, L. R. Sykes, and Y. P. Aggarwal, "Earthquake prediction: a physical basis," *Science*, vol. 181, no. 4102, pp. 803–810, 1973.

[17] T. Ishido, "Experimental and theoretical basis of electrokinetic phenomena in rock-water systems and its applications to geophysics," *Journal of Geophysical Research*, vol. 86, no. 3, pp. 1763–1775, 1981.

[18] F. D. Morgan, E. R. Williams, and T. R. Madden, "Streaming potential properties of Westerly granite with applications," *Journal of Geophysical Research*, vol. 94, no. 9, pp. 12449–12461, 1989.

[19] L. Jouniaux and J. P. Pozzi, "Permeability dependence of streaming potential in rocks for various fluid conductivities," *Geophysical Research Letters*, vol. 22, no. 4, pp. 485–488, 1995.

[20] B. Lorne, F. Perrier, and J. P. Avouac, "Streaming potential measurements: 1. Properties of the electrical double layer from crushed rock samples," *Journal of Geophysical Research B*, vol. 104, no. 8, pp. 17857–17877, 1999.

[21] F. Perrier and T. Froidefond, "Electrical conductivity and streaming potential coefficient in a moderately alkaline lava series," *Earth and Planetary Science Letters*, vol. 210, no. 1-2, pp. 351–363, 2003.

[22] L. Jouniaux, S. Lallemant, and J. P. Pozzi, "Changes in the permeability, streaming potential and resistivity of a claystone from the Nankai prism under stress," *Geophysical Research Letters*, vol. 21, no. 2, pp. 149–152, 1994.

[23] L. Jouniaux and J. P. Pozzi, "Streaming potential and permeability of saturated sandstones under triaxial stress: consequences for electrotelluric anomalies prior to earthquakes," *Journal of Geophysical Research*, vol. 100, no. 6, pp. 10197–10209, 1995.

[24] S. Yoshida, O. C. Clint, and P. R. Sammonds, "Electric potential changes prior to shear fracture in dry and saturated rocks," *Geophysical Research Letters*, vol. 25, no. 10, pp. 1577–1580, 1998.

[25] B. Lorne, F. Perrier, and J. P. Avouac, "Streaming potential measurements 2. Relationship between electrical and hydraulic flow patterns from rock samples during deformation," *Journal of Geophysical Research B*, vol. 104, no. 8, pp. 17879–17896, 1999.

[26] L. Jouniaux, M. L. Bernard, M. Zamora, and J. P. Pozzi, "Streaming potential in volcanic rocks from Mount Pelée," *Journal of Geophysical Research B*, vol. 105, no. 4, pp. 8391–8401, 2000.

[27] S. Yoshida, "Convection current generated prior to rupture in saturated rocks," *Journal of Geophysical Research B*, vol. 106, no. 2, pp. 2103–2120, 2001.

[28] P. M. Reppert and F. D. Morgan, "Temperature-dependent streaming potentials: 2. Laboratory," *Journal of Geophysical Research B*, vol. 108, no. 11, pp. 4–13, 2003.

[29] T. Tosha, N. Matsushima, and T. Ishido, "Zeta potential measured for an intact granite sample at temperatures to 200°C," *Geophysical Research Letters*, vol. 30, no. 6, pp. 28–1, 2003.

[30] S. R. de Groot and P. Mazur, *Non-Equilibrium Thermodynamics*, North-Holland, Amsterdam, The Netherlands; Interscience Publishers, New York, NY, USA, 1962.

[31] S. Pride, "Governing equations for the coupled electromagnetics and acoustics of porous media," *Physical Review B*, vol. 50, no. 21, pp. 15678–15696, 1994.

[32] J. B. Walsh and W. F. Brace, "The effect of pressure on porosity and the transport properties of rock," *Journal of Geophysical Research*, vol. 89, no. 11, pp. 9425–9431, 1984.

[33] O. Nishizawa, "New multi-wire type and co-axial type feedthroughs for an oil pressure medium vessel," *Bulletin of the Geological Survey of Japan*, vol. 48, pp. 431–438, 1997.

[34] T. Ishido and O. Nishizawa, "Effects of Zeta potential on microcrack growth in rock under relatively low uniaxial compression," *Journal of Geophysical Research*, vol. 89, no. 6, pp. 4153–4159, 1984.

[35] M. Takahashi, "Permeability change during experimental fault smearing," *Journal of Geophysical Research B*, vol. 108, no. 5, pp. 1–15, 2003.

[36] R. L. Kranz, J. S. Saltzman, and J. D. Blacic, "Hydraulic diffusivity measurements on laboratory rock samples using an oscillating pore pressure method," *International Journal of Rock Mechanics and Mining Sciences*, vol. 27, no. 5, pp. 345–352, 1990.

[37] G. J. Fischer, "The determination of permeability and storage capacity: pore pressure oscillation method," in *Fault Mechanics and Transport Properties of Rocks*, B. Evan and T. F. Wong, Eds., pp. 187–211, Academic Press, San Diego, Calif, USA, 1992.

[38] G. J. Fischer and M. S. Paterson, "Measurement of permeability and storage capacity in rocks during deformation at high temoerature and pressure," in *Fault Mechanics and Transport Properties of Rocks*, B. Evan and T. F. Wong, Eds., pp. 213–252, Academic Press, San Diego, Calif, USA, 1992.

[39] I. Song and J. Renner, "Analysis of oscillatory fluid flow through rock samples," *Geophysical Journal International*, vol. 170, no. 1, pp. 195–204, 2007.

[40] R. G. Packard, "Streaming potentials across glass capillaries for sinusoidal pressure," *The Journal of Chemical Physics*, vol. 21, no. 2, pp. 303–307, 1953.

[41] P. M. Reppert, F. D. Morgan, D. P. Lesmes, and L. Jouniaux, "Frequency-dependent streaming potentials," *Journal of Colloid and Interface Science*, vol. 234, no. 1, pp. 194–203, 2001.

[42] P. M. Reppert and F. D. Morgan, "Streaming potential collection and data processing techniques," *Journal of Colloid and Interface Science*, vol. 233, no. 2, pp. 348–355, 2001.

[43] W. F. Brace, J. B. Walsh, and W. T. Frangos, "Permeability of granite under high pressure," *Journal of Geophysical Research*, vol. 73, no. 6, pp. 2225–2236, 1968.

[44] Y. Bernabé, U. Mok, and B. Evans, "Permeability-porosity relationships in rocks subjected to various evolution processes," *Pure and Applied Geophysics*, vol. 160, no. 5-6, pp. 937–960, 2003.

[45] W. Zhu, L. G. J. Montési, and T. F. Wong, "A probabilistic damage model of stress-induced permeability anisotropy during cataclastic flow," *Journal of Geophysical Research B*, vol. 112, no. 10, Article ID B10207, 2007.

[46] X. Guichet, L. Jouniaux, and J. P. Pozzi, "Streaming potential of a sand column in partial saturation conditions," *Journal of Geophysical Research B*, vol. 108, no. 3, article 2141, p. 12, 2003.

[47] N. Linde and A. Revil, "Inverting self-potential data for redox potentials of contaminant plumes," *Geophysical Research Letters*, vol. 34, no. 14, Article ID L14302, 2007.

[48] A. Revil, N. Linde, A. Cerepi, D. Jougnot, S. Matthäi, and S. Finsterle, "Electrokinetic coupling in unsaturated porous media," *Journal of Colloid and Interface Science*, vol. 313, no. 1, pp. 315–327, 2007.

[49] T. Ishido, "Electrokinetic mechanisms for the "W"-shaped self-potential profile on volcanoes," *Geophysical Research Letters*, vol. 31, no. 15, article L15616, p. 5, 2004.

[50] S. Onizawa, N. Matsushima, T. Ishido, H. Hase, S. Takakura, and Y. Nishi, "Self-potential distribution on active volcano controlled by three-dimensional resistivity structure in Izu-Oshima, Japan," *Geophysical Journal International*, vol. 178, no. 2, pp. 1164–1181, 2009.

Permissions

The contributors of this book come from diverse backgrounds, making this book a truly international effort. This book will bring forth new frontiers with its revolutionizing research information and detailed analysis of the nascent developments around the world.

We would like to thank all the contributing authors for lending their expertise to make the book truly unique. They have played a crucial role in the development of this book. Without their invaluable contributions this book wouldn't have been possible. They have made vital efforts to compile up to date information on the varied aspects of this subject to make this book a valuable addition to the collection of many professionals and students.

This book was conceptualized with the vision of imparting up-to-date information and advanced data in this field. To ensure the same, a matchless editorial board was set up. Every individual on the board went through rigorous rounds of assessment to prove their worth. After which they invested a large part of their time researching and compiling the most relevant data for our readers. Conferences and sessions were held from time to time between the editorial board and the contributing authors to present the data in the most comprehensible form. The editorial team has worked tirelessly to provide valuable and valid information to help people across the globe.

Every chapter published in this book has been scrutinized by our experts. Their significance has been extensively debated. The topics covered herein carry significant findings which will fuel the growth of the discipline. They may even be implemented as practical applications or may be referred to as a beginning point for another development. Chapters in this book were first published by Hindawi Publishing Corporation; hereby published with permission under the Creative Commons Attribution License or equivalent.

The editorial board has been involved in producing this book since its inception. They have spent rigorous hours researching and exploring the diverse topics which have resulted in the successful publishing of this book. They have passed on their knowledge of decades through this book. To expedite this challenging task, the publisher supported the team at every step. A small team of assistant editors was also appointed to further simplify the editing procedure and attain best results for the readers.

Our editorial team has been hand-picked from every corner of the world. Their multi-ethnicity adds dynamic inputs to the discussions which result in innovative outcomes. These outcomes are then further discussed with the researchers and contributors who give their valuable feedback and opinion regarding the same. The feedback is then collaborated with the researches and they are edited in a comprehensive manner to aid the understanding of the subject.

Apart from the editorial board, the designing team has also invested a significant amount of their time in understanding the subject and creating the most relevant covers. They scrutinized every image to scout for the most suitable representation of the subject and create an appropriate cover for the book.

The publishing team has been involved in this book since its early stages. They were actively engaged in every process, be it collecting the data, connecting with the contributors or procuring relevant information. The team has been an ardent support to the editorial, designing and production team. Their endless efforts to recruit the best for this project, has resulted in the accomplishment of this book. They are a veteran in the field of academics and their pool of knowledge is as vast as their experience in printing. Their expertise and guidance has proved useful at every step. Their uncompromising quality standards have made this book an exceptional effort. Their encouragement from time to time has been an inspiration for everyone.

The publisher and the editorial board hope that this book will prove to be a valuable piece of knowledge for researchers, students, practitioners and scholars across the globe.

List of Contributors

Shohei Minato and Toshifumi Matsuoka
Graduate School of Engineering, Kyoto University, C1 Kyoto-Daigaku Katsura, Nishikyo-ku, Kyoto 6158540, Japan

Takeshi Tsuji
Graduate School of Engineering, Kyoto University, C1 Kyoto-Daigaku Katsura, Nishikyo-ku, Kyoto 6158540, Japan
International Institute for Carbon-Neutral Energy Research (I2CNER), Kyushu University, Ito Campus, 744 Motooka, Nishi-Ku, Fukuoka 819-0395, Japan

Koichiro Obana
Japan Agency for Marine-Earth Science and Technology, Institute for Research on Earth Evolution, 3173-25 Showa-machi, Kanazawa-ku, Yokohama 236-0001, Japan

Charles S. Carrano, Cesar E. Valladares and Keith M. Groves
Boston College, Chestnut Hill, MA 02467, USA

Igor Mingalev and Victor Mingalev
Kola Scientific Center of the Russian Academy of Sciences, Polar Geophysical Institute, Fersman Street 14, Apatity 184209, Russia

Sabine Muller and Ernst Niederleithinger
Division 8.2 of Non-destructive Damage Assessment and Environmental Measurement Methods, BAM Federal Institute for Materials Research and Testing, Unter den Eichen 87, 12205 Berlin, Germany

Thomas Bohlen
Geophysical Institute, KIT Karlsruhe Institute of Technology, Hertzstr. 16, 76187 Karlsruhe, Germany

Sergey Pulinets
Space Research Institute, Russian Academy of Sciences, Moscow 117997, Russia

A. D'Alterio and R. Solimene
Dipartimento di Ingegneria dell'Informazione, Seconda Universita di Napoli, Via Roma 29, 81031 Aversa, Italy

Tao Zhu, Jian-Guo Zhou and Jin-Qi Hao
Institute of Geophysics, China Earthquake Administration, Beijing 100081, China

Yuji Nishi and Tsuneo Ishido
GSJ, National Institute of Advanced Industrial Science and Technology (AIST), Tsukuba Central 7, Tsukuba, Ibaraki 305-8567, Japan

T. L. Battle
Program in Atmospheric Sciences, Howard University, Washington, DC 20059, USA

J. W. Smith
Program in Atmospheric Sciences, Howard University, Washington, DC 20059, USA
Atmospheric Science Group, Department of Geosciences, Texas Tech University, Lubbuck, TX 79409, USA

A. S. Pratt
Earth Resources Technology, Inc., Laurel, MD 20707, USA

S. Salack, A. Diop, T. Fall and A. Gaye
Laboratory of Atmospheric Physics—Simeon Fongong, Cheikh Anta Diop University, Dakar, Senegal

B. Klotz
Rosenstiel School of Marine and Atmospheric Science, Cooperative Institute for Marine and Atmospheric Science Studies, University of Miami, Miami, FL 33149, USA

D. Grant
Environmental Protection Agency, Region 4, Atlanta, GA 30303, USA

D. Robertson
Department of Mechanical Engineering, Howard University, Washington, DC 20059, USA

M. S. De Longe
Department of Environmental Science, Policy and Management, University of California Berkeley, Berkeley, CA 94720, USA

S. Chan
Department of Environmental Sciences, University of Virginia, Charlottesville, VA 22903, USA

A. E. Reynolds
NASA Goddard Space Flight Center, Greenbelt, MD 20771, USA
Atmospheric Science Group, Department of Geosciences, Texas Tech University, Lubbuck, TX 79409, USA

Valery Grigoryev and Evgeniy Tereshchenko
Kola Science Centre, Polar Geophysical Institute, Russian Academy of Science, Murmansk, 15 Khalturina Street, Murmansk 183010, Russia

Igor Trofimov
Geo electromagnetic Research Centre of Schmidt Institute of Physics of the Earth, Russian Academy of Sciences, P.O. Box 30, Troitsk, Moscow Region 142190, Russia

Darya Orekhova and Yury Scshors
Kurchatov Institute, Moscow, 1 Akademika Kurchatova Square, Moscow 123182, Russia

Pavel Tereshchenko
Institute of Terrestrial Magnetism, Ionosphere and Radiowave Propagation, Russian Academy of Sciences, St. Petersburg Branch, 1 Mendeleevskaya Linia, St. Petersburg 199034, Russia

Sergey Korotaev
Geo electromagnetic Research Centre of Schmidt Institute of Physics of the Earth, Russian Academy of Sciences, P.O. Box 30, Troitsk, Moscow Region 142190, Russia
Kurchatov Institute, Moscow, 1 Akademika Kurchatova Square, Moscow 123182, Russia

Mikhail Kruglyakov
Kurchatov Institute, Moscow, 1 Akademika Kurchatova Square, Moscow 123182, Russia
Moscow State University, Moscow, GSP-1 Leninskie Gory, Moscow 119991, Russia

Raffaele Luongo
DIFA, Universita della Basilicata, Contrado Macchia Romana, 85100 Potenza, Italy
IMAA-CNR, Contrada Santa Loja Zona Industriale, Tito Scalo, 85055 Potenza, Italy

Angela Perrone, Sabatino Piscitelli and Vincenzo Lapenna
IMAA-CNR, Contrada Santa Loja Zona Industriale, Tito Scalo, 85055 Potenza, Italy

Kumar Ramachandran and Bryan Tapp
Department of Geosciences, The University of Tulsa, Tulsa, OK 74104-9700, USA

Tayler Rigsby
Department of Geosciences, The University of Tulsa, Tulsa, OK 74104-9700, USA
Colorado School of Mines, CO 80401-1893, USA

Erin Lewallen
Department of Geosciences, The University of Tulsa, Tulsa, OK 74104-9700, USA
Chevron Corporation, Houston, TX 77002, USA

Said Keshta
Geology Department, Faculty of Education, Suez Canal University, Arish, Egypt

Farouk J. Metwalli
Geology Department, Faculty of Science, Helwan University, Cairo, Egypt

H. S. Al Arabi
Geology Department, Faculty of Science, Suez Canal University, Ismailia, Egypt

G. Bayanjargal
Research Center for Astronomy and Geophysics, Mongolian Academy of Sciences (MAS), Ulaanbaatar, Mongolia

Naser Tamimi and Thomas L. Davis
Reservoir Characterization Project, Department of Geophysics, Colorado School of Mines, Golden, CO 80401, USA

Antonio Satriani, Antonio Loperte, Vito Imbrenda and Vincenzo Lapenna
Institute of Methodologies for Environmental Analysis, CNR, 85050 Tito, Italy

Akihiro Takeuchi
Earthquake Prediction Research Center, Institute of Oceanic Research and Development, Tokai University, 3-20-1 Orido, Shimizu-ku, Shizuoka 424-8610, Japan

Kan Okubo
Division of Information and Communications Systems Engineering, Tokyo Metropolitan University, 6-6 Asahigaoka, Hino 191-0065, Japan

Nobunao Takeuchi
Research Center for Prediction of Earthquakes and Volcanic Eruptions, Graduate School of Science, Tohoku University, 6-6 Aza-aoba, Aramaki, Aoba-ku, Sendai 980-8578, Japan

Osamu Kuwano and Shingo Yoshida
Earthquake Research Institute, University of Tokyo, 1-1-1 Yayoi, Bunkyo-ku, Tokyo 113-0032, Japan